Microbial Source Tracking

Emerging Issues in Food Safety
SERIES EDITOR, Michael P. Doyle

Microbiology of Fresh Produce
Edited by Karl R. Matthews

Microbial Source Tracking
Edited by Jorge W. Santo Domingo and Michael J. Sadowsky

ALSO IN THIS SERIES

Microbial Risk Analysis in Foods (2007)
Edited by Donald W. Schaffner

***Enterobacter sakazakii* (2007)**
Edited by Jeffrey M. Farber and Stephen Forsythe

Safety of Imported Foods: Microbiological Issues and Challenges (2007)
Edited by Michael P. Doyle and M. C. Erickson

Microbial Source Tracking

EDITED BY

Jorge W. Santo Domingo
ORD/NRMRL/WSWRD/MCCB
U.S. Environmental Protection Agency
Cincinnati, Ohio

AND

Michael J. Sadowsky
Department of Soil, Water, and Climate
and BioTechnology Institute
University of Minnesota
St. Paul, Minnesota

WASHINGTON, D.C.

Address editorial correspondence to ASM Press, 1752 N St., N.W., Washington, DC 20036-2904, USA

Send orders to ASM Press, P.O. Box 605, Herndon, VA 20172, USA
Phone: 800-546-2416; 703-661-1593
Fax: 703-661-1501
E-mail: books@asmusa.org
Online: http://estore.asm.org

Library of Congress Cataloging-in-Publication Data

Microbial source tracking / edited by Jorge W. Santo Domingo and Michael J. Sadowsky.
p. ; cm.—(Emerging issues in food safety)
Includes bibliographical references and index.
ISBN-13: 978-1-55581-374-1 (hardcover)
ISBN-10: 1-55581-374-7 (hardcover)
1. Diagnostic microbiology. 2. Pathogenic microorganisms —Detection. 3. Microbial toxins. 4. Foodborne diseases. 5. Microorganisms. I. Santo Domingo, Jorge W. II. Sadowsky, M. J. (Michael J.) III. Series.
[DNLM: 1. Food Microbiology. 2. Disease Outbreaks —prevention & control. 3. Environmental Monitoring—methods. 4. Feces—microbiology. 5. Food Contamination—prevention & control. 6. Microbiological Techniques—methods. 7. Water Microbiology. QW 85 M62619 2007]
QR67.M53 2007
616.9′0475—dc22

2006024176

10 9 8 7 6 5 4 3 2 1

Printed in the United States of America

Cover illustration: Microbial water quality biochip.
Courtesy of Jorge W. Santo Domingo (see Chapter 1, Figure 5).

Contents

Contributors

TERESA M. BERGHOLZ
National Food Safety and Toxicology Center, 165 Food Safety and Toxicology Building, Michigan State University, East Lansing, MI 48824

DAVID BOYLE
Washington State Dept. of Health, 1610 N.E. 150th Street, Public Health Laboratories, Shoreline, WA 98155-7224

DOUGLAS R. CALL
Department of Veterinary Microbiology & Pathology, Washington State University, Pullman, WA 99164-7040

VALERIE J. HARWOOD
Department of Biology, SCA 110, University of South Florida, 4202 E. Fowler Ave., Tampa, FL 33620

JOHN SCOTT MESCHKE
School of Public Health and Community Medicine, University of Washington, Seattle, WA 98105-6099

CINDY H. NAKATSU
Department of Agronomy, Purdue University, West Lafayette, IN 47907-2054

PETER T. PESENTI
Environmental Microbiology, Department of Homeland Security, Directorate for Science and Technology, Biological Countermeasures, Washington, DC 20528

Suresh D. Pillai
Departments of Nutrition and Food Science & Poultry Science, Texas A&M University, College Station, Texas 77843-2472

Albert Rhodes
Homeland Security Institute, 2900 S. Quincy St., Ste. 800, Arlington, VA 22206

Michael J. Sadowsky
Department of Soil, Water & Climate and BioTechnology Institute, University of Minnesota, 1991 Upper Buford Circle, 439 BorH, St. Paul, MN 55108

Jorge W. Santo Domingo
U.S. Environmental Protection Agency, ORD/NRMRL/WSWRD/MCCB, 26 W. Martin Luther King Dr., MS 387, Cincinnati, OH 45268

Jill R. Stewart
National Oceanic and Atmospheric Administration, 219 Ft. Johnson Rd., Charleston, SC 29412-9110

Everardo Vega
Centers for Disease Control and Prevention, Atlanta, GA 30333

Timothy J. Wade
U.S. Environmental Protection Agency, ORD/NHEERL/HSD/EBB, MD 58C, Research Triangle Park, NC 27711

John E. Whitlock
Division of Mathematics and Sciences, Hillsborough Community College, Tampa, FL 33614

Thomas S. Whittam
National Food Safety and Toxicology Center, 165 Food Safety and Toxicology Building, Michigan State University, East Lansing, MI 48824

Jayson D. Wilbur
Department of Mathematical Sciences, Worcester Polytechnic Institute, Worcester, MA 01609-2280

Series Editor's Foreword

The ability to trace microorganisms, including food-borne pathogens, to the point of origin has major ramifications for the food industry, food regulatory agencies, and public health. Such information would enable food producers and processors to better understand sources of contamination and thereby take corrective actions to prevent pathogen transmission. It would assist regulatory agencies in identifying food producers responsible for supplying foods involved in human illnesses. It would also provide public health investigators the opportunity to track food-borne disease outbreaks to their original source and prevent future occurrences.

Microbial source tracking, which currently is largely focused on determining sources of fecal contamination in waterways, is providing the scientific community tools that may be adapted to broader applications, including the tracking of food-borne pathogens. Some day in the near future such techniques will likely enable the traceback of pathogens, like *Salmonella* and *E. coli* O157:H7, in foods to their origin. As an example, *Salmonella enterica* serovar Poona has been responsible for outbreaks of salmonellosis associated with cantaloupe consumption and continues to be a nemesis for the melon industry. Using the tools of microbial source tracking, investigators would be able to trace an outbreak-associated strain of serovar Poona from patients to cantaloupes from a food service or retail outlet, to the field in which the cantaloupes were produced, to irrigation water which was obtained from a river contaminated with cattle manure, to the cattle farm from which serovar Poona-contaminated manure polluted the river. Whether microbial pathogens contaminate foods by intentional or unintentional routes, the tools of microbial source tracking would enable identification of their origin for attribution of their source. The ultimate beneficiary of this technology will be the consumer, as the tools of microbial source tracking will provide greater public health protection to our water and food supplies.

This book provides the current state of the science for microbial source tracking, especially as it relates to identifying sources of waterborne microbes of fecal origin. It not only addresses current methods used for identification of sources of fecal contamination but also provides perspectives on the assumptions and limitations of tracking approaches and the challenges faced by scientists conducting tracking studies. In addition, there is an attempt to gaze into the future, providing a look at how emerging molecular and genomic techniques will advance the field of microbial source tracking. This is a "must have" resource for anyone interested in tracking microbes to their origin.

I thank and commend Jorge Santo Domingo, Michael Sadowsky, and their colleagues for preparing this exceptional treatise on an extremely important topic of public health significance.

MICHAEL P. DOYLE, Series Editor
Emerging Issues in Food Safety

Preface

Microbial Source Tracking (MST) is a term that has recently been used to define a variety of phenotypic and genotypic methods that are used to determine sources of fecal contamination in waterways. While the idea of discriminating between human and animal fecal contamination dates back to the 1960s, recent progress in molecular biology has accelerated the development of more rapid and sensitive typing methods that can be used to classify microbial isolates into different fecal source groups. For the MST concept to work, differences in gut conditions (influenced by diet and digestive systems) must serve as the driver for natural selection. This, in turn, leads to the systematic development of a microbial community dominated by populations that are better adapted to compete for space and nutrients that are specific to a particular gut environment. That is, some microbes must display a degree of host specificity in order for MST methods to be of value in field applications. Like those found in any habitat, however, some gut microbial populations are transitory, while others are commonly found in the gut ecosystems of all warm-blooded animals, due to a lack of preference for colonization, retention, or the supply of specific nutrients. Within this background, those performing source-tracking methods have the challenging task of finding host-specific populations that can survive long enough outside of the gut to be identified and quantified and that provide monitoring and risk assessment value to regulators.

While strict host-specific distribution has only been proven for a few microbial groups, several studies have successfully identified primary sources of fecal pollution in environmental settings using some MST methods. This suggests that some fecal microbial populations indeed tend to favor particular gut environments. Many of the same techniques that have been used for assessing fecal sources in contaminated watersheds can also be used to track sources of fecal pollution in contaminated foods, or they can be used for epidemiological analyses in disease outbreaks. Regardless of the matrix, MST methods can help those in the water and food industries in the evaluation of pollution control, and in the development of remediation practices and more accurate microbial risk assessment models.

The primary goal of this book is to provide the reader with some of the most recent information regarding the following: (i) methods used for the identification of fecal contamination sources in water and food matrices; (ii) the assumptions and limitations of MST approaches; (iii) the challenges that people performing source tracking studies are currently facing; and (iv) the ways in which the newest tools in genomics and biotechnology will help in method development and in environmental monitoring. In this book we do not present any of the approaches that measure chemical surrogates used to trace fecal pollution sources (e.g., stanols, sterols, caffeine, and fluorescent brighteners). Rather, we decided to focus on microbe-based markers. However, recent studies have presented evidence for the use of chemical surrogates, particularly when tracing human pollution. In many cases, these methods can be used as a first line of evidence, particularly when monitoring for the presence of human pollution.

This book is organized into eight chapters that are intended to complement each other. Aside from the first few chapters, which provide an introduction to MST methods and assumptions and limitations associated with their use, there is no specific order in which the chapters should be read, as they cover a wide spectrum of topics that relate MST to environmental monitoring, public health, population biology, and microbial ecology. Whenever possible, the link between MST and the water and food industries is established from both theoretical and applied standpoints. For example, in chapters 1, 2, 3, and 8, the authors provide scenarios in which MST methods can be useful for the prevention, identification, and remediation of water- and food-related contamination events. Similarly, the chapter by Whittam and Large (chapter 4) discusses how molecular typing technologies that are used in water, food, and clinical settings can be useful in issues relating to the transmission of food-borne illness and food safety, and that by Meschke and Boyle (chapter 5) focuses on the shellfish industry and how methods developed for microbial water-quality applications can benefit this industry.

There are a number of assumptions and limitations associated with each MST approach, and these can have a significant impact on which method is ultimately selected for real-life applications and on how the data is interpreted. Harwood (chapter 2) provides the reader with insights on how to plan and assess the value of MST data in light of the basic assumptions and the many limitations associated with MST approaches. Like any other scientific endeavor, the validity of data and method application strongly depends on adequate statistical analysis. Hence, the entire chapter by Wilbur and Whitlock (chapter 6) is devoted to discussion of many of the statistical approaches that can be used in MST studies, and some of the issues pertaining to experimental design.

While most of the chapters in this book describe methods developed for microbial water quality, Nakatsu et al. (chapter 7) discuss how recent developments in homeland security research can be used, not only to identify biological threats, but also in fecal source identification studies. The authors discuss these developments considering some of the challenges related to the diversity, abundance, and distribution of microorganisms in their natural reservoirs. Other recent technological advances relevant to MST and the food industry, and the technical challenges facing those interested in sensitive microbial detection, are discussed in chapter 3 by by Pillai and Vega. Finally, Sadowsky et al. (chapter 8) discuss the future of MST, taking into consideration emerging molecular and genomic methodologies and the ways in which newly developed research tools will likely improve marker discovery and the simultaneous detection of biomarkers relevant to public health.

As we hope the reader will appreciate, the field of MST goes beyond just the simple identification of pollution sources and their relationship to public health and environmental protection. Many of the methods used in source tracking are also widely used in clinical microbiological studies, although in most cases the studies focus on specific pathogens. The methods and approaches discussed herein can also be used to answer critical questions related to microbial ecology, evolutionary biology, population biology, microbial diversity, and microbial transport. Consequently, we hope that this book will be useful not only to those performing MST, but also to the applied and environmental microbiology community in general.

We are very grateful to Mike Doyle (University of Georgia), who invited us to write an interdisciplinary MST book, and to Eleanor Riemer (ASM Press), for her advice and guidance on issues related to the editorial process. We also acknowledge the help provided by Luis Tenorio (Colorado School of Mines), who reviewed chapter 4, and John Ferguson (University of Minnesota), who helped sort through the many references cited in this book. Special thanks to all the authors for your contributions and to Sally Gutierrez (U.S. EPA) and Donald Reasoner for early discussions, guidance, and support.

JORGE W. SANTO DOMINGO
MICHAEL J. SADOWSKY

Microbial Source Tracking
Edited by Jorge W. Santo Domingo and Michael J. Sadowsky

Fecal Pollution, Public Health, and Microbial Source Tracking

1

Jill R. Stewart, Jorge W. Santo Domingo, and Timothy J. Wade

Fecal contamination of water and food is a common and enduring problem, resulting each year in closed beaches and shellfish beds, tainted meat products, and polluted waterways. These problems are common to all nations regardless of economic status, although the levels of fecal pollution as well as the primary agents of disease differ among different regions. While problems associated with fecal pollution could probably be traced through all civilizations, the link between fecal pollution, water quality, and disease only became clear in the mid-1800s. Around this time the germ theory was gaining popular acceptance among the scientific and medical communities, due to the theoretical postulates and pioneering research of scientists like Henle, Lister, Pasteur, and Koch. The body of work compiled by these scientists provided the means to combat microorganisms before and after exposure of humans to pathogens. But it was the work of John Snow, in 1854, that first implicated fecally impacted drinking water as the source of an enteric disease (i.e., cholera) (26, 120). A few decades after Snow's findings, Escherich reported that "*Bacillus coli*" (now known as *Escherichia coli*) was a predominant bacterium in feces. The concept of indicator organisms was then born, and shortly after, sanitation officials began to routinely use the densities of coliform bacteria as an indication of the levels of fecal pollution in water.

Today, assays for the detection and enumeration of fecal coliforms, *E. coli*, and *Enterococcus* are typically used to identify fecal contamination in surface waters. Bacterial indicators are used to detect fecal contamination in environmental

Jill R. Stewart, National Oceanic and Atmospheric Administration, 219 Ft. Johnson Rd., Charleston, SC 29412-9110. Jorge W. Santo Domingo, U.S. Environmental Protection Agency, ORD/NRMRL/WSWRD/MCCB, 26 W. Martin Luther King Dr., MS 387, Cincinnati, OH 45268. Timothy J. Wade, U.S. Environmental Protection Agency, ORD/NHEERL/HSD/EBB, MD 58C, Research Triangle Park, NC 27711.

waters because it is not yet practical to directly identify and enumerate the hundreds of pathogenic microorganisms associated with fecal pollution. Indicator bacteria are not necessarily pathogens themselves, but may be associated with fecal contamination. Ideally, indicators and pathogens should share common fecal origins; however, indicators should be present in higher numbers than and survive as long as and be easier to detect and quantify than pathogens (12).

Coliform bacteria, the most commonly employed microbial indicator group, have been used to assess the safety of drinking water for nearly a century (76). In water microbiology, the coliform group includes aerobic and facultatively anaerobic, gram-negative, non-spore-forming, rod-shaped bacteria. Total coliforms further ferment lactose with gas formation within 48 hours at 35°C, or they develop red colonies with a metallic (golden) sheen within 24 h at 35°C on an Endo-type medium containing lactose (2). Fecal or thermotolerant coliforms ferment lactose as detectable using modified tests with elevated temperatures. These are operational rather than taxonomic definitions, and the coliform group includes a variety of organisms. The fecal coliform group was established in an effort to detect organisms exclusively of fecal origin, and it is the most frequently used microbial indicator for surface-water monitoring programs. More recently, *E. coli* has been used as a more specific indicator of fecal pollution.

Enterococcus is a recently recognized genus which was originally considered a subgroup of the streptococcal group. Currently, there are over 20 recognized enterococcal species, although *Enterococcus faecalis*, *Enterococcus faecium*, *Enterococcus durans*, and *Enterococcus hirae* are the species primarily associated with feces (48). The enterococci are differentiated from the streptococci by their ability to grow in 6.5% sodium chloride, pH 9.6, at 10°C and 45°C, with resistance to 60°C for 30 min, and by their ability to reduce 0.1% methylene blue (2). Several studies have suggested that enterococci may be better indicator organisms than the fecal coliforms because they survive longer in marine environments (15, 71) and through water treatment processes (94). Furthermore, their numbers in recreational waters correlate with the risk of gastrointestinal illness, while those of coliforms do not (15). Enterococci are now recommended by the U.S. Environmental Protection Agency (124) as the indicators of choice for marine recreational waters.

While the use of bacterial indicators to monitor microbial water quality has been criticized by many (3), others have argued that the use of current bacterial indicators continues to protect human health at relatively modest costs (37). Some of the main arguments against the use of bacterial indicators relate to the differences in survival and transport rates between these indicators and human pathogens, particularly enteric viruses and protozoa. In addition, the recently documented occurrence of *E. coli* (130) and enterococci (96) in nonintestinal environments suggests that some subpopulations of indicator bacteria might not be of recent fecal origin. Hence, alternate indicators might be needed to more accurately monitor fecal pollution in natural environments.

Another problem with the current use of indicators is that simply detecting the presence of fecal bacteria does not provide information about sources of contamination. Without knowledge of sources, it is difficult to accurately conduct risk assessments, choose effective remediation strategies, or bring chronically polluted waters into compliance with regulatory policies. Microbial source tracking (MST) seeks to provide solutions to these problems. With the tools being developed for fecal source identification, host-specific characteristics of microorganisms can be identified and used to distinguish sources of pollution. The methods and principles used to identify fecal contamination sources in water and food are similar, although in the food industry, pathogens instead of bacterial indicators are more often used as the target organisms. The goal of this chapter is to define the problems associated with fecal pollution (with special attention to gastroenteritis), describe some of the different uses of source tracking tools, briefly describe the methods commonly used in source identification, establish the link between source tracking and food-borne and waterborne illnesses, and discuss the impact accurate identification has on microbial risk assessment and in the implementation of risk management strategies.

Health Impacts of Fecal Pollution

Fecal contamination of waters can be assessed in public health terms (i.e., illness and death cases) and economic terms (i.e., decrease of tourism or increased use of sick leave). The Centers for Disease Control and Prevention estimate that 76 million people suffer food-borne illnesses in the United States each year, accounting for 325,000 hospitalizations and more than 5,000 deaths (19). In 2000, the Economic Research Service of the U.S. Department of Agriculture estimated the annual cost of five common bacterial food-borne pathogens at $6.9 billion (38). Waterborne diseases can be even more devastating, particularly in developing countries. Globally, the World Health Organization estimates that 3.4 million people die as a result of water-related diseases, making them the leading cause of disease and death around the world (34).

Although the frequency and level of severity are higher in developing countries, waterborne outbreaks are relatively common in all countries of the world. Most of the bacterial, viral, and protozoan pathogens associated with waterborne outbreaks are commonly found in the feces of higher mammals

(77); hence, preventing mammalian fecal contamination of source water and recreational waters is critical to human health. In addition, due to potential health risks associated with avian species, watershed protection programs are also implementing risk management practices to minimize the contribution of waterfowl to water pollution. Indeed, avian species can be important sources of human and animal pathogens, as migratory birds are known to be vectors of pathogens like *Cryptosporidium parvum*, *Giardia*, and *Campylobacter* (49, 70, 102). Exposure to fecally polluted water can cause gastroenteritis as well as respiratory and eye-, ear-, and skin-related illnesses. Waterborne pathogens frequently implicated in gastrointestinal diseases include the enteric bacteria, enteric viruses, and protozoan parasites (Table 1). A brief discussion of waterborne and food-borne outbreaks will follow to highlight the magnitude and the diversity of agents responsible for reported cases of gastroenteritis. The reader is encouraged to consult previously published reports for more comprehensive reviews and surveillance data analysis (10, 20, 29, 34, 43, 47, 59, 113, 117, 122).

One of the most severe waterborne outbreaks recorded in the United States occurred in Milwaukee, Wisconsin in 1993. Over 400,000 people became ill with gastroenteritis (18), over 90 million dollars was associated with total medical costs and productivity losses (27), and more than 100 fatalities were also associated with this outbreak. Unusual weather conditions and a failure in the treatment process were found to contribute to water contamination (84). The primary pathogen implicated in this outbreak was *Cryptosporidium.* Molecular typing methods have suggested that the main fecal contamination source was of human origin (134), although runoff from dairy farms might have also been a potential source (30). It should be noted that identifying fecal sources of protozoa is a challenging task, as some protozoan species are present in very different types of hosts. For example, *Cryptosporidium* infections have been reported for 79 mammalian species, 30 bird species, 57 reptilian species, 9 fish species, and 2 amphibian species (100). *Cryptosporidium parvum*, which is the most identified human pathogen of the genus, is often isolated from multiple hosts. In addition, a limited number of environmental isolates are often obtained as part of waterborne and food-borne outbreaks, limiting the amount of source typing that is performed (107). Hence, the need for developing tools that can accurately identify the primary source of protozoan outbreaks is relevant to both risk assessors and risk managers. Other outbreaks involving *Cryptosporidium* have been reviewed by Lisle and Rose (82).

Another waterborne outbreak occurred in the city of Walkerton, Ontario, Canada, in May 2000. Approximately 2,300 individuals experienced gastroenteritis, and 7 people died. A period of heavy rainfall contributed to the

Table 1 Selected list of etiologic agents associated with fecal contamination

Agent
Bacteria
Enterobacteriaceae[a]
Escherichia
Klebsiella
Enterobacter
Salmonella
Shigella
Citrobacter
Providencia
Proteus
Yersinia
Campylobacter
Aeromonas
Clostridium
Viruses
Enteroviruses
Poliovirus
Coxsackievirus A
Coxsackievirus B
Echovirus
Other enteroviruses
Hepatitis A virus (HAV)
Adenovirus
Reovirus
Rotavirus
Calicivirus
Astrovirus
Coronavirus
Protozoan parasites
Cryptosporidium
Giardia
Entamoeba
Microsporidia
Toxoplasma
Naegleria
Acanthamoeba
Cyclospora

[a]Many species in the *Enterobacteriaceae* are inhabitants of the gastrointestinal (GI) tract. Normally, they do not cause GI disease, but they can cause disease in other organ systems and sites in the body. Some *Enterobacteriaceae*, including enterotoxigenic and enterohemorrhagic *E. coli*, can also cause disease in the GI tract if they possess virulence factors.

contamination of groundwater wells (4). As in the Milwaukee outbreak, cow manure was implicated as the primary source of fecal contamination (7). However, the bacterial pathogens *E. coli* O157:H7 and *Campylobacter jejuni* were believed to have been primarily responsible for the outbreak. These bacteria were also associated with one of the largest bacterial waterborne outbreaks in the United States (13). Other enteric bacteria have also been implicated in waterborne outbreaks. For example, *Salmonella enterica* serovar Typhimurium was found to be the etiological agent associated with the Gideon, Missouri outbreak (23). In this particular case, contamination from birds (pigeons) was determined as the most likely fecal source. This and the studies described in the previous paragraph illustrate the impact that zoonotic vectors have on the microbial quality of drinking water and the importance of discriminating human and animal sources of pollution (40, 123).

Enteric viruses also impact the microbial quality of water. Many of the reported waterborne outbreaks worldwide have implicated human caliciviruses (1, 11, 61, 69, 99). As a group, enteric viruses are also believed to be responsible for a significant percentage of waterborne outbreaks in which the etiological agent is unknown. For years, viral outbreaks have been difficult to study, in part due to the relatively low densities of viruses in environmental samples and the lack of cell culture systems to grow many of the viruses that cause gastroenteritis (e.g., noroviruses). The advent of PCR-based methods has greatly increased viral detection, and the application of these methods in environmental monitoring studies has in turn further demonstrated the importance of viruses in microbial water quality (52).

Many waterborne outbreaks are associated with water ingestion during recreational activities. Indeed, several studies have shown that there is a correlation between swimming-associated gastrointestinal illness and the quality of the bathing water (14, 90). These studies have been used to establish the microbial water quality guidelines for recreational waters. In the United States, the recommended guidelines are monthly geometric means of ≤33 CFU/100 ml for enterococci or ≤126 CFU/100 ml for *Escherichia coli* for freshwater recreational waters and a monthly geometric mean of ≤35 CFU/100 ml for enterococci in marine waters. In spite of these guidelines, the number of waterborne-disease outbreaks associated with recreational water in the United States has increased in recent years, with an all-time high total of 65 outbreaks reported during 2001 and 2002 (133). During this period, *Cryptosporidium* was the most commonly identified etiologic agent, linked to 36.7% of cases. Acute gastroenteritis illnesses of unknown etiology accounted for nearly 25% of the cases. In most cases of unknown etiology, pollution sources were not identified. Moreover, 73% of nearly 20,000 U.S. beach closings and swimming advisory days reported in 2004 were due to elevated bacterial levels (http://www .nrdc.org/water/oceans/qttw.asp). The source of contamination associated with these beach closures was unknown in an estimated 43 to 52 percent of cases (http://www.epa.gov/waterscience/beaches/aboutsurvey.html). The inability to link outbreaks with particular sources of pollution often impedes implementation of effective management practices.

Food-borne illnesses are also associated with pathogens of fecal origin. In fact, many of the etiological agents are the same for food-borne and waterborne outbreaks. Food-borne disease surveillance programs, like the FoodNet in the United States (20) (http://www.cdc.gov/foodnet) and the OzFoodNet in Australia (http://www.ozfoodnet.org.au), monitor the trends and potential sources of food-borne diseases and provide information for those interested in learning relevant historical trends. The FoodNet surveillance program monitors seven bacterial pathogens (*Campylobacter*, *E. coli* O157, *Listeria*, *Salmonella*, *Shigella*, *Vibrio*, and *Yersinia*) and two parasitic pathogens (*Cryptosporidium* and *Cyclospora*) in selected counties of 10 states (California, Colorado, Connecticut, Georgia, Maryland, Minnesota, New Mexico, New York, Oregon, and Tennessee).

One common theme in food-borne outbreak data is that while the transmission of pathogens can often be attributed to a limited number of sources (i.e., a particular food type), food contamination can occur in different ways. For example, food-borne outbreaks attributed to consumption of raw shellfish sometimes occur due to the fact that shellfish can concentrate pathogens via their feeding behavior. Improper food handling and storage can alternatively lead to shellfish contamination and associated outbreaks. Food handling has also been implicated in several food-borne outbreaks (103), and in some cases the source has been traced back to a particular food handler (111). Food processing can also be a source of fecal pollution, particularly when equipment is not maintained in proper sanitary conditions. For this reason, many food-borne outbreaks can be prevented by food handlers using proper sanitary practices, in addition to the use of management practices that enforce proper handling, storage, and cooking of foods.

Practically all foods can be sources of food-borne illnesses, especially if they are manually handled and improperly cooked. Vegetables, fruits, juices, egg-containing foods, shellfish, beef, poultry, swine, and dairy products have all been associated with food-borne illnesses. In the United States, *Salmonella*, *Campylobacter*, and *Shigella* have been reported to be associated with the greatest number of food-borne outbreaks (Table 2), although an overall decline in the incidence of infection (estimated per 100,000 persons) has been observed in recent years (101, 126). *Salmonella* and *Campylobacter* are also among the leading bacterial food-borne pathogens in European countries (http://www.bfr.bund.de/internet/8threport/8threp_fr.htm). Brucellosis was

Table 2 Infections caused by pathogens[a]

Etiological agent	No. of infections by yr 2004	2003	2002	2001	2000	1999	1998	1997	Total per pathogen
Campylobacter	5,665	5,273	5,059	4,751	4,713	3,884	4,025	3,974	37,344
E. coli O157	401	444	638	560	626	510	500	340	4,019
Listeria	120	139	98	94	105	114	112	77	859
Salmonella	6,464	6,040	6,150	5,240	4,330	4,488	2,839	2,205	37,756
Shigella	2,231	3,041	4,113	2,219	2,355	1,040	1,480	1,273	17,752
Vibrio	124	110	104	79	54	48	50	51	620
Yersinia	173	162	169	144	133	164	181	139	1,265
Cryptosporidium	613	481	531	575	535	457	566	468	4,226
Cyclospora	15	15	42	32	22	12	9	49	196
Total per yr	15,806	15,705	16,904	13,694	12,873	10,717	9,762	8,576	

[a]Data for 1997 to 2003 were obtained from annual FoodNet reports (http://www.cdc.gov/foodnet/reports.htm). Data for 2004 were obtained from a preliminary report (20). The numbers of states used for the annual surveillance were 10, 9, 9, 9, 8, 7, 7, and 5 for 2004, 2003, 2002, 2001, 2000, 1999, 1998, and 1997, respectively. A total of 104,037 cases were reported.

reported to be a relatively important food-borne illness in Greece, Italy, Portugal, and Turkey. In northern Taiwan, *Vibrio parahaemolyticus* and *Staphylococcus aureus* were the most relevant food-borne pathogens during 1995 to 2001 (121).

While enteric bacteria account for a large number of the food-borne outbreaks, enteric viruses are increasingly being implicated in a significant number of outbreaks. In one study, noroviruses (Norwalk-like viruses) were estimated to be associated with approximately 67% of the illnesses attributable to food-borne transmission (91). This is in sharp contrast with a surveillance summary for 1993 to 1997 that showed that viruses were implicated in only 6% of the outbreaks for which the etiology was determined. Since viruses are not currently monitored by the FoodNet, it is difficult to establish the actual number of cases linked to enteric viruses in the United States. Results from a recent cohort study implicated viral pathogens as the leading cause of gastroenteritis in the Netherlands (32), although the percentage of viral food-borne illnesses was much lower (21%) than the number estimated by Mead et al. (91). Despite these contrasting findings, noroviruses continue to be reported among the most common causes of outbreaks of nonbacterial gastroenteritis worldwide (22, 39, 74, 85).

The source of a food- or waterborne outbreak is typically determined during an ensuing epidemiological investigation, when methods such as pulsed-field gel electrophoresis are used to match genetic fingerprints of clinical isolates implicated in the outbreak with sources. In general, MST borrows from the basic premise of characterizing microorganisms in order to determine sources of microbial contamination. However, the aim of MST is to apply these techniques before a disease outbreak (Fig. 1).

Methods for MST

Public health and environmental protection officials have recognized the importance of discriminating between sources of fecal pollution for several decades. An early approach involved use of the ratio of fecal coliforms to fecal streptococci or FC-FS (46). Water samples containing an FC-FS ratio of greater than or equal to 4.0 were considered to be impacted by human feces, while ratios below 0.7 were believed to be associated with animal feces. However, the use of these ratios was eventually deemed unreliable due to variable survival rates of the bacterial species involved and the variability of FC-FS ratios in different animals. In fact, one study showed that *E. coli* densities were higher than fecal streptococci and enterococci in 80% of the samples, regardless of fecal origin (108). More-recent efforts have focused on the development of methods that use phenotypic and genotypic fingerprints of particular microorganisms to classify environmental isolates. Collectively, the field is known as Microbial Source Tracking, although the terms Bacterial Source

Figure 1 Different approaches for epidemiological and MST investigations.

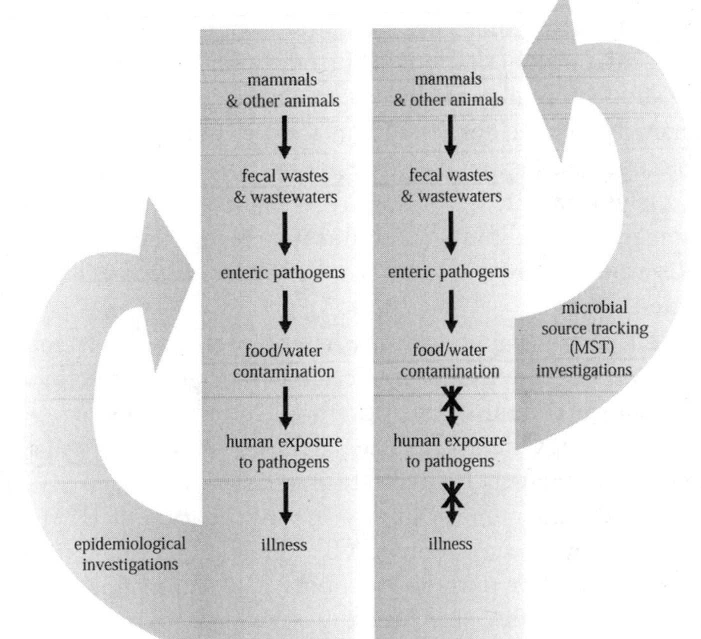

Tracking and Fecal Source Identification might be more appropriate depending on the method used and the goals of the study.

Approaches to MST are commonly classified as library-dependent methods (LDMs) or library-independent methods (LIMs). In MST terms, a library is a database of characteristics (e.g., genetic fingerprints or antibiotic resistance profiles) of microorganisms from known sources (i.e., isolates from animal feces). Characteristics of isolates from contaminated waters are compared to the library to find matches, thereby identifying the source of contamination. Both phenotypic and genotypic characteristics have been used in library-based MST studies. Tested phenotypes have included antibiotic resistance profiles (54, 57, 104, 131), carbon utilization profiles (55), and whole-cell fatty acids (36, 58). Tested genotypic methods have included the fingerprints generated from ribotyping (16, 105, 115), pulsed-field gel electrophoresis (118), and rep-PCR (17, 33, 68). Each of these methods will be discussed in much greater detail in the next chapter. We will introduce some of the key elements pertaining to MST methods and briefly discuss some of the limitations.

Recent studies have suggested that MST libraries must contain characteristics from thousands of bacterial isolates to be representative of those found in watersheds suspected to be impacted by multiple sources (65). As a result, even for techniques that do not require a significant amount of resources and technical expertise, the development of comprehensive libraries may become relatively expensive and time consuming (although the use of robotics could help circumvent these limitations). Furthermore, libraries tend to be geographically and temporally specific, necessitating construction of a library for each watershed intended for study and then requiring maintenance of the library to account for temporal changes in microbial populations. Lastly, the statistical analysis required for identifying sources of fecal contamination may be cumbersome or difficult, and thus may require experience in database analysis. Regardless of these limitations, LDMs have been widely used in the United States.

Library-independent approaches, whereby users of the technology look for a known host-specific characteristic, show promise for freeing practitioners of the resource commitment and complications inherent in building a representative library. Often, sequence databases are necessary to aid in identifying host-specific targets to ascertain potential cross-reactivity of selected markers during development of library-independent methods and to develop quantitative assays (i.e., real-time PCR). Validation of targets against fecal samples is also necessary during development of library-independent host-specific assays and possibly over time to gauge genetic drift. However, practitioners are not required to build and maintain pure culture-based libraries in order to apply these methods. Some of the most critical limitations of current LIMs relate to targeting only one gene, the fact that targets are found in low numbers, the lack of molecular (i.e., sequencing) databases, and the fact that many of the currently targeted genes have very little to do with microbial interactions.

Several MST LIMs have relied on identification of host-specific microbial species. In terms of using bacterial species, researchers have explored whether host specificity of *Enterococcus, Streptococcus, Bifidobacterium*, and *Rhodococcus* species could aid in source identification (45, 87, 97, 129). Coliphages and phages of specific strains of *Bacteroides* species have also been suggested as useful in discriminating between human and animal sources of fecal pollution (24, 110). Similarly, detection of human-associated enteroviruses or adenoviruses using PCR-based assays has been used to identify human source water contamination (66, 79, 98). Bovine enteroviruses and porcine teschoviruses have been used to identify pollution originating from particular animal species (67, 80, 86). The host-specific nature of enteric viruses makes them good targets for the development of MST markers. While direct detection of enteric viruses can be used to identify fecal sources of pollution impacting surface waters, concentration steps are necessary in order to increase the sensitivity of viral detection methods (50).

In addition to detection of host-specific microbes, direct detection of host-specific genetic markers has been proposed as a means of identifying sources of fecal contamination in environmental waters. Some of these approaches require a cultivation step prior to detection, while others do not. Proposed targets include host-specific sequences of toxin genes in *E. coli* and *Enterococcus faecium* (72, 73, 116) and of 16S ribosomal RNA genes in *Bacteroides* (6). This latter approach has been reported to provide relatively accurate results for identification of human and ruminant contamination (41). Eukaryotic genes are another source of host-specific markers. For example, Martellini et al. (88) developed mitochondrial assays to differentiate between human, bovine, porcine, and ovine sources in fecally contaminated surface water. The survival of eukaryotic genes in natural waters plus the potential presence of eukaryotic genes in some of the PCR reagents represent some of the potential problems when using eukaryotic-based assays. Thus far, most MST LIMs target only one gene from one bacterial group. In some cases, the targets are found in low numbers and have been evaluated in only a limited number of studies. Another potential problem is that many of the targeted genes have little to do with host-microbe interactions. Regardless of these limitations, library-independent methods hold promise for routine application, although more targets need to be identified, the methods need to be validated by multiple laboratories, and efforts need to be made to make these approaches quantitative.

Researchers have discovered advantages and disadvantages associated with various MST approaches. Thus far, no single method has emerged as superior, and the choice of MST methods for a particular application will depend on a number of parameters (125). Considerations include the necessary level of specificity and discriminatory power, geographic scale, number and type of potential contributing sources, time available for analysis, need for presence/absence or quantitative identification, environmental persistence and fate of chosen microorganisms, and association with risk. Practicality, cost, and ease of analysis are also important factors.

Many of the challenges intrinsic to MST studies relate to critical issues in microbial ecology and population biology. Some of the other issues relate to the relatively limited empirical data both at the bench and field scale levels, poor experimental design, marker stability, and the lack of standard protocols (Fig. 2). One of the key challenges is to identify host-specific characteristics of microorganisms. In simple terms, it is possible to hypothesize three different scenarios in regard to the distribution of fecal markers in different hosts. The markers are (i) universally present, (ii) preferentially present in a limited number of hosts, or (iii) truly host specific (Fig. 3). Unfortunately, the levels of host distribution in most, if not all, of the host-specific markers available in the scientific literature are not known. This information is critical to the success of an MST study, as the level of host distribution of any marker will greatly impact the validity of the results and hence the success of any predictive model and any remediation strategy.

Figure 2 Some of the potential issues associated with library-dependent methods (LDMs) and library-independent methods (LIMs) currently used in microbial source tracking studies.

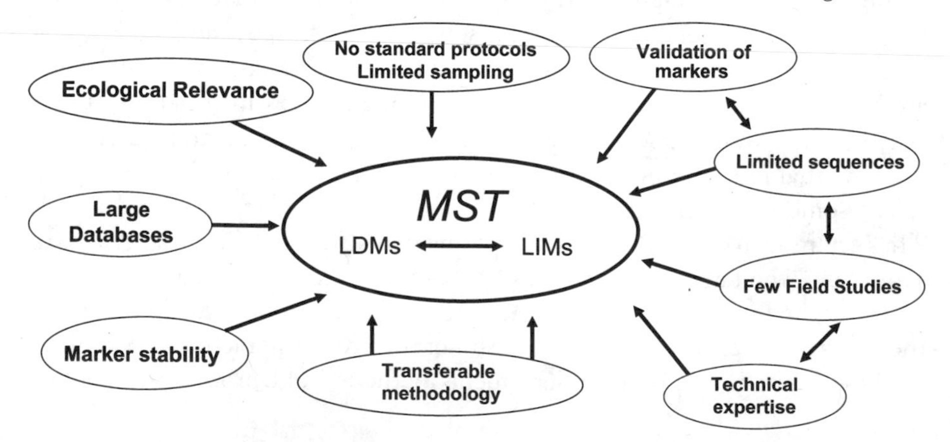

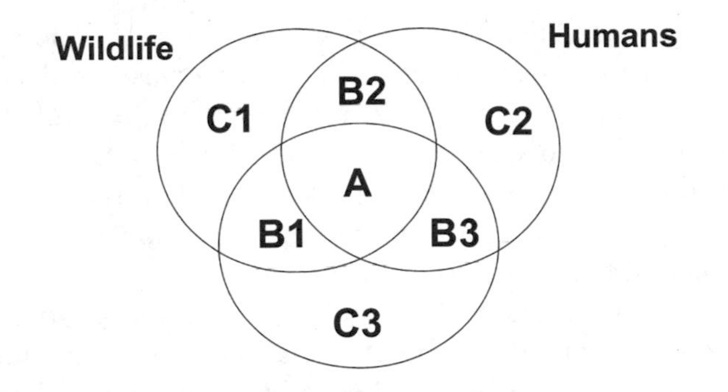

Figure 3 Types of host distribution patterns associated with microorganisms and genetic markers. Microbes as markers can be universally present (A), preferentially present in a limited number of hosts (B), or truly host specific (C).

Many MST approaches assume coevolution between host species and their gut symbionts. While evidence of cospeciation between symbiotic bacteria and their hosts has been reported (44, 60), the validity of this assumption is not clear for many of the facultatively anaerobic bacteria colonizing mammalian guts, including the indicators of fecal pollution like fecal coliforms and enterococci. The latter organisms are opportunistic colonizers and can survive long enough in secondary environments to be able to take up residence in the guts of more than one host species. Analogous genetic fingerprints observed between humans and dogs living in the same households (L. Webster, unpublished results) are evidence of this possible transmission. Prolonged survival of microbial targets in secondary habitats might also significantly disrupt the clonal composition of strains of fecal origin, diminishing their value when trying to quantify the contribution of specific sources of pollution (5). Therefore, for microbial targets that could be opportunistic colonizers, it might be necessary for researchers to adopt a population ecology perspective. This perspective allows that a studied characteristic or property may not be exclusive to a source but that it can be predominantly source specific. Identification of the characteristic, then, could still be useful for MST, and conclusions could be drawn on the basis of probability. In the meantime, the search for sensitive assays exhibiting high-fidelity host-specific properties of microorganisms continues.

MST Applications

Applications for MST are widespread and, to some extent, have dictated method development. Some of the main questions are as follows. Which microbes should be targeted? What host sources are most important to identify? Do we need to be quantitative or is presence/absence data sufficient? Conflicting goals of practitioners may be addressed in the future by a simple, cost-effective method to quantify sources of fecal contamination. In the method development phase where we presently find ourselves, however, it is important to understand the goals of potential practitioners in order to place the current state of MST into context.

Some resource managers wish to use MST as a means to comply with Total Maximum Daily Load (TMDL) regulations required by the U.S. EPA. In the United States, approximately 13% of assessed river and stream miles and 15% of assessed estuarine waters are out of compliance with fecal pollution standards, as measured by indicator bacteria (124). The U.S. EPA issued regulations in 1985 and 1992 that implemented section 303(d) of the Clean Water Act, including the TMDL provisions. TMDL provisions require states to develop a loading estimate for each pollutant for each watershed out of compliance with water quality standards. Since 1996, the total number of impaired waters listed in the United States has exceeded the 34,000 mark (http://oaspub.epa.gov/waters/national_rept.control), a number that could significantly grow due to anthropogenic as well as natural factors.

Pathogens of fecal origin represent the second leading category of impairments in the United States. Other general categories of pollutants responsible for the impairment of water bodies and watersheds are metals, nutrients, sedimentation and siltation, low dissolved oxygen, pH, and turbidity. Pathogen impairments are normally assessed by measuring fecal coliform levels. A geometric mean of 200 CFU/100 ml is the most common threshold value for contact recreation, although this number might vary in different states depending on water usage (recreational versus source water) and season (summer versus winter months). If a watershed is found to exceed the regulatory levels, it must then be listed as impaired, after which the state has a finite number of years to develop a TMDL. However, it is difficult to assign TMDLs for coliforms without knowledge of the primary contamination sources. Identification of primary sources of pollution is a challenging task considering the lack of simple and accurate tools to monitor nonpoint sources (e.g., wildlife, pet wastes, and faulty septic systems), which in a significant number of cases are primarily responsible for exceeding of fecal loads. While the U.S. EPA has provided general guidance on TMDLs, most states and the U.S. EPA have yet to develop TMDLs for each impaired watershed in the United States. Only 14,871 TMDLs out of approximately 34,000 have been approved as of 2005. Citizen groups have brought legal actions against the U.S. EPA, seeking listing of waters and development of TMDLs in compliance with section 303(d) regulations, and the U.S. EPA is under court order or consent decree in many states to ensure that TMDLs are established.

Source tracking practitioners whose primary goal is to compute TMDLs need to quantify loadings of indicator bacteria from various sources. They have a need to target *E. coli* or *Enterococcus*, depending on what indicator was used to determine that the water was out of compliance, and they need to quantify the relative contribution of pollution sources. An evaluation that takes weeks or months is typically acceptable when developing a TMDL, so rapid or real-time methodology is not critical. Those interested in using MST to develop TMDLs have been a major driving force for the technological advances that have been made over the past few years, as these managers are under pressure to comply with the regulations. Many of these practitioners do not have time to wait until research scientists develop a more ideal method before making source assessments.

In other cases, MST is being sought as a tool to conserve resources while mitigating contamination in an area where the cause of pollution is unknown. In these cases, identification of sources of fecal pollution could suggest different management strategies. Water quality management options for point sources of fecal pollution (e.g., wastewater discharges) generally include treatment options and permit reductions. Options are also available for human and animal fecal pollution entering waters through nonpoint sources. Alternatives include reduction of impervious surfaces and other major hydrologic changes within developed and developing areas as well as strategies to intercept or reduce urban storm water runoff. Pet waste cleanup campaigns are also options, as is reduction of waste dumping from boats through education and stricter enforcement of existing laws. There are also management options for reduction of wildlife sources of fecal bacteria. Storage tank modifications and scarecrow adaptations can be used to discourage overpopulation of birds in some areas. Population control strategies may also be adopted for urban wildlife (raccoons, for example) that may be a nuisance or threat to human populations. Without MST, however, it can be difficult for resource managers to know which remediation option to apply for effective reductions of fecal contamination. For this reason, limited funds can end up being spent implementing well-intentioned but ineffective management strategies.

Environmental professionals whose primary goal is to identify effective remediation strategies typically have flexibility in choosing an MST approach. There is oftentimes an interest or preference in using standard bacterial indicators as method targets, primarily due to their use in environmental monitoring of fecal pollution. However, there are no national mandates to use indicators in choosing management options. While the success of management practices can be measured using conventional bacterial enumeration methods, novel MST tools can be used to evaluate the efficiency of the targeted management practices. Time is also not critical, as a study design that would take days or months to determine sources of pollution is typically acceptable.

Another application for MST is to directly relate contamination to public health risks. Hundreds of food- and waterborne diseases have been described, with symptoms varying depending on etiological agent, immunocompetency of the host, and other factors. Practitioners interested primarily in relating MST data directly to the protection of public health tend to be more interested in identifying human or other particular sources perceived to have high human health risks. Presence/absence data may be acceptable to these practitioners, but rapid detection methods are typically needed. Use of regulated indicators tends to be less critical for these applications, and incorporation of particular pathogens or virulence factors while tracing sources of contamination can be of interest.

Implications for Risk Assessment

MST methods have the potential to greatly improve the understanding of risk associated with exposure to fecal contamination. While it is debatable that human sources of fecal contamination are a more important cause of illnesses than contamination from animals, it is likely that human and animal sources of contamination differ in the types of risks they pose to humans. This would be expected, because pathogens transmitted via the fecal-oral route generally are the cause of different illnesses compared to pathogens transmitted from animals to humans (zoonotic transmission). In terms of human illnesses, variation in the source of contamination could result in a different risk with regard to the frequency or severity of illness or both. As a hypothetical example, human fecal contamination of a recreational water body could result in a higher proportion of ill swimmers than animal fecal contamination of a water body. Although fewer swimmers may become ill following exposure to animal-associated fecal contamination, the severity of illness could be more significant and result in a larger public health burden. It is hypotheses like these that could be tested and evaluated by combining MST methods with epidemiological studies. The incorporation of MST data into microbial risk assessment models, as well as human epidemiological studies, would allow the development of distinct exposure-response relationships and would allow a broader range of hypotheses to be addressed. These abilities would result in greater accuracy in the understanding of human risk and, once the risk resulting from specific exposures is better understood, could also result in the implementation of better, more specific control measures and regulations.

MST would provide better information regarding risk associated with pathogens and fecal indicator organisms. Current bacterial indicators are used to evaluate public health risk associated with waterborne pathogens, although in most cases all they indicate is the previous occurrence of fecal contamination events. The occurrence of certain densities of these fecal organisms is used to establish regulations or guidelines based on risk assessments or epidemiological studies. Currently, regulations do not distinguish between animal and human sources of contamination. If there are truly distinct differences in the exposure response relationships between animal and human sources of contamination, the incorporation of MST data would allow the development of better, more specific regulations.

While several studies have evaluated the ability of various MST methods to accurately predict sources of contamination, few, if any, risk assessments or epidemiological studies have separately quantified the risk associated with human versus animal sources of contamination. In the simplest form, epidemiological studies could incorporate knowledge about the source of a particular form of contamination and evaluate whether this improved or changed the prediction of health effects. For example, one important hypothesis that could easily be tested would be the contention that contamination from human sources represents a greater risk than contamination from animal sources. Epidemiological studies could be designed to evaluate the effectiveness of various MST techniques in predicting illness as well as in evaluating the difference in risk posed by different sources of contamination.

A recent study of exposure to fecal contamination among swimmers conducted in Mission Bay, Calif., is an example where knowledge of the source of contamination was vital to the interpretation of the results (25). The current recreational water quality standards in the United States were developed after epidemiological approaches linked levels of fecal pollution (as monitored by observing the densities of fecal bacteria) with incidence of symptoms of gastrointestinal illness. While numerous studies have confirmed this association (109, 127), the vast majority of these were conducted in waters with a known human source of fecal contamination. Historically, the beaches at Mission Bay have had poor water quality. The entire bay was listed as an impaired water body in 1998 for bacterial exceedences under the Clean Water Act. In efforts to improve water quality and reduce human exposure to pathogens, improvements and repairs were made to the sanitary sewerage system and storm drains were diverted. Often, however, water quality remained in exceedence of current standards.

An epidemiological study, similar to those conducted by the U.S. EPA in the 1970s and 1980s (15, 35), was conducted during the summer of 2004. In this study over 8,000 beachgoers were enrolled, and water quality was measured using bacterial indicators of fecal contamination (*Enterococcus,* fecal coliforms, total coliforms), using both traditional methods and more rapid molecular methods (i.e., quantitative PCR). Bacteriophage, adenovirus, and Norwalk-like viruses, nonbacterial indicators of fecal contamination, were also measured. Beachgoers were interviewed 10 to 14 days after water exposure and were asked about the occurrence of illness after their beach visit. This study, unlike most previous studies, found no association between the level of fecal contamination and the occurrence of gastrointestinal or other illness. Furthermore, the proportion of ill swimmers on days when traditional indicators exceeded current regulations was no different from the proportion of ill swimmers on days when current regulations were met.

To put the results of the Mission Bay study in context, knowledge about the type or source of the fecal contamination is important. An in-depth study of the sources of bacterial contamination was conducted in Mission Bay between 2002 and 2004, immediately prior to the epidemiology study (53). A main goal of the study was to use MST to determine the source of the bacterial contamination. Ribotyping, in conjunction with PCR, was used to determine the source of the contamination in the bay. The ribotyping analysis concluded that 95% of the contamination was from nonhuman sources and implicated storm drains as the major source of the bacterial contamination. The majority of the bacterial contamination was determined to be from avian sources (67%).

The Mission Bay study was the first large-scale study conducted at a beach with poor water quality but with no distinct source of human contamination. Microbial source tracking data confirmed and identified that the main source of contamination was nonhuman. The results of this study indicate the importance of understanding sources of fecal contamination in evaluating the risks associated with recreational water exposure. The study results imply there may be separate and distinct exposure-response relationships between illnesses following recreational water exposure, depending on the source of fecal contamination. Such information can be taken into account when developing risk assessment models and regulations for environmental exposures that rely on bacterial fecal indicators or other indicators of fecal contamination. In addition to recreational water contact, the evaluation of risks related to contamination of shellfish-growing waters and drinking water sources could benefit from MST methods.

Combining MST with risk assessment has proven to be a powerful tool for protection of public health in at least one instance. In Denmark, as part of an ongoing *Salmonella* control program, isolates from human infections were compared with isolates from various sources: pigs, chicken, pork, or eggs. By comparing the isolates from human infections with a large library of subtypes from each source generated through intensive and continuous monitoring, the number of illnesses caused by each source was estimated. This method of intensive surveillance has allowed a more complete understanding of the dynamics of *Salmonella* infection in the region. Moreover, it has allowed the direct evaluation of the benefits of control measures for broiler chickens and eggs. The number of infections attributable to broiler chickens and eggs declined following the implementation of control measures specifically designed to reduce or eliminate exposures from identified sources (128).

The assessment of risk associated with *Cryptosporidium* is another example of a case where MST methodologies can be applied to dramatically improve the understanding of human risk associated with a particular exposure event. *Cryptosporidium* oocysts, an important cause of human illness, have been identified in mice, pigs, marsupials, cattle, sheep, and other mammals (114). Among this host-range diversity, the human *Cryptosporidium hominis* and *C. parvum* genotypes are responsible for the vast majority of human infections and illnesses. It is uncertain whether the noncattle genotypes of *C. parvum* can cause human infections and illness. Moreover, within the *C. parvum* species, different isolates resulted in different dose-response relationships in human infectivity studies (21). There are important geographical differences in the dynamics of *Cryptosporidium* infection. In Australia and North America, most cases are due to the *C. hominis* human genotype, whereas in Europe, the *C. parvum* cattle genotype appears to be more common (123). The issue is even more complicated in environmental samples. As an example, of 12 genotypes identified from a stream in New York, 8 were genotypes whose host was unknown (114). MST techniques which can reliably and accurately characterize the type of *Cryptosporidium* could dramatically change the reaction of a regulatory or public health agency in response to the identification of *Cryptosporidium* in a clinical setting, an outbreak, or an environmental sample.

As the above examples indicate, identification of sources of fecal pollution will suggest different management and prevention strategies. Additional epidemiological studies are needed to quantify the relative risk to human health posed by fecal contamination from different sources. The results of these studies, combined with MST data, would allow more accurate risk assessments within food and water microbiology. Moreover, MST methods are of great value to those addressing key areas related to the fecal pollution continuum (Fig. 4), like pathogen survival and risk exposure levels. Ultimately, successful application of an MST toolbox will require integration of all the key elements within the fecal pollution continuum in order to link sources with potential health effects and health outcomes.

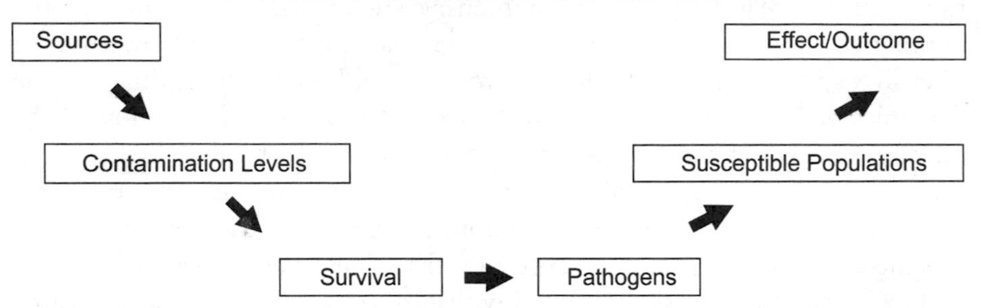

Figure 4 Potential linkages across the fecal-source-to-outcome continuum. Each arrow represents a research area in which MST tools can provide critical information.

Links between Water, Fecal Pollution, the Food Industry, and MST

As mentioned previously, most waterborne and food-borne outbreaks are caused by similar etiological agents. The public sporadically becomes aware of these pathogens when news of outbreaks splashes across their television sets or newspapers. For example, *E. coli* in ground beef or apple juice, hepatitis A in green onion, and *Cryptosporidium* in drinking water all have reached the public consciousness through the popular media. Tracking contamination sources after an outbreak is typically limited to a relatively small number of isolates of a particular pathogenic species, which in most cases is overtly present in the contamination source. This approach minimizes the complexity of the analysis, as it decreases considerably the diversity of clones examined. In contrast, most individuals involved in MST have targeted more diverse populations of fecal microorganisms under varying ecological conditions (Fig. 1). Many MST scientists use indicators of fecal contamination that are commonly present in the animal gut system but which might survive in secondary habitats. In such cases where entire bacterial genera (e.g., enterococci) are used for source identification, the level of complexity increases dramatically, as different species might exhibit different host distribution and different survival rates. It must also be noted that indicators are surrogate organisms and their detection might or might not correlate with the presence of the pathogen(s) of interest. Thus, as they are currently used in MST, results with some host-specific assays do not provide enough information for microbial risk assessment. This is in contrast with the food-related investigations that target pathogens and use this information in prediction models.

Detecting sources of fecal contamination by MST has practical implications for the food industry, particularly in reference to the water used in food preparation and in the irrigation of land used for agriculture and pasture

(63). Since irrigation is necessary for maintaining growing crops, it is in many cases the first time food may have contact with fecal pollution. For example, contamination of lettuce with *E. coli* O157:H7 and *S. enterica* via irrigation has been demonstrated in several studies (64). The problem is more significant in developing countries, where the use of sewage effluent for irrigation is common (92) despite the fact that many countries outlaw the practice. Irrigation using reclaimed water and application of biosolids to grasslands are another potential link between water, food, and fecal pollution. While reclaimed water and biosolids are pretreated, they typically contain fecal indicator bacteria and pathogens that could be taken up by pasturing animals. In turn, these animals could become carriers of human pathogens. Incidentally, this process could complicate source tracking analyses that use host-specific assays dependent on targeting microbial groups with a cosmopolitan lifestyle. The use of reclaimed water is increasingly becoming a water conservation practice in southern and western cities of the United States due to the dwindling capacity of water sources. The need for MST tools that can discriminate between reclaimed water as a source of pollution and other sources of human pollution represents a difficult challenge that has yet to be tackled by individual source tracking technologies.

Composted animal manure is another potential source of pathogens that can contaminate foods and watersheds (8). This is a significant problem due to the fact that even healthy animals can harbor relevant human pathogens, like *E. coli* O157:H7 (75, 81), which is responsible for causing over 70,000 illnesses in the United States annually (112). Hence, proper manure-composting practices are necessary in order to reduce pathogen numbers to acceptable limits (83). Manure application can also have an impact on pathogen persistence in soil, possibly due to the increase in available nutrients (64). Once manures are land applied, the fecally associated organisms can enter natural water systems after rainfall events. Similarly, grazing animals can have a significant impact on the microbial water quality of bodies of water used in recreational activities and as sources of drinking water (62).

Animal drinking water represents a potential source of fecal pathogens that can be transmitted to domesticated animals and subsequently lead to human illnesses. Waterborne *C. jejuni* was implicated in the contamination of a chicken broiler farm, which in turn was linked to a *Campylobacter* outbreak in Bournemouth, United Kingdom (106). *Campylobacter* can also be transmitted to cattle by drinking of contaminated water (56). Cattle water troughs are thought to be good reservoirs of *E. coli* O157:H7 and a source of livestock infection (78). These studies show a direct link between animal water pollution and food-borne outbreaks. Consequently, accurate identification of animal fecal sources contaminating water can provide valuable information leading to the understanding of how particular pathogens contaminate food. This is especially important for zoonotic pathogens, as some bacterial and protozoan agents can inhabit multiple hosts (89).

Reports suggest that adenoviruses and enteroviruses can potentially link water pollution with food-borne illness and enteric disease in animals. Human enteric viruses have been detected in shellfish harvesting areas (42), suggesting that prevention of water pollution from nearby human fecal sources is necessary to minimize shellfish contamination. Water is also a vector for animal viruses, as is evident from the fact that bovine enteroviruses have been detected in river water collected downstream from cattle pasturing fields as reported by Ley et al. (80). In that study, bovine viruses were also detected in oysters receiving contaminated water. Porcine enteroviruses have also been isolated from surface water (31) and drinking water (51). Although the frequency of viral infections of animals via contaminated water is unknown, it is possible that these could have significant economic implications to meat and milk producers.

Another important link between water, food, and fecal pollution relates to the emergence of new or uncommon pathogenic forms associated with food-borne illnesses. The global increase in the emergence of new food-borne diseases may be explained by the increase in international travel and trade, microbial adaptation, changes in food production, and changes in human demographics and lifestyle. Multiple-antibiotic-resistant bacteria are among the most relevant emerging pathogenic groups in both the water and food industries. The excessive uses of antibiotics in clinical settings and for non-therapeutic use in animal husbandry are likely to be responsible for the increase in the numbers of multiple-antibiotic-resistant bacteria in the last decades (119, 132). Antibiotic resistance genes can be transferred between phylogenetically distant bacteria via self-transferable plasmids or other mobile genetic elements (28), increasing the chances for pathogens to acquire resistance and possibly explaining the incidence of multidrug-resistant bacterial outbreaks associated with water and food consumption (93, 95).

Not all emerging food-borne pathogens have also been associated with waterborne illnesses or fecal pollution. A good example is *Enterobacter sakazakii*, a pathogen responsible for neonatal meningitis and necrotizing meningoencephalitis acquired from powdered infant formula (9). Interestingly, this organism has been isolated from fecal samples in the past; however, the extent to which fecal contamination can be directly implicated in any of the *E. sakazakii* food-borne outbreaks is not known. This bacterium has not been associated with any waterborne outbreak, so thus far its primary role in human health is through food-borne outbreaks. Methods that can link emerging food- and waterborne pathogens to specific sources will be relevant tools to reduce the threat of these types of diseases in the future.

CONCLUDING REMARKS

Microbial source tracking is a young field with a simple goal: to develop a tool(s) that discerns the host origins of enteric microorganisms found in environmental samples. Successful application of these tools will constitute a significant advance in environmental health microbiology. This technology will allow us not only to detect fecal contamination using indicators but also to simultaneously determine the source of contamination and pathogens present in both fecal and water samples (Fig. 5). These tools will have widespread applications, including use for regulatory compliance, pollution remediation, and risk assessments. These tools will reduce the incidence of illness associated with food and water and will improve public health and well-being. Potential applications of MST have been driving methods development, and we currently find ourselves with multiple library-dependent and library-independent approaches in various stages of development and validation. No one method has emerged as being superior enough to be adopted as a standard. However, advances in molecular methods and our understanding of the microbial ecology, microbial genetics, and microbial population dy-

Figure 5 Microbial water quality biochip. The main goal of this simulated biochip is to simultaneously determine the presence of indicators of fecal pollution, pathogen-associated genes, and MST markers. This type of information will greatly enhance our current environmental monitoring capabilities and provide data critical for improving risk assessment models. (Modified from the *Journal of Microbiological Methods* [132a] with permission from Elsevier.)

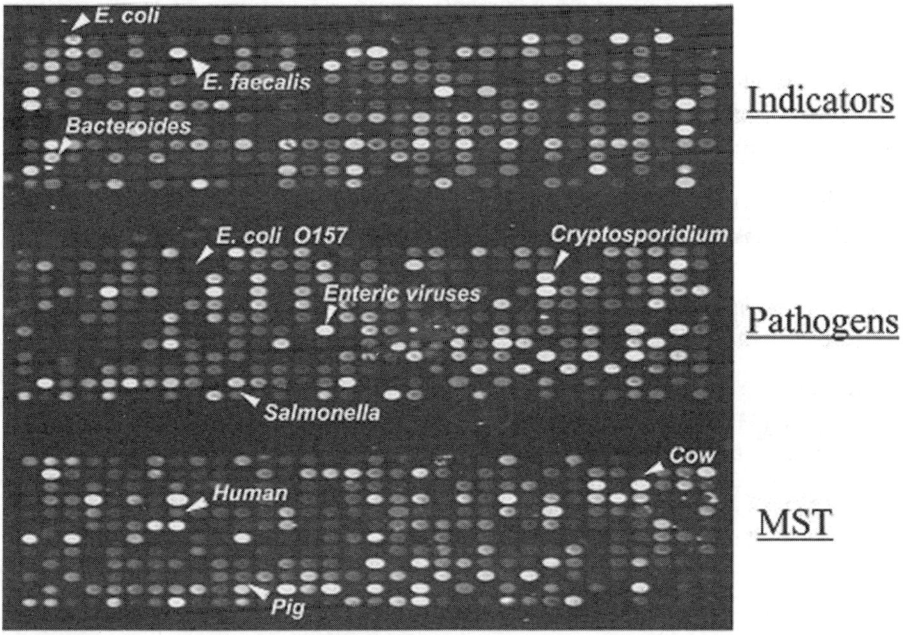

namics of targeted species are likely to allow us to increase the accuracy and ease of microbial source tracking.

REFERENCES

1. **Anderson, A. D., A. G. Heryford, J. P. Sarisky, C. Higgins, S. S. Monroe, R. S. Beard, C. M. Newport, J. L. Cashdollar, G. S. Fout, D. E. Robbins, S. A. Seys, K. J. Musgrave, C. Medus, J. Vinjé, J. S. Bresee, H. M. Mainzer, and R. I. Glass.** 2003. A waterborne outbreak of Norwalk-like virus among snowmobilers—Wyoming, 2001. *J. Infect. Dis.* **187**:303–306.
2. **APHA (American Public Health Association).** 1998. *Standard Methods for the Examination of Water and Wastewater*, 20th ed. American Public Health Association, Washington, D.C.
3. **Ashbolt, N. J., W. O. K. Grabow, and M. Snozzi.** 2001. Indicators of microbial water quality, p. 289–315. *In* L. Fewtrell and J. Bartram (ed.), *Water Quality: Guidelines, Standards and Health.* IWA Publishing, London, United Kingdom.
4. **Auld, H., D. MacIver, and J. Klaassen.** 2004. Heavy rainfall and waterborne disease outbreaks: the Walkerton example. *J. Toxicol. Environ. Health A.* **67**:1879–1887.
5. **Barnes, B., and D. M. Gordon.** 2004. Coliform dynamics and the implications for source tracking. *Environ. Microbiol.* **6**:501–509.
6. **Bernhard, A. E., T. Goyard, M. T. Simonich, and K. G. Field.** 2003. Application of a rapid method for identifying fecal pollution sources in a multi-use estuary. *Water Res.* **37**: 909–913.
7. **BGOSHU (Bruce-Grey-Owen Sound Health Unit).** 2000. Waterborne outbreak of gastroenteritis associated with a contaminated municipal water supply, Walkerton, Ontario, May–June 2000. *Can. Commun. Dis. Rep.* **20**:170–172.
8. **Bicudo, J. R., and S. M. Goyal.** 2003. Pathogens and manure management systems: a review. *Environ. Technol.* **24**:115–130.
9. **Biering, G., S. Karlsson, N. C. Clark, K. E. Jónsdóttir, P. Lúdvígsson, and O. Steingrímsson.** 1989. Three cases of neonatal meningitis caused by *Enterobacter sakazakii* in powdered milk. *J. Clin. Microbiol.* **27**:2054–2056.
10. **Blackburn, B. G., G. F. Craun, J. S. Yoder, V. Hill, R. L. Calderon, N. Chen, S. H. Lee, D. A. Levy, and M. Beach.** 2004. Surveillance for waterborne-disease outbreaks associated with drinking water—United States, 2001–2002. *Morb. Mortal. Wkly. Rep.* **53**:23–45.
11. **Boccia, D., A. E. Tozzi, B. Cotter, C. Rizzo, T. Russo, G. Buttinelli, A. Caprioli, M. L. Marziano, and F. M. Ruggeri.** 2002. Waterborne outbreak of Norwalk-like virus gastroenteritis at a tourist resort, Italy. *Emer. Infect. Dis.* **8**:563–568.
12. **Bonde, G. J.** 1977. Bacterial indicators of water pollution, p. 273–364. *In* M. Droop and H. W. Jannasch (ed.), *Advances in Aquatic Microbiology.* Academic Press, London, United Kingdom.
13. **Bopp, D. J., B. D. Sauders, A. L. Waring, J. Ackelsberg, N. Dumas, E. Braun-Howland, D. Dziewulski, B. J. Wallace, M. Kelly, T. Halse, K. A. Musser, P. F. Smith, D. L. Morse, and R. J. Limberger.** 2003. Detection, isolation, and molecular subtyping of *Escherichia coli* O157:H7 and *Campylobacter jejuni* associated with a large waterborne outbreak. *J. Clin. Microbiol.* **41**:174–180.

14. **Cabelli, V.** 1983. *Health Effects Criteria for Marine Recreational Waters.* U.S. EPA report number EPA-600/1-80-031. U.S. Environmental Protection Agency, Cincinnati, Ohio.
15. **Cabelli, V. J., A. P. Dufour, L. J. McCabe, and M. A. Levin.** 1982. Swimming-associated gastroenteritis and water quality. *Am. J. Epidemiol.* **115**:606–616.
16. **Carson, C. A., B. L. Shear, M. R. Ellersieck, and A. Asfaw.** 2001. Identification of fecal *Escherichia coli* from humans and animals by ribotyping. *Appl. Environ. Microbiol.* **67**: 1503–1507.
17. **Carson, C. A., B. L. Shear, M. R. Ellersieck, and J. D. Schnell.** 2003. Comparison of ribotyping and repetitive extragenic palindromic-PCR for identification of fecal *Escherichia coli* from humans and animals. *Appl. Environ. Microbiol.* **69**:1836–1839.
18. **Centers for Disease Control and Prevention.** 2001. Updated guidelines for evaluating public health surveillance systems. *Morb. Mortal. Wkly. Rep.* **50**:1–35.
19. **Centers for Disease Control and Prevention.** 2005. *Foodborne Illness: Frequently Asked Questions.* [Online.] http://www.cdc.gov/ncidod/dbmd/diseaseinfo/files/foodborne_illness _FAQ.pdf.
20. **Centers for Disease Control and Prevention.** 2005. Preliminary FoodNet data on the incidence of infection with pathogens transmitted commonly through food—10 sites, United States, 2004. *Morb. Mortal. Wkly. Rep.* **54**:352–356.
21. **Chappel, C. L., P. C. Okhuysen, H. L. DuPont, C. R. Sterling, and S. Tzipori.** 2003. *Cryptosporidium parvum* volunteer study: infectivity and immunity, p. 79–81. *In* R. A. Thompson, A. Armson, and U. M. Ryan (ed.), *Cryptosporidium: from Molecules to Disease.* Elsevier, Amsterdam, The Netherlands.
22. **Cheng, P. K. C., D. K. K. Wong, T. W. H. Chung, and W. W. L. Lim.** 2005. Norovirus contamination found in oysters worldwide. *J. Med. Virol.* **76**:593–597.
23. **Clark, R. M., E. E. Geldreich, K. R. Fox, E. W. Rice, C. H. Johnson, J. A. Goodrich, J. A. Barnick, F. Abdesaken, J. E. Hill, and F. J. Angulo.** 1996. A waterborne *Salmonella typhimurium* outbreak in Gideon, Missouri: results from a field investigation. *Int. J. Environ. Health Res.* **6**:187–193.
24. **Cole, D., S. C. Long, and M. D. Sobsey.** 2003. Evaluation of F^+ RNA and DNA coliphages as source-specific indicators of fecal contamination in surface waters. *Appl. Environ. Microbiol.* **69**:6507–6514.
25. **Colford, J. M., T. J. Wade, K. C. Schiff, C. Wright, J. F. Griffith, S. K.. Sukhminder, and S. B. Weisberg.** 2005. *Recreational Water Contact and Illness in Mission Bay, California. Southern California Coastal Water Research Project.* [Online.] http://www.environmental-expert.com/files/19961/articles/4564/4564.pdf.
26. **Colwell, R. R.** 1996. Global climate and infectious disease: the cholera paradigm. *Science* **274**:2025–2031.
27. **Corso, P. S., M. H. Kramer, K. A. Blair, D. G. Addiss, J. P. Davis, and A. C. Haddix.** 2003. Cost of illness in the 1993 waterborne *Cryptosporidium* outbreak, Milwaukee, Wisconsin. *Emerg. Infect. Dis.* **9**:426–431.
28. **Courvalin, P.** 1994. Transfer of antibiotic resistance genes between gram-positive and gram-negative bacteria. *Antimicrob. Agents Chemother.* **38**:1447–1451.
29. **Craun, G. F., R. L. Calderon, and M. F. Craun.** 2004. Waterborne outbreaks caused by zoonotic pathogens in the U.S.A. *In* J. A. Cotruvo, A. Dufour, G. Rees, J. Bartram,

R. Carr, D. O. Cliver, G. F. Craun, R. Fayer, and V. P. J. Gannon (ed.), *Waterborne Zoonoses*. IWA Publishing, London, United Kingdom.

30. **Dawson, D.** 2003. *Foodborne Protozoan Pathogens*. International Life Sciences Institute Press, Washington, D.C.
31. **Derbyshire, J. B., and E. G. Brown.** 1978. Isolation of animal viruses from farm livestock waste, soil and water. *J. Hyg.* (London) **81:**295–302.
32. **de Wit, M. A., M. P. Koopmans, L. M. Kortbeek, W. J. Wannet, J. Vinje, F. van Leusden, A. I. Bartelds, and Y. T. van Duynhoven.** 2001. Sensor, a population-based cohort study on gastroenteritis in The Netherlands: incidence and etiology. *Am. J. Epidemiol.* **154:** 666–674.
33. **Dombek, P. E., L. K. Johnson, S. T. Zimmerley, and M. J. Sadowsky.** 2000. Use of repetitive DNA sequences and the PCR to differentiate *Escherichia coli* isolates from human and animal sources. *Appl. Environ. Microbiol.* **66:**2572–2577.
34. **Dufour, A., M. Snozzi, W. Koster, J. Bartram, E. Ronchi, and L. Fewtrell (ed.).** 2003. *Assessing Microbial Safety of Drinking Water: Improving Approaches and Methods.* World Health Organization. IWA Publishing, London, United Kingdom.
35. **Dufour, A. P.** 1984. *Health Effects Criteria for Fresh Recreational Waters*. U.S. Environmental Protection Agency publication no. EPA-600/1-84-004. Office of Research and Development, U.S. Environmental Protection Agency, Washington, D.C.
36. **Duran, M., B. Z. Haznedaroğlu, and D. H. Zitomer.** 2006. Microbial source tracking using host specific FAME profiles of fecal coliforms. *Water Res.* **40:**67–74.
37. **Edberg, S. C., E. W. Rice, R. J. Karlin, and M. J. Allen.** 2000. *Escherichia coli*: the best biological drinking water indicator for public health protection. *Symp. Ser. Soc. Appl. Microbiol.* **29:**106S–116S.
38. **ERS (Economic Research Service).** 2004. *Economics of Foodborne Disease.* USDA Economic Research Service, Washington, D.C. [Online.] http://www.ers.usda.gov/briefing/FoodborneDisease/features.htm.
39. **Fankhauser, R. L., J. S. Noel, S. S. Monroe, T. Ando, and R. I. Glass.** 1998. Molecular epidemiology of "Norwalk-like viruses" in outbreaks of gastroenteritis in the United States. *J. Infect. Dis.* **178:**1571–1578.
40. **Field, K. G.** 2004. Fecal source identification, p. 19–26. *In* J. A. Cotruvo, A. Dufour, G. Rees, J. Bartram, R. Carr, D. O. Cliver, G. F. Craun, R. Frayer, and V. P. J. Gannon (ed.), *Waterborne Zoonoses: Identification, Causes and Control.* World Health Organization. IWA Publishing, London, United Kingdom.
41. **Field, K. G., E. C. Chern, L. K. Dick, J. Fuhrman, J. Griffith, P. A. Holden, M. G. LaMontagne, J. Le, B. Olson, and M. T. Simonich.** 2003. A comparative study of culture-independent, library-independent genotypic methods of fecal source tracking. *J. Water Health* **1:**181–194.
42. **Fong, T.-T., D. W. Griffin, and E. K. Lipp.** 2005. Molecular assays for targeting human and bovine enteric viruses in coastal waters and their application for library-independent source tracking. *Appl. Environ. Microbiol.* **71:**2070–2078.
43. **Frost, F. J., G. F. Craun, and R. L. Calderon.** 1996. Waterborne disease surveillance. *J. Am. Water Works Assoc.* **88:**66–75.
44. **Funk, D. J., L. Helbling, J. J. Wernegreen, and N. A. Moran.** 2000. Intraspecific phylogenetic congruence among multiple symbiont genomes. *Proc. Biol. Sci.* **267:**2517–2521.

45. **Geldreich, E. E.** 1978. Bacterial populations and indicator concepts in feces, sewage, stormwater and solid wastes, p. 51–97. *In* G. Berg (ed.), *Indicators of Viruses in Water and Food.* Ann Arbor Science Publishers, Inc., Ann Arbor, Mich.

46. **Geldreich, E. E., and B. A. Kenner.** 1969. Concepts of fecal streptococci in stream pollution. *J. Water Pollut. Control Fed.* **41**(Suppl.):R336–R352.

47. **Gibson, C. J., III, C. N. Haas, and J. B. Rose.** 1999. Risk assessment of waterborne protozoa: current status and future trends. *Parasitology* **117**:205–212.

48. **Godfree, A. F., D. Kay, and M. D. Wyer.** 1997. Faecal streptococci as indicators of faecal contamination in water. *J. Appl. Microbiol.* **83**:110S–119S.

49. **Graczyk, T. K., R. Fayer, J. M. Trout, E. J. Lewis, C. A. Farley, I. Sulaiman, and A. A. Lal.** 1998. *Giardia* sp. cysts and infectious *Cryptosporidium parvum* oocysts in the feces of migratory Canada geese (*Branta canadensis*). *Appl. Environ. Microbiol.* **64**:2736–2738.

50. **Graczyk, T. K., A. S. Girouard, L. Tamang, S. P. Nappier, and K. J. Schwab.** 2006. Recovery, bioaccumulation, and inactivation of human waterborne pathogens by the Chesapeake Bay nonnative oyster, *Crassostrea ariakensis*. *Appl. Environ. Microbiol.* **72**: 3390–3395.

51. **Gratacap-Cavallier, B., O. Genoulaz, K. Brengel-Pesce, H. Soule, P. Innocenti-Francillard, M. Bost, L. Gofti, D. Zmirou, and J. M. Seigneurin.** 2000. Detection of human and animal rotavirus sequences in drinking water. *Appl. Environ. Microbiol.* **66**: 2690–2692.

52. **Griffin, D. W., C. J. Gibson III, E. K. Lipp, K. Riley, J. H. Paul III, and J. B. Rose.** 1999. Detection of viral pathogens by reverse transcriptase PCR and of microbial indicators by standard methods in the canals of the Florida Keys. *Appl. Environ. Microbiol.* **65**:4118–4125.

53. **Gruber, S. J., L. M. Kay, R. Kolb, and K. Henry.** 2005. Mission Bay bacterial source identification study. *Stormwater* [Online.] http://www.stormh2o.com/sw_0505_mission.html.

54. **Hagedorn, C., S. L. Robinson, J. R. Filtz, S. M. Grubbs, T. A. Angier, and R. B. Reneau, Jr.** 1999. Determining sources of fecal pollution in a rural Virginia watershed with antibiotic resistance patterns in fecal streptococci. *Appl. Environ. Microbiol.* **65**:5522–5531.

55. **Hagedorn, C., J. B. Crozier, K. A. Mentz, A. M. Booth, A. K. Graves, N. J. Nelson, and R. B. Reneau, Jr.** 2003. Carbon source utilization profiles as a method to identify sources of faecal pollution in water. *J. Appl. Microbiol.* **94**:792–799.

56. **Hanninen, M. L., M. Niskanen, and L. Korhonen.** 1998. Water as a reservoir for *Campylobacter jejuni* infection in cows studied by serotyping and pulsed-field gel electrophoresis (PFGE). *Zentralbl. Veterinarmed. B* **45**:37–42.

57. **Harwood, V. J., J. Whitlock, and V. Withington.** 2000. Classification of antibiotic resistance patterns of indicator bacteria by discriminant analysis: use in predicting the source of fecal contamination in subtropical waters. *Appl. Environ. Microbiol.* **66**:3698–3704.

58. **Haznedaroglu, B. Z., D. H. Zitomer, G. B. Hughes-Strange, and M. Duran.** 2005. Whole-cell fatty acid composition of total coliforms to predict sources of fecal contamination. *J. Environ. Eng.* **131**:1426–1432.

59. **Hedberg, C. W., and M. T. Osterholm.** 1993. Outbreaks of food-borne and waterborne viral gastroenteritis. *Clin. Microbiol. Rev.* **6**:199–210.

60. **Hedlund, B. P., and J. T. Staley.** 2002. Phylogeny of the genus *Simonsiella* and other members of the *Neisseriaceae*. *Int. J. Syst. Evol. Microbiol.* **52**:1377–1382.

61. **Hoebe, C. J. P. A., H. Vennema, A. M. de Roda Husman, and Y. T. H. P. van Duynhoven.** 2004. Norovirus outbreak among primary schoolchildren who had played in a recreational water fountain. *J. Infect. Dis.* **189:**699–705.
62. **Hubbard, R. K., G. L. Newton, and G. M. Hill.** 2004. Water quality and the grazing animal. *J. Anim Sci.* **82:**E255–E263.
63. **Islam, M., J. Morgan, M. P. Doyle, and X. Jiang.** 2004. Fate of *Escherichia coli* O157:H7 in manure compost-amended soil and on carrots and onions grown in an environmentally controlled growth chamber. *J. Food Prot.* **67:**574–578.
64. **Islam, M., M. P. Doyle, S. C. Phatak, P. Millner, and X. Jiang.** 2004. Persistence of enterohemorrhagic *Escherichia coli* O157:H7 in soil and on leaf lettuce and parsley grown in fields treated with contaminated manure composts or irrigation water. *J. Food Prot.* **67:** 1365–1370.
65. **Jenkins, M. B., P. G. Hartel, T. J. Olexa, and J. A. Stuedemann.** 2003. Putative temporal variability of *Escherichia coli* ribotypes from yearling steers. *J. Environ. Qual.* **32:**305–309.
66. **Jiang, S., R. Noble, and W. Chu.** 2001. Human adenoviruses and coliphages in urban runoff-impacted coastal waters of Southern California. *Appl. Environ. Microbiol.* **67:** 179–184.
67. **Jimenez-Clavero, M. A., C. Fernandez, J. A. Ortiz, J. Pro, G. Carbonell, J. V. Tarazona, N. Roblas, and V. Ley.** 2003. Teschoviruses as indicators of porcine fecal contamination of surface water. *Appl. Environ. Microbiol.* **69:**6311–6315.
68. **Johnson, L. K., M. B. Brown, E. A. Carruthers, J. A. Ferguson, P. E. Dombek, and M. J. Sadowsky.** 2004. Sample size, library composition, and genotypic diversity among natural populations of *Escherichia coli* from different animals influence accuracy of determining sources of fecal pollution. *Appl. Environ. Microbiol.* **70:**4478–4485.
69. **Kageyama, T., M. Shinohara, K. Uchida, S. Fukushi, F. B. Hoshino, S. Kojima, R. Takai, T. Oka, N. Takeda, and K. Katayama.** 2004. Coexistence of multiple genotypes, including newly identified genotypes, in outbreaks of gastroenteritis due to *Norovirus* in Japan. *J. Clin. Microbiol.* **42:**2988–2995.
70. **Kassa, H., B. J. Harrington, and M. S. Bisesi.** 2004. Cryptosporidiosis: a brief literature review and update regarding *Cryptosporidium* in feces of Canada geese (*Branta canadensis*). *J. Environ. Health* **66:**34–40.
71. **Kay, D., J. M. Fleisher, R. L. Salmon, F. Jones, M. D. Wyer, A. F. Godfree, Z. Zelenauch-Jacquotte, and R. Shore.** 1994. Predicting likelihood of gastroenteritis from sea bathing: results from randomised exposure. *Lancet* **344:**905–909.
72. **Khatib, L. A., Y. L. Tsai, and B. H. Olson.** 2002. A biomarker for the identification of cattle fecal pollution in water using the LTIIa toxin gene from enterotoxigenic *Escherichia coli*. *Appl. Microbiol. Biotechnol.* **59:**97–104.
73. **Khatib, L. A., Y. L. Tsai, and B. H. Olson.** 2003. A biomarker for the identification of swine fecal pollution in water, using the STII toxin gene from enterotoxigenic *Escherichia coli*. *Appl. Microbiol. Biotechnol.* **63:**231–238.
74. **Kirkwood, C. D., R. Clark, N. Bogdanovic-Sakran, and R. F. Bishop.** 2005. A 5-year study of the prevalence and genetic diversity of human caliciviruses associated with sporadic cases of acute gastroenteritis in young children admitted to hospital in Melbourne, Australia (1998–2002). *J. Med. Virol.* **77:**96–101.
75. **Kudva, I. T., P. G. Hatfield, and C. J. Hovde.** 1997. Characterization of *Escherichia coli* O157:H7 and other Shiga toxin-producing *E. coli* serotypes isolated from sheep. *J. Clin. Microbiol.* **35:**892–899.

76. **Leclerc, H., D. A. Mossel, S. C. Edberg, and C. B. Struijk.** 2001. Advances in the bacteriology of the coliform group: their suitability as markers of microbial water safety. *Annu. Rev. Microbiol.* **55**:201–234.

77. **Leclerc, H., L. Schwartzbrod, and E. Dei-Cas.** 2002. Microbial agents associated with waterborne diseases. *Crit. Rev. Microbiol.* **28**:371–409.

78. **LeJeune, J. T., T. E. Besser, and D. D. Hancock.** 2001. Cattle water troughs as reservoirs of *Escherichia coli* O157. *Appl. Environ. Microbiol.* **67**:3053–3057.

79. **Lewis, G. D., M. W. Loutit, and F. J. Austin.** 1985. A method for detecting human enteroviruses in aquatic sediments. *J. Virol. Methods* **10**:153–162.

80. **Ley, V., J. Higgins, and R. Fayer.** 2002. Bovine enteroviruses as indicators of fecal contamination. *Appl. Environ. Microbiol.* **68**:3455–3461.

81. **Liebana, E., R. P. Smith, M. Batchelor, I. McLaren, C. Cassar, F. A. Clifton-Hadley, and G. A. Paiba.** 2005. Persistence of *Escherichia coli* O157 isolates on bovine farms in England and Wales. *J. Clin. Microbiol.* **43**:898–902.

82. **Lisle, J. T., and J. B. Rose.** 1995. *Cryptosporidium* contamination of water in the U.S.A. and U.K.: a mini-review. *J. Water Supply Res. Technol. - Aqua* **44**:103–117.

83. **Lung, A. J., C.-M. Lin, J. M. Kim, M. R. Marshall, R. Nordstedt, N. P. Thompson, and C. I. Wei.** 2001. Destruction of *Escherichia coli* O157:H7 and *Salmonella enteritidis* in cow manure composting. *J. Food Prot.* **64**:1309–1314.

84. **Mac Kenzie, W. R., N. J. Hoxie, M. E. Proctor, M. S. Gradus, K. A. Blair, D. E. Peterson, J. J. Kazmierczak, D. G. Addiss, K. R. Fox, J. B. Rose, and J. P. Davis.** 1994. A massive outbreak in Milwaukee of *Cryptosporidium* infection transmitted through the public water supply. *N. Engl. J. Med.* **331**:161–167.

85. **Maguire, A. J., J. Green, D. W. Brown, U. Desselberger, and J. J. Gray.** 1999. Molecular epidemiology of outbreaks of gastroenteritis associated with small round-structured viruses in East Anglia, United Kingdom, during the 1996–1997 season. *J. Clin. Microbiol.* **37**:81–89.

86. **Maluquer de Motes, C., P. Clemente-Casares, A. Hundesa, M. Martin, and R. Girones.** 2004. Detection of bovine and porcine adenoviruses for tracing the source of fecal contamination. *Appl. Environ. Microbiol.* **70**:1448–1454.

87. **Mara, D. D., and J. I. Oragui.** 1981. Occurrence of *Rhodococcus coprophilus* and associated actinomycetes in feces, sewage, and freshwater. *Appl. Environ. Microbiol.* **42**:1037–1042.

88. **Martellini, A., P. Payment, and R. Villemur.** 2005. Use of eukaryotic mitochondrial DNA to differentiate human, bovine, porcine and ovine sources in fecally contaminated surface water. *Water Res.* **39**:541–548.

89. **Mathis, A., R. Weber, and P. Deplazes.** 2005. Zoonotic potential of the microsporidia. *Clin. Microbiol. Rev.* **18**:423–445.

90. **McBride, G. B., C. E. Salmondo, D. R. Bandaranayake, S. J. Turner, G. D. Lewis, and D. G. Till.** 1998. Health effects of marine bathing in New Zealand. *Int. J. Environ. Health Res.* **8**:173–189.

91. **Mead, P. S., L. Slutsker, V. Dietz, L. F. McCaig, J. S. Bresee, C. Shapiro, P. M. Griffin, and R. V. Tauxe.** 1999. Food-related illness and death in the United States. *Emerg. Infect. Dis.* **5**:607–625.

92. **Melloul, A. A. V., L. V. Hassani, and L. V. Rafouk.** 2001. *Salmonella* contamination of vegetables irrigated with untreated wastewater. *World J. Microbiol. Biotechnol.* **17**: 207–209.

93. **Mermin, J. H., R. Villar, J. Carpenter, L. Roberts, A. Samaridden, L. Gasanova, S. Lomakina, C. Bopp, L. Hutwagner, P. Mead, B. Ross, and E. D. Mintz.** 1999. A massive epidemic of multidrug-resistant typhoid fever in Tajikistan associated with consumption of municipal water. *J. Infect. Dis.* **179**:1416–1422.

94. **Miescier, J. J., and V. J. Cabelli.** 1982. Enterococci and other microbial indicators in municipal wastewater effluents. *J. Water Pollut. Control Fed.* **41**:164–168.

95. **Molbak, K., D. L. Baggesen, F. M. Aarestrup, J. M. Ebbesen, J. Engberg, K. Frydendahl, P. Gerner-Smidt, A. M. Petersen, and H. C. Wegener.** 1999. An outbreak of multidrug-resistant, quinolone-resistant *Salmonella enterica* serotype typhimurium DT104. *N. Engl. J. Med.* **341**:1420–1425.

96. **Muller, T., A. Ulrich, E. M. Ott, and M. Muller.** 2001. Identification of plant-associated enterococci. *J. Appl. Microbiol.* **91**:268–278.

97. **Nebra, Y., X. Bonjoch, and A. R. Blanch.** 2003. Use of *Bifidobacterium dentium* as an indicator of the origin of fecal water pollution. *Appl. Environ. Microbiol.* **69**:2651–2656.

98. **Noble, R. T., and J. A. Fuhrman.** 2001. Enteroviruses detected by reverse transcriptase polymerase chain reaction from the coastal waters of Santa Monica Bay, California: low correlation to bacterial indicators. *Hydrobiologia* **460**:175–184.

99. **Nygard, K., L. Vold, E. Halvorsen, E. Bringeland, J. A. Rottingen, and P. Aavitsland.** 2004. Waterborne outbreak of gastroenteritis in a religious summer camp in Norway, 2002. *Epidemiol. Infect.* **132**:223–229.

100. **O'Donoghue, P. J.** 1995. *Cryptosporidium* and cryptosporidiosis in man and animals. *Int. J. Parasitol.* **25**:139–195.

101. **Olsen, S. J., L. C. MacKinnon, J. S. Goulding, N. H. Bean, and L. Slutsker.** 2000. Surveillance for foodborne-disease outbreaks—United States, 1993–1997. *Morb. Mortal. Wkly. Rep.* **49**:1–51.

102. **Pacha, R. E., G. W. Clark, E. A. Williams, and A. M. Carter.** 1988. Migratory birds of central Washington as reservoirs of *Campylobacter jejuni*. *Can. J. Microbiol.* **34**:80–82.

103. **Parashar, U. D., L. Dow, R. L. Fankhauser, C. D. Humphrey, J. Miller, T. Ando, K. S. William, C. R. Eddy, J. S. Noel, T. Ingram, J. S. Bresee, S. S. Monroe, and R. I. Glass.** 1998. An outbreak of viral gastroenteritis associated with consumption of sandwiches: implications for the control of transmission by food handlers. *Epidemiol. Infect.* **121**: 615–621.

104. **Parveen, S., R. L. Murphree, L. Edmiston, C. W. Kaspar, K. M. Portier, and M. L. Tamplin.** 1997. Association of multiple-antibiotic-resistance profiles with point and nonpoint sources of *Escherichia coli* in Apalachicola Bay. *Appl. Environ. Microbiol.* **63**: 2607–2612.

105. **Parveen, S., K. M. Portier, K. Robinson, L. Edmiston, and M. L. Tamplin.** 1999. Discriminant analysis of ribotype profiles of *Escherichia coli* for differentiating human and nonhuman sources of fecal pollution. *Appl. Environ. Microbiol.* **65**: 3142–3147.

106. **Pearson, A. D., M. Greenwood, T. D. Healing, D. Rollins, M. Shahamat, J. Donaldson, and R. R. Colwell.** 1993. Colonization of broiler chickens by waterborne *Campylobacter jejuni*. *Appl. Environ. Microbiol.* **59**:987–996.

107. **Peng, M. M., L. Xiao, A. R. Freeman, M. J. Arrowood, A. A. Escalante, A. C. Weltman, C. S. L. Ong, W. R. Mac Kenzie, A. A. Lal, and C. B. Beard.** 1997. Genetic polymorphism among *Cryptosporidium parvum* isolates: evidence of two distinct human transmission cycles. *Emerg. Infect. Dis.* **3**:567–573.

108. **Pourcher, A. M., L. A. Devriese, J. F. Hernandez, and J. M. Delattre.** 1991. Enumeration by a miniaturized method of *Escherichia coli*, *Streptococcus bovis* and enterococci as indicators of the origin of faecal pollution of waters. *J. Appl. Bacteriol.* **70:** 525–530.

109. **Pruss, A.** 1998. Review of epidemiological studies on health effects from exposure to recreational water. *Int. J. Epidemiol.* **27**:1–9.

110. **Puig, A., N. Queralt, J. Jofre, and R. Araujo.** 1999. Diversity of *Bacteroides fragilis* strains in their capacity to recover phages from human and animal wastes and from fecally polluted wastewater. *Appl. Environ. Microbiol.* **65**:1772–1776.

111. **Quiroz, E. S., C. Bern, J. R. MacArthur, L. Xiao, M. Fletcher, M. J. Arrowood, D. K. Shay, M. E. Levy, R. I. Glass, and A. Lal.** 2000. An outbreak of cryptosporidiosis linked to a foodhandler. *J. Infect. Dis.* **181**:695–700.

112. **Rangel, J. M., P. H. Crowe, C. Sparling, P. M. Griffin, and D. L. Swerdlow.** 2005. Epidemiology of *Escherichia coli* O157:H7 outbreaks, United States, 1982–2002. *Emerg. Infect. Dis.* **11**:603–609.

113. **Rose, J. B., and T. R. Slifko.** 1999. *Giardia*, *Cryptosporidium*, and *Cyclospora* and their impact on foods: a review. *J. Food Prot.* **62**:1059–1070.

114. **Ryan, U. M.** 2003. Molecular characterization and taxonomy of *Cryptosporidium*, p. 147–160. *In* R. A. Thompson, A. Armson, and U. M. Ryan (ed.), *Cryptosporidium: from Molecules to Disease*. Elsevier, Amsterdam, The Netherlands.

115. **Scott, T. M., S. Parveen, K. M. Portier, J. B. Rose, M. L. Tamplin, S. R. Farrah, A. Koo, and J. Lukasik.** 2003. Geographical variation in ribotype profiles of *Escherichia coli* isolates from humans, swine, poultry, beef, and dairy cattle in Florida. *Appl. Environ. Microbiol.* **69**:1089–1092.

116. **Scott, T. M., T. M. Jenkins, J. Lukasik, and J. B. Rose.** 2005. Potential use of a host associated molecular marker in *Enterococcus faecium* as an index of human fecal pollution. *Environ. Sci. Technol.* **19**:145–152.

117. **Sharma, S., P. Sachdeva, and J. S. Virdi.** 2003. Emerging water-borne pathogens. *Appl. Microbiol. Biotechnol.* **61**:424–428.

118. **Simmons, G. M., Jr., D. F. Waye, S. Herbein, S. Myers, and E. Walker.** 2000. *Estimating Nonpoint Fecal Coliform Sources in Northern Virginia's Four Mile Run Watershed*. Virginia Department of Environmental Quality, Richmond, Va.

119. **Simonsen, G. S., L. Smabrekke, D. L. Monnet, T. L. Sorensen, J. K. Moller, K. G. Kristinsson, A. Lagerqvist-Widh, E. Torell, A. Digranes, S. Harthug, and A. Sundsfjord.** 2003. Prevalence of resistance to ampicillin, gentamicin and vancomycin in *Enterococcus faecalis* and *Enterococcus faecium* isolates from clinical specimens and use of antimicrobials in five Nordic hospitals. *J. Antimicrob. Chemother.* **51**:323–331.

120. **Snow, J.** 1855. *On the Mode of Communication of Cholera*. John Churchill, London, United Kingdom.

121. **Su, H. P., S. I. Chiu, J. L. Tsai, C. L. Lee, and T. M. Pan.** 2005. Bacterial food-borne illness outbreaks in northern Taiwan, 1995–2001. *J. Infect. Chemother.* **11**:146–151.

122. **Tauxe, R. V.** 1997. Emerging foodborne diseases: an evolving public health challenge. *Emerg. Infect. Dis.* **3**:425–434.

123. **Thompson, R. A.** 2003. The zoonotic potential of *Cryptosporidium*, p. 113–119. *In* R. A. Thompson, A. Armson, and U. M. Ryan (ed.), *Cryptosporidium: from Molecules to Disease*. Elsevier, Amsterdam, The Netherlands.

124. **U.S. Environmental Protection Agency.** 2000. *National Water Quality Inventory.* Report number EPA-841-R-02-001. U.S. EPA Office of Water, Washington, D.C.

125. **U.S. Environmental Protection Agency.** 2005. *Microbial Source Tracking Guide Document.* Document number EPA-600/R-05/064. U.S. EPA Office of Research and Development, Washington, D.C.

126. **Vugia, D., A. Cronquist, J. Hadler, M. Tobin-D'Angelo, D. Blythe, K. Smith, K. Thornton, D. Morse, P. Cieslak, T. Jones, R. Varghese, J. Guzewich, F. Angulo, P. Griffin, R. Tauxe, and J. Dunn.** 2005. Preliminary FoodNet data on the incidence of infection with pathogens commonly transmitted through food—10 sites, United States, 2004. *Morb. Mortal. Wkly. Rep.* **54**:352–356.

127. **Wade, T. J., N. Pai, J. N. Eisenberg, and J. M. J. Colford.** 2003. Do U.S. Environmental Protection Agency water quality guidelines for recreational waters prevent gastrointestinal illness? A systematic review and meta-analysis. *Environ. Health Perspect.* **111**:1102–1109.

128. **Wegener, H. C., T. Hald, D. Lo Fo Wong, M. Madsen, H. Korsgaard, F. Bager, P. Gerner-Smidt, and K. Mølbak.** 2003. *Salmonella* control programs in Denmark. *Emerg. Infect. Dis.* **9**:774–780.

129. **Wheeler, A. L., P. G. Hartel, D. G. Godfrey, J. L. Hill, and W. I. Segars.** 2002. Potential of *Enterococcus faecalis* as a human fecal indicator for microbial source tracking. *J. Environ. Qual.* **31**:1286–1293.

130. **Whitman, R. L., and M. B. Nevers.** 2003. Foreshore sand as a source of *Escherichia coli* in nearshore water of a Lake Michigan beach. *Appl. Environ. Microbiol.* **69**:5555–5562.

131. **Wiggins, B. A., R. W. Andrews, R. A. Conway, C. L. Corr, E. J. Dobratz, D. P. Dougherty, J. R. Eppard, S. R. Knupp, M. C. Limjoco, J. M. Mettenburg, J. M. Rinehardt, J. Sonsino, R. L. Torrijos, and M. E. Zimmerman.** 1999. Use of antibiotic resistance analysis to identify nonpoint sources of fecal pollution. *Appl. Environ. Microbiol.* **65**:3483–3486.

132. **Witte, W.** 2000. Selective pressure by antibiotic use in livestock. *Int. J. Antimicrob. Agents* **16**:S19–S24.

132a. **Ye, R. W., T. Wang, L. Bedzyk, and K. M. Croker.** 2001. Applications of DNA microarrays in microbial systems. *J. Microbiol. Methods* **47**:257–272.

133. **Yoder, J. S., B. G. Blackburn, G. F. Craun, V. Hill, D. A. Levy, N. Chen, S. H. Lee, R. L. Calderon, and M. J. Beach.** 2004. Surveillance for waterborne-disease outbreaks associated with recreational water—United States, 2001–2002. *Morb. Mortal. Wkly. Rep.* **53**: 1–22.

134. **Zhou, L., A. Singh, J. Jiang, and L. Xiao.** 2003. Molecular surveillance of *Cryptosporidium* spp. in raw wastewater in Milwaukee: implications for understanding outbreak occurrence and transmission dynamics. *J. Clin. Microbiol.* **41**:5254–5257.

Microbial Source Tracking
Edited by Jorge W. Santo Domingo and Michael J. Sadowsky

Assumptions and Limitations Associated with Microbial Source Tracking Methods

2

Valerie J. Harwood

OVERVIEW

The goal of this chapter is to provide a critical evaluation of microbial source tracking (MST) methods, including an analysis of the current expectations of MST tools. The ultimate goal of all microbial source tracking studies is to link fecal contamination with its host source, whether contamination is a concern in water or food. Characteristics considered desirable for an "ideal" MST tool are presented in this chapter, followed by a discussion of the assumptions made about tools currently used, research studies that have addressed these assumptions, and known limitations of the tools. The focus is on biological markers (i.e., organisms and their genes); however, chemical markers for fecal sources may well be a useful addition to the source tracking "toolbox" approach (C. Hagedorn, http://filebox.vt.edu/users/chagedor/biol_4684/BST/BST.html). While no attempts will be made to cover the extensive body of literature on food-borne disease and epidemiology, this chapter will emphasize studies that attempt to link fecal contamination of water and food to its probable source(s). Finally, recommendations will be made for testing assumptions in future studies, which may provide some guidance for scientists, resource managers, and other personnel for planning and assessing the value of current and future MST studies.

The various targets of MST studies will be grouped under the umbrella of "source identifiers," each of which may be a chemical, a virus, a bacterium, a group of bacteria, or another microorganism (Table 1). In many MST applications, the source identifier (SI) is subtyped ("fingerprinted") in order to discriminate between particular species, strains, or variants (Fig. 1). The

VALERIE J. HARWOOD, Department of Biology, SCA 110, University of South Florida, 4202 E. Fowler Ave., Tampa, FL 33620.

Table 1 Examples of source identifiers and SPMs used in MST studies, with example references

Source identifier	SPMs	Reference(s)
E. coli	Antibiotic resistance pattern	45, 75
	Ribotype	76, 85
	Rep-PCR pattern (including BOX-PCR)	17, 52
	PFGE pattern	64, 71, 97
	RAPD pattern[a]	4
Genus *Enterococcus*	Antibiotic resistance pattern	45, 108
	Ribotype	24, 96
	Rep-PCR pattern (including BOX-PCR)	15
	esp gene (enterococcal surface protein)	84
Bacteroidetes	16S rRNA gene	7
Bifidobacterium spp.	16S rRNA gene (PCR) of *B. adolescentis* and *B. dentium*	12
Campylobacter spp.	RAPD pattern	113
Salmonella spp.	PFGE pattern	25
Listeria monocytogenes	Virulence genes and ribotype	53
F-specific RNA coliphage	Serotype	21
	Genotype	49, 82
Enterovirus	Reverse transcriptase PCR	73
Adenovirus	Nested PCR	73
Human polyomavirus	PCR	68
Yeast[b]	RAPD pattern and microsatellite analysis	77

[a]RAPD, randomly amplified polymorphic DNA.
[b]*Saccharomyces cerevisiae.*

overriding hypothesis of MST methods is that some of the SI variants are associated with a particular host species or host group. Fingerprints (patterns) may be created by phenotypic methods (e.g., antibiotic resistance or carbon source utilization) or genotypic methods (i.e., ribotyping, repetitive extragenic palindromic [rep-PCR], or pulsed-field gel electrophoresis [PFGE]). Other discriminatory characteristics of SIs used in MST include specific genes or DNA sequences (genetic markers), which are generally detected by PCR amplification. These discriminatory characteristics will be grouped under the term species/pattern/marker (SPM). Characteristics of ideal SIs and SPMs are compared below with the more realistic expectations for good or useful SIs and SPMs (Table 2).

SPECIFIC PARAMETERS THAT INFLUENCE METHOD PERFORMANCE

Host Specificity

Completely unambiguous identification of fecal contamination from a given host species would require absolute host specificity of an SPM, i.e., that it be

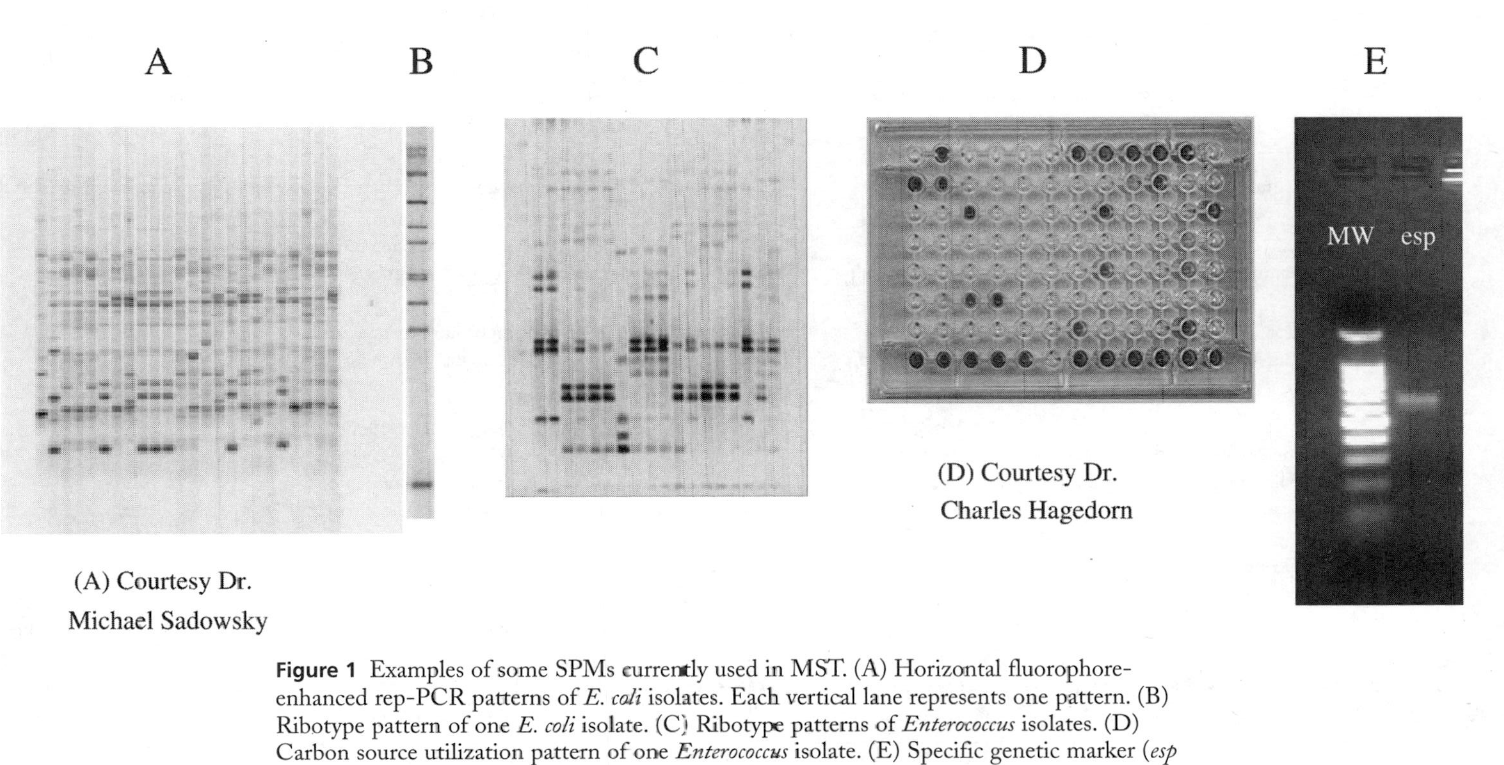

Figure 1 Examples of some SPMs currently used in MST. (A) Horizontal fluorophore-enhanced rep-PCR patterns of *E. coli* isolates. Each vertical lane represents one pattern. (B) Ribotype pattern of one *E. coli* isolate. (C) Ribotype patterns of *Enterococcus* isolates. (D) Carbon source utilization pattern of one *Enterococcus* isolate. (E) Specific genetic marker (*esp* of *E. faecium*) amplified by PCR of *Enterococcus* DNA. MW, molecular weight.

Table 2 Characteristics of an ideal source identifier (SI) compared to those of a useful SI

Characteristic	Ideal SI	Useful SI
Host specificity	SPM found only in one host species	SPM is differentially distributed among host species
Distribution in host population	Found in all members of all populations of a host species	Consistently found in fecal material of host species that could impact the target sites
Stability of SPM[a]	Not subject to mutation or methodological variability that would preclude detection	Rarely subject to mutation; methodology as defined reproducibility[b]
Temporal stability in host	Frequency in host individuals and populations does not change over time	If frequency in individual hosts varies, it is balanced by lack of variation in host populations
Geographic range/stability	The frequencies of SPMs in geographically separated host populations are similar	SPM associated with a particular host can be consistently identified across the geographic area to be studied
Diversity	The diversity of the SI can be represented by a small sample size	The diversity of the SI can be represented by a reasonable sample size
Survival in water and food		
Rate of decay	Consistent decay rate in various matrices and habitats; no growth under any conditions	Predictable decay rate in various matrices and habitats; no growth under ambient conditions; all SPMs decay at the same rate after leaving host
Abundance in 1° vs 2° habitat	The distributions of SPMs in source material, i.e., feces, remain constant upon delivery to water or food (target sites)	The distributions of SPMs in target sites are similar to those found in contaminating fecal material
Quantitative assessment	The relative and absolute contribution of each host to contamination can be assessed	Accurately indicates presence/absence of contamination source
Relevance to regulatory tools	The SI itself is also a regulatory parameter	The SI is correlated with a regulatory parameter
Relevance to health risk	The SI itself constitutes a health risk	The SI is correlated with health risk

[a]Strain/pattern/marker is abbreviated SPM.
[b]Methodological reproducibility refers to the ability to generate the same pattern (i.e., a DNA or phenotypic profile) or result (i.e., PCR +/−) from independent assays.

found in only one host species. Host specificity of commensal and mutualistic microflora of the gastrointestinal (GI) tract may, however, be the exception rather than the rule, particularly among the culturable fraction of the inhabitants of the GI tract. MST methods as related to food are also concerned with direct detection and tracking of the source of food-borne pathogens, some of which are zoonotic and are easily transmitted from animal hosts to humans. When the source of pathogens must be determined in food, MST may follow the route of epidemiological studies, in which identification of one pathogen subtype is sought. In contrast, the safety of food preparations can also benefit from MST methods utilizing population-based approaches, in which a more general picture of the source(s) of fecal contamination is desired.

A central assumption of MST methodologies is that certain SIs are represented by some SPMs that are host specific, or are at least differentially distributed in host species (1, 46). The term differential distribution recognizes that certain SPMs are found at higher frequency, and possibly at a greater density in certain hosts, but does not imply absolute host specificity. Strains that inhabit multiple host types have been termed "transient" (46, 71), a term borrowed from earlier work on *Escherichia coli* population dynamics; however, in the population dynamics literature the term "transient" has a different meaning, as it describes SPMs that were sampled only once or infrequently from a host individual (reviewed in reference 43). Others have utilized the term "cosmopolitan" to describe the ability of a bacterial strain to reside in more than one host type (27, 105), thus referring to the organism's ability to inhabit various host species. Because the term cosmopolitan implies nothing about the length of the habitation in the host and does not conflict with previous uses in the literature, it may be preferred over the term "transient" when describing SPMs that inhabit multiple host types. It should be noted that apparent lack of host specificity (observation of an SPM in more than one host group) could be due to insufficient discrimination by the typing method used, as less-discriminatory typing methods may group SPMs that are actually different. However, even highly discriminatory typing methods, such as PFGE, have identified cosmopolitan *E. coli* strains (97), and sequence analysis of *Bacteroidales* 16S rRNA genes revealed many cosmopolitan phylotypes among the genera *Bacteroides* and *Prevotella* (22).

The cosmopolitan distribution of many SPMs can represent a confounding factor for MST applications, as they do not contribute to identification of fecal source and may cause inaccurate source assignments. As mentioned above, the ability to discriminate between SPMs depends on the method utilized for subtyping (39, 52); however, studies using several different methods have documented cosmopolitan distribution of *E. coli* subtypes. Multilocus enzyme electrophoresis of *E. coli* isolates revealed that 24 of 270 electrophoretic types were found in more than one (up to seven) distinct hosts (74). Genotyping by rep-PCR revealed some identical *E. coli* subtypes in gull feces and sewage (67). Twenty-two percent of all distinct *E. coli* ribotypes (typed by the two-enzyme method) found in analysis of strains from cattle, chickens, horses, and swine were shared by some combination of host species (41), representing 66% of all isolates typed. SPM sharing among host species has been compared using ribotyping and antibiotic resistance analysis (ARA) of *E. coli* (2). Antibiotic resistance patterns (n = 857) were more frequently shared among different host species than ribotypes (n = 222), but 22.5% of all *E. coli* ribotypes were shared between cattle and horses, and 6.5% of ribotypes were shared between cattle, horses, and human hosts. These data suggest that the cosmopolitan distribution of many SPMs defined using these methods may be the norm rather than the exception.

The F-specific RNA coliphages also display cosmopolitan distribution. Three serotypes (types I, II, and III) were found in municipal wastewater (21), although groups II and III were the dominant serotypes. Members of groups I, II, and III were also found in animal feces, and group IV coliphages were found only in animal feces. Similar results were obtained by genetic typing of F-specific RNA coliphages (83). Genotypes II and III have been reported to be associated with human feces, while types I and IV are associated with animal species, but the specificity is not absolute. In Florida, F-specific RNA coliphages isolated from sewage were predominantly genotype II, although genotypes I and III were also isolated (37).

Escherichia coli, *Enterococcus* spp., and coliphages are commensal fecal microorganisms that are broadly distributed in feces and are widely used by the regulatory and MST community. Consequently, it is clear that a better understanding of the cosmopolitan distribution of these organisms in host species is needed if they are to be successfully used in MST methods. Furthermore, as SI/SPMs are proposed for new methods, their host specificity and host range should be fully explored.

Many typing methods used in source tracking studies rely on DNA sequences with unknown function (e.g., rep-PCR) or ubiquitous functionality that is shared among all organisms (e.g., small-subunit rRNA). However, it has been hypothesized that markers that contribute to the specific interaction between host and fecal microbe may have a greater probability of being host specific than do "housekeeping" genes (92). Candidate markers include the genes that code for microbial appendages such as pili and adhesins, which mediate attachment to cells of the host gastrointestinal tract. One method capitalizing on this approach is PCR amplification of an *Enterococcus faecium* surface protein gene (*esp*), which may be human specific (or found at comparatively high frequency in enterococci from human hosts) (84).

Enteric viruses, which rely on specific cell surface receptors to bind to host cells, have the advantage of being inherently species specific and have been used to assess the presence of human fecal contamination in environmental waters and MST studies (36, 51). Recently, reverse transcriptase PCR, and/or nested PCR, was used to assess the presence of human adenoviruses, human enteric viruses, and bovine enteric viruses in river water in Georgia (29). No testing for specificity of the methods on fecal or waste samples was performed, although limited testing against two nontarget viral groups was carried out. A real-time PCR method for adenoviruses, with a detection limit of 100 plaque-forming units (PFU), has been developed, which was also tested on sewage and seawater samples (47). A possible drawback of using a genetic marker associated with a pathogen is that the prevalence of many pathogens in a host population may be very low. Such markers may be frequently detected in sewage samples that represent a large population, but not in feces from individual humans (27, 73).

Distribution in Host Populations

It is tempting to assume that the observed distribution of SPMs in host populations sampled in a given study is applicable to other host populations. This notion is central to efforts to form libraries of data (e.g., PCR fingerprints, ribotypes, genotypes, or phenotypes) that extend across broad geographic regions (e.g., statewide, nationwide). Questions about the distribution of SPMs in host populations apply both to the patterns that make up the libraries and to species or markers that are used for library-independent MST methods. Testing the assumptions that are made about these distributions is crucial to the future success of MST methods. This discussion will focus on assumptions about SPM distribution at the following levels:

- Genes of pathogens and other SPMs that are found at low frequency in host populations are useful for MST.
- SPM distribution in a given host population is consistent across wide geographic areas.
- The host specificity of SPMs does not vary across geographic areas.
- The temporal variability of SPM distribution in hosts is limited.

An MST tool that is adopted for beach water quality monitoring, total maximum daily load (TMDL) assessment, and food quality programs throughout the United States should meet the last three assumptions listed; otherwise, extensive regional testing and updating of methods will be required, which may make the methods so expensive as to be nonadoptable.

The use of relatively rare SPMs for MST studies must take into account SPM distribution, both when the study is designed and when data are interpreted. Pathogenic microorganisms are not expected to be as widely distributed in host populations as commensal organisms; therefore, markers associated with them may suffer from drawbacks due to limited distribution. Such markers may be frequently detected in sewage samples that represent a large population, but not in feces from individual humans (27, 38, 73). A multilaboratory comparison of MST methods found that nested PCR amplification of *E. coli* toxin genes (the heat-stable enterotoxin STIa and the heat-labile toxin LTII) was able to identify water samples seeded with sewage, but the toxin genes were not frequently detected in samples containing feces from human individuals (27). Similar results were found in the same study for adenoviruses and enteroviruses (38, 73). However, some studies have provided evidence that the genes of certain pathogens have a sufficiently frequent distribution, at least on a herd or population level, to be useful for MST applications. Assays on the majority of river water samples in a study done in Georgia (29) were positive for at least one human viral group. In a study conducted in Spain (65), pooled fecal samples from 7 pig and 2 cattle farms and 11 of 12 sewage samples were positive for porcine adenovirus, bovine adenovirus, and human adenovirus, respectively. No cross-amplification by primers designed for viruses of a given host species on fecal material from another host species was observed, although specificity testing was limited to porcine, bovine, and human fecal material. The prevalence of animal-specific *E. coli* toxin genes LTIIa and STII was measured in animal waste material from farms in several states (57, 58). More than 93% of samples from cattle waste lagoons were positive for the cattle-specific LTIIa marker when $<10^3$ *E. coli* bacteria were screened, and the frequency of positive results rose to 100% when $>10^5$ *E. coli* bacteria were screened (57). The swine-specific STII marker was found in 100% of samples when 35 *E. coli* colonies were screened (58).

The hypothesis that SPMs of commensal SIs have widespread distribution in the gastrointestinal tracts of their respective hosts must be rigorously tested. In Europe a bacteriophage that infects *Bacteroides fragilis* HSP40 was found only in human sewage and in sewage-contaminated waters (98). While this bacteriophage was considered a promising candidate as a human-specific fecal marker, its limited distribution in sewage (86) and the relative difficulty of the method (62) have curtailed its use in the United States. F-specific coliphages are common in sewage, but it has been estimated that only ~3% of humans carry this type of phage (reviewed in reference 62). In the multilaboratory study mentioned above, F-specific coliphage genotyping was able to identify water contaminated with sewage, but not water contaminated with feces from individual humans (38, 73).

Ideally, host-specific SPMs should be present at about the same density in separate populations of a given host species, which would provide greater confidence that sampling effort was adequate when using standardized protocols. Furthermore, it would be advantageous if host-specific SPMs were found at about the same density in various individual animals within a host population, which would facilitate accurate quantification. For the majority of MST methods, very little is known about these concerns.

Geographic variability in *E. coli* populations isolated from various hosts in three regions of Florida was explored by ribotyping using a one-enzyme method (85). The authors concluded that geographic variability had a negative effect on correct classification of library isolates; however, they did not specifically assess the effect of geographic separation on SPM distribution. When waste lagoons at cattle and pig farms were tested for the presence of *E. coli* carrying virulence factors (genes for STII and LTIIa and *stxI*, *stxII*, and *eaeA*) (19), some combination of these genes could always be detected by nested PCR, although the prevalence of the genes suggested that the majority of individual animals did not shed *E. coli* carrying these virulence factors. Studies conducted in the United States (21) and in South Africa and Spain (83) have shown that genotypes II and III are predominantly associated with human feces and that genotypes I and IV are predominantly associated with animal feces. Thus, the distribution of F-specific RNA coliphage genotypes appears to be consistent across geographic ranges.

Temporal Stability in Host Populations

The ideal SI and its associated SPM(s) would be highly stable within individual host animals and within host populations over time, i.e., it would be sampled at a consistent frequency from the host population, even when samples were taken years apart. On the other hand, temporal variability of SPMs in a host population may have many contributing factors, including the following: (i) a naturally rapid turnover rate; (ii) changes in microbial populations due to environmental perturbations such as dietary changes, exposure of the host to a novel environment or a new group (herd), or exposure of the host to antimicrobial agents; and (iii) changes in host genotype or phenotype that alter their ability to support certain SPMs (gain or loss).

Of the commensal and mutualistic microflora of the mature GI tract, *E. coli* population dynamics are the most completely explored. The temporal variability of *E. coli* populations observed in the feces of one human individual established the concept of transient versus resident types, initially using multilocus enzyme electrophoresis (MLEE) as a typing method. Over an 11-month study, Caugant et al. (18) defined a "transient" population as one observed to occur once or rarely within the individual, while a "resident" population was one observed multiple times over an extended period. Most *E. coli* types observed were transient, and diversity was high (53 electrophoretic types from 550 isolates). The authors concluded that the observed diversity was very likely to be due to successive invasions of new electrophoretic types and unlikely to be due to genetic exchange among residents. The considerable evidence for the transient nature of most *E. coli* subtypes was reviewed over 20 years ago (43), and more recently by Gordon (32). The number of MLEE subtypes in natural populations of *E. coli* has been estimated to be 100 to 1,000 per host species (87) and was placed at ~55 in feral house mice in Australia (31). Work is needed to determine whether such high subtype diversity is comparable in other SIs. Molecular methods have provided supporting evidence on the transience of *E. coli* subtypes in host species such as humans, cattle and horses. A study on temporal stability of *E. coli* in humans, cattle, and horses defined a "persistent" SPM as one that was sampled from an individual in at least two sampling events (2). The SPMs analyzed were ribotypes (one-enzyme method) and antibiotic resistance patterns. At least one persistent ribotype was observed per human, although only 4 of 36 (11%) of the ribotypes observed in the three humans were persistent. *E. coli* populations in horses and cattle tended to display higher diversity (more subtypes per host) than those in humans; however, they followed a similar trend in that most of the *E. coli* subtypes (both ribotypes and antibiotic resistance patterns) observed were not persistent (2). Restriction fragment length polymorphism-PCR of the *fliC* gene also revealed temporal variability in the *E. coli* population from an individual cow (5). These studies indicate a high probability that the *E. coli* subtype(s) obtained from a single host at a given instant in time is not representative of the *E. coli* population in the animal's feces over longer time periods. Such a limitation has major repercussions in the establishment of host origin libraries, which may require continuous updating in order for a particular MST methodology to be able to track the host species over an extended period of time (50).

Several lines of evidence demonstrate that *E. coli* SPMs are not temporally stable within host individuals, and therefore *E. coli* does not meet the temporal-stability criterion for an ideal SI. However, temporal stability at the level of the host population is a criterion for a useful SI. In a recent study on the temporal stability of *E. coli* ribotypes in cattle, individual animals were sampled at random over four sample events (50). The "resident" *E. coli* ribotypes (observed in more than one sample event) represented only 8.3% of 240 ribotypes. Among the 20 resident ribotypes, none was found at all four sampling times or in all of the steers sampled. Although many *E. coli* isolates were analyzed per cow (about 11 to 25), individual cattle were not resampled throughout the study. Thus, it can be argued that the observed variability was as likely due to undersampling of individuals in the herd as to temporal variability. However, in support of the above results are data from an 8-month study of three beef cattle from one herd (2, 3) that were repeatedly sampled. Variability in *E. coli* ribotypes of strains isolated from fecal material from these animals was high, sharing between herd members was low, and temporal variability in the dominant ribotypes within each animal was consistently noted. Evidence of temporal variability of *E. coli* populations in other host species has also been observed in humans and horses (2, 3). In contrast, results from a large-scale study on *Enterococcus* spp. were more encouraging, as the temporal stability of a large library of *Enterococcus* spp. subtyped by antibiotic resistance analysis was demonstrated for up to a year (109). However, this method may not have been able to discriminate subtypes to the same level as other methods used.

There is evidence that dietary changes can affect the population structure of *E. coli* in host GI tracts. Serotyping of *E. coli* isolates from cattle that were fed different diets revealed that (i) diet impacted *E. coli* diversity and (ii) that serotypes of *E. coli* isolated from cattle whose diets included roughage were clustered more closely than those of cattle fed only grain (9). The diversity and distribution of *E. coli* ribotypes differed in captive versus wild deer (42), which was attributed to diet.

Most MST methods assume that mutations in host individuals that can alter the specificity of an SPM are very rare. An individual could lose the ability to support the SPM if, for example, a receptor in the gastrointestinal tract experienced decreased affinity for the SPM. Conversely, an individual from a different species might acquire the ability to support the SPM by mutation or horizontal gene transfer. While a recent mutation in a host population would not be of major concern, because few individuals would carry the mutation, over generations it could pose a problem, particularly in isolated host populations.

Stability of the SPM

A useful SPM must have a "signal" that is stable in water or food, whether that signal is a phenotypic pattern, a genetic pattern, or a PCR amplicon. In this case, "stability" of the SPM is used in a narrow sense, i.e., does the antibiotic resistance pattern, ribotype, or genetic marker remain associated with the organism, and are its characteristics (e.g., fingerprint, size) consistent after the SI spends some time outside the gastrointestinal tract? The assumption that genetic patterns/markers represent more stable signals than phenotypic patterns has appeared frequently in MST literature (76, 86, 92). The lines of reasoning behind this assumption include that (i) bacterial phenotypes are influenced by environmental conditions as well as the genetic makeup of the organisms, and (ii) traits such as antibiotic resistance are strongly influenced by selective pressure. Although both lines of reasoning have some validity, the dominant assumption they evoke, that phenotypic SPMs are more likely to be influenced by environmental parameters than genotypic SPMs, has not been systematically tested. It is important to note that questions about environmental influences on phenotypic SPMs are not concerned with differences in gene expression, since the assays are conducted in the laboratory under controlled conditions. Rather, the relevant questions about stability of the signal apply to the maintenance of the genes that encode the antibiotic-resistant phenotype.

The stability of antibiotic resistance phenotypes is important on several levels that have yet to be adequately addressed. While it is known that *E. coli* strains can lose plasmid-encoded resistance to antibiotics when selective pressure is removed (30, 104), it is unknown whether this phenomenon occurs often enough to significantly impact the accuracy of MST methods such as antibiotic resistance analysis. Similarly, the consequences of transfer of antibiotic resistance genes from one SI to another have not been established for MST purposes. A study on the stability of antibiotic resistance phenotypes in *E. coli* found that the antibiotic resistance pattern of only 1 of 35 isolates (2.9%) changed upon subculture (72). Vancomycin-resistant enterococci persisted in livestock populations in Europe long after application of the selective agent (pressure), avoparcin, was discontinued (13). Resistance to vancomycin and erythromycin was maintained in *Enterococcus* populations in raw and treated sewage (102). Some argue that the change from antibiotic-resistant to -susceptible phenotype is slowed by many factors that contribute maintenance of antibiotic resistance (48, 80). The cost of fitness (reduced competitiveness) associated with some modes of antibiotic resistance may be decreased or eliminated by compensatory mutations in other genes (10, 11).

The gene(s) that encodes ribosomal RNA (rRNA) is frequently targeted for MST studies (17, 76), in part because significant mutations in this gene(s) influence growth and survival and thus are likely to be fatal to bacteria in which they occur. The low mutation rates of rRNA genes do contribute to the stability of many types of fingerprints; in fact, sequencing of rRNA genes within a species such as *E. coli* generally results in very little strain discrimination (39). Ribotyping, as it is used as an MST method, should be clarified as "genomic ribotyping", which is not dependent solely on variability within rRNA operons but also detects changes within the genome. Genomic ribotyping begins with purification of chromosomal DNA, which is cut with one or more restriction enzymes. Chromosomal fragments are separated by electrophoresis, and probes consisting of labeled fragments of the rRNA gene(s) are hybridized to the chromosomal DNA fragments. This method can be quite discriminatory, even within a species, because much of the variability in patterns is due to variation outside the conserved rRNA operons. One study on *E. coli* found that ribotypes were stable over 1 week of incubation in pond water (1). Ribotypes probably represent a very stable SPM; however, no comparisons with phenotypic or other genotypic methods have been published.

The stability of other genotypic SPMs has been explored. The PFGE pattern of one *E. coli* O157:H7 isolate was monitored over 110 subcultures and did not change (101). BOX-PCR and 16-to-23S rRNA intergenic spacer region PCR fingerprints of *E. coli* isolated from cows and dogs were stable over a 150-day incubation period (89). In contrast, the PFGE pattern of one *E. coli* isolate did change slightly over an 8-week incubation period in irrigation water (64). The persistence of *Bacteroides distasonis* DNA was also influenced by temperature, as persistence of up to 5 days was noted in cooler 14°C water. However, the PCR assay was only able to detect *B. distasonis* for 1 day when incubation was carried out at 30°C (60). Stability of the human-specific *Bacteroides* signal in fresh water was assessed by real-time PCR (88). When mesocosms inoculated with sewage were incubated at 4°C, 12°C, and 28°C, the marker was detected for up to 8 days at 28°C, and for up to 24 days at the higher temperatures.

Much less is known about the population biology of SIs other than *E. coli*. The population biology of antibiotic-resistant *Enterococcus* spp. (44, 90, 110, 112) has received some attention, but most aspects of these studies are not directly pertinent to this discussion, as they focus on a very minor component of the *Enterococcus* population in host species.

Influence of Geography on SI/SPM Distribution

An ideal SI and associated SPMs should be broadly and consistently distributed across geographic distances. Although it has been argued that geographic variability is desirable for MST (32), this assertion assumes that discrimination between geographically separate populations of host animals is a goal. If this level of discrimination were attainable it would be useful; however, most would agree that discrimination between geographically separate populations is not a practical goal at this stage of development in the MST field. Several pertinent assumptions that apply to the geographic distribution of an ideal SI can be identified: (i) SPMs sampled from one population of a host species are similar to SPMs sampled from another population of the same host species, and (ii) SPMs sampled from host populations separated by broad geographic distances are highly similar.

A small but significant amount of the variability in *E. coli* populations isolated from humans can be attributed to geographic separation (18, 107). This limited geographic effect may be due, in part, to the mobility of human populations (32). Caugant et al. (18) reported that little geographic structure was observed in *E. coli* populations isolated from families living in one city; 6% of the genetic variability was explained by geographic distance, and only 1% was explained by geographic distance for families living in different cities. Geographic structure accounted for only a small percentage (2%) of the variability in *E. coli* subtypes in mice (31). While studies that have compared *E. coli* population structure in various animals have found significant contributions to diversity from both geographic location and host species (35, 94), only a small percentage of the variability (<20%) was accounted for by these factors. Geographic structure in *E. coli* populations of cattle and horses was observed in one U.S. study, and more ribotypes were shared in host populations in closer geographic proximity. However, no geographic structure was observed for *E. coli* from chickens and swine (41). An Australian study on the effect of geography on the prevalence of *E. coli* in animal feces reported that mammals living in desert regions were least likely to harbor *E. coli,* followed by mammals in tropical regions. *E. coli* was most prevalent in mammals residing in semiarid or temperate regions (34).

If library-based methods are to be widely applicable in the United States, databases should be applicable across broad geographic ranges. Thus, in an ideal case, host populations in all parts of the United States would share similar SIs and SPMs. In spite of the limited contribution of geographic separation to *E. coli* population structure noted above, studies completed to date suggest that this ideal will not be met, particularly in the case of library-based methods. In a study performed in Florida, *E. coli* isolates from beef, dairy, poultry, swine, and human hosts that were geographically separated were ribotyped after HindIII digestion (85). Although the method accurately differentiated *E. coli* of human versus nonhuman origin, it could not differentiate the source of isolates from the various nonhuman host species. The diets of host animals may differ significantly by geographic region, providing one of the drivers for geographic variability of commensal bacterial populations within a host species. Libraries made from *E. coli* and *Enterococcus* strains from three geographic regions were assessed for broad geographic applicability (24) using ARA, ribotyping (one-enzyme), and PFGE subtyping methods. The regional sublibraries (Florida; Shenandoah Valley, Virginia; and southwest Virginia) identified isolates collected from within the region significantly more accurately than they identified isolates from outside the region. A three-region merged library identified the source of isolates much less accurately than each of the regional libraries, and this generalization held true for each of the methods and SIs.

The predictive accuracy of an *Enterococcus* ARA library over geographic ranges was broadened by increasing library size and representation of isolates from a number of watersheds in the Shenandoah Valley region of Virginia (109). Six watershed-specific libraries were merged to produce a library of 6,587 isolates, which had an internal accuracy (average rate of correct classification) of 57% when sources were grouped into three categories (human, domesticated animals, and wild animals). The geographic range of the merged library was limited, as it identified isolates from southwest Virginia and Florida significantly less accurately than it did isolates collected within the region.

Representative Sampling

In addition to the aspects of sampling design such as sample number, sample size, frequency of sampling, and site placement that influence the success of any study, one of the most important assumptions of library-based MST methods is that the SI population can be adequately sampled so that all (or most) SPMs are represented. The assumption of representative sampling is extremely important with respect to sampling of both host fecal material and SPMs in water samples. Many factors impose limits on the amount of material or isolates that can be analyzed, including cost and time. Undersampling of SI populations in fecal sources leads to nonrepresentative libraries, which may have high correct-classification rates (internal accuracy) but low predictive accuracy for isolates that are not included in the library (105, 109). Furthermore, nonrepresentative libraries are almost certain to be highly variable from both temporal and geographic standpoints. Sampling design is also a crucial component of non-library-based methods, and the factors to be considered are similar to those mentioned above that are important for any environmental or ecological study.

Various estimates of *E. coli* subtype diversity within host populations have been advanced, and these generally fall between 50 and 1,000 subtypes (69, 87). Feces from individual gulls at Lake Michigan beaches contained a median of 8 unique *E. coli* subtypes per 10 isolates typed by rep-PCR (28). *E. coli* populations in gull feces, sewage, and beach water in Wisconsin were assessed by PFGE and found to be highly diverse (66). Biochemical patterns (Phene Plate) of *E. coli* isolated from pig feces (sows and piglets) demonstrated high diversity, as up to 10 unique phenotypes out of 11 isolates were identified in one animal (54). The diversity of the dominant *E. coli* populations from horses, cattle, and humans, as assessed by ribotyping, varied according to host species (2). *E. coli* populations in horses displayed the highest diversity, while those in humans had the lowest diversity.

Rarefaction analysis of an *E. coli* rep-PCR library determined that a library size of 1,535 isolates from humans and 12 animal species was not close to saturation (52). Nearly 59% of rep-PCR types in the library were sampled only once, which demonstrates the great diversity in *E. coli* genotypes. High-diversity *E. coli* populations were identified by PFGE subtyping in water and sediments of irrigation water in the Rio Grande river in Texas (64). Complicating the issue is the fact that different host species and sample types (for example, human feces versus sewage) contain *E. coli* populations of differing richness (97), indicating that sampling efforts should be adjusted based on host species and sample type. The apparently low frequency of "resident" *E. coli* subtypes compared to "transients" may be due in part to error that is generated by undersampling the *E. coli* population of individuals (50). Achieving representative sampling of *E. coli* populations in environmental waters will be affected by similar concerns. Chivukula and Harwood reported that highly diverse *E. coli* populations were found in both pristine waters and in water impacted by anthropogenic activities (20). A 2:1 ratio of isolates analyzed to estimated subtype richness has been suggested as a minimal requirement for capturing population diversity (50, 76). Thus, if this recommendation is followed, then the sampling effort for making a representative *E. coli* library in a small watershed will be very great.

Enterococcus diversity has also been investigated in fecal sources, in water and in sewage (15, 61, 96). A European study using a phenotypic SPM (Phene Plate) found diverse *Enterococcus* populations in individual animals and humans and highly diverse populations in sewage and surface waters (61, 102). Similar results were obtained in a U.S. study using a genotypic SPM (ribotyping with two enzymes), which demonstrated that sewage samples tended to contain highly diverse *Enterococcus* populations (96). Diversity in deer, dairy cattle, and dogs was particularly great, and species accumulation curves were not close to saturation at seven isolates per host individual. In contrast, diversity in enterococci was substantially lower in humans, where two isolates per host individual adequately represented diversity. *Enterococcus* diversity in a sample from a sewage lift station is demonstrated by BOX-PCR subtyping in Fig. 2 (V. J. Harwood and M. J. Brownell, unpublished data). Great differences in *Enterococcus* genotype richness (BOX-PCR) were observed in various water types in Florida (15), as demonstrated here by species accumulation curves (Fig. 3A). *Enterococcus* populations were most diverse in storm water, followed by sewage and then surface waters impacted by storm water (Siesta Key). In contrast, the relatively nonimpacted waters of the Myakka River harbored a low-diversity *Enterococcus* population. Comparison of *E. coli* and *Enterococcus* diversity (Fig. 3B) on two sampling dates, the first of which was during a rain event and the second of which was

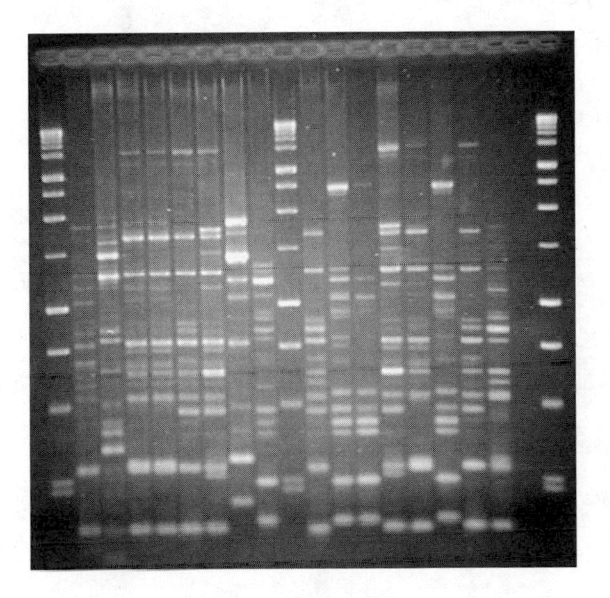

Figure 2 BOX-PCR subtyping of *Enterococcus* spp. isolated from a lift station (central sewer) shows diversity of enterococci.

during a dry period, reveals that diversity was strongly affected by sampling date, probably due to the rain event. Both *E. coli* and *Enterococcus* populations were very diverse in samples collected during the rain event, and the species accumulation curves for *E. coli* compared to *Enterococcus* populations are very similar. In contrast, the species accumulation curves show less-diverse *E. coli* and *Enterococcus* populations during the dry period, and richness is nearing saturation at 20 isolates per sample (Fig. 3B). In contrast to the low-diversity *Enterococcus* population observed in Myakka River waters, *Enterococcus* diversity in sediments from a Gulf of Mexico (saltwater) beach was very high (Fig. 4). These data indicate that the SPM diversity in both source (fecal) samples and water samples can have a great impact on whether sampling is representative and that preliminary data about population diversity is essential before planning a library-based MST study.

Persistence of SIs and SPMs in Secondary Habitats

If the gastrointestinal tract is considered to be the primary habitat of microbial indicator organisms, then water, soil, sediments, sewage treatment systems, and other habitats they occupy after being excreted from their hosts can be considered secondary habitats (31, 81). Food can also be considered a secondary habitat of indicator organisms. Gordon (32) pointed out that one of the most crucial assumptions of MST is that "the clonal composition of the species isolated from soil and water [the secondary habitat] represents the clonal composition of the species in the host populations responsible for the fecal inputs [primary habitat] to the environment." This assumption

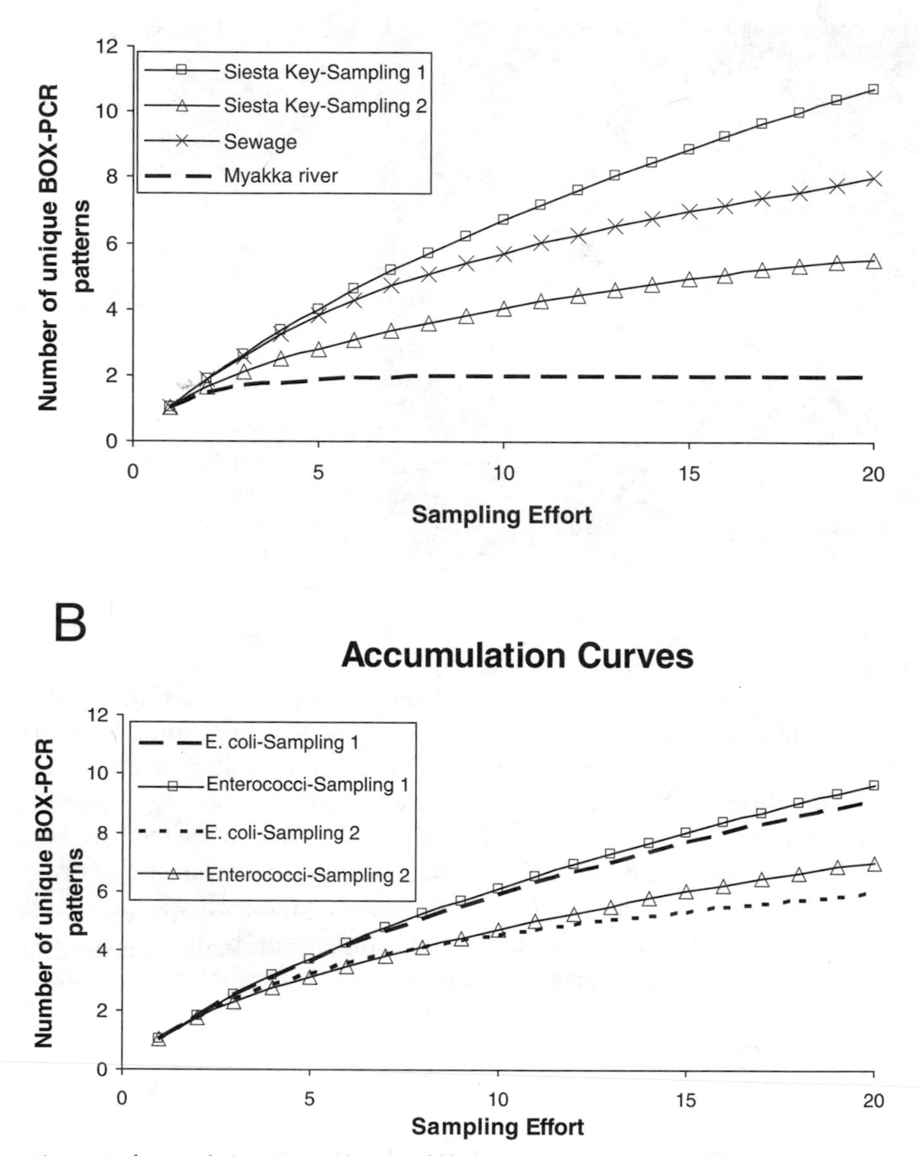

Figure 3 Accumulation curves showing (A) the number of unique BOX-PCR subtypes of *Enterococcus* spp. as a function of sampling effort in various types of water and (B) the number of unique BOX-PCR subtypes of *E. coli* and *Enterococcus* spp. as a function of sampling effort in a storm water system. For Siesta Key samples the accumulation curves from several sites in a storm drain system were averaged: storm pipe water, beach water, ditch water, and vault water. Sampling 1 occurred during a rain event, while sampling 2 was dry. The Myakka River samples represent a relatively unimpacted site in a state park. Accumulation curves were averaged from the following samples: storm pipe water, vault water, ditch water, ditch sediment, beach water, and beach sediment. Sampling 1 occurred during a rain event, while sampling 2 was dry.

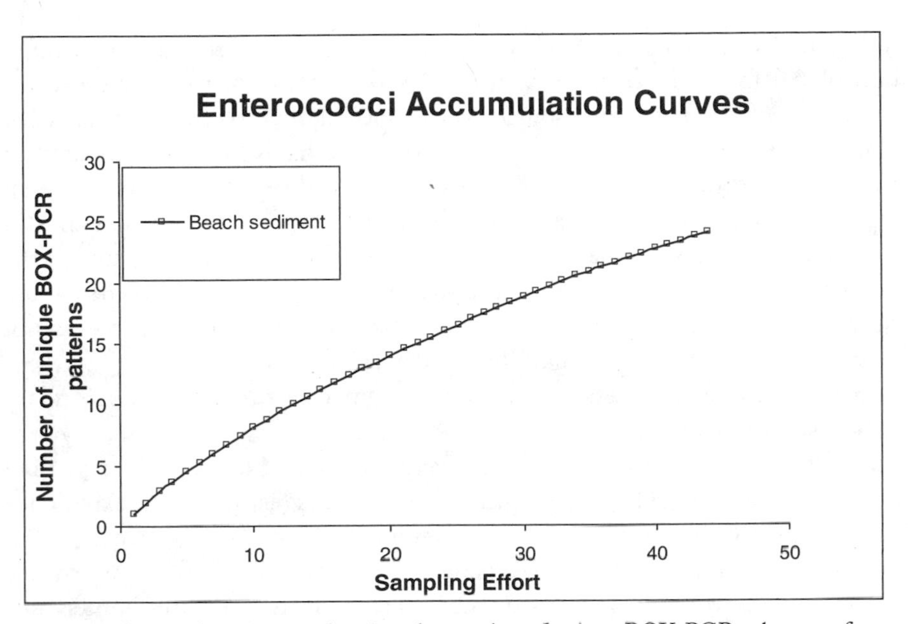

Figure 4 Accumulation curve showing the number of unique BOX-PCR subtypes of *Enterococcus* spp. as a function of sampling effort in Gulf of Mexico beach sediments.

must be validated if any MST method is to generate an accurate quantitative or semiquantitative estimate of source contributions to microbial loading in water or food. If the fecal SI/SPM persists poorly in the secondary habitat relative to other SPMs, the source-specific fecal signal will rapidly disappear. Conversely, if certain SIs/SPMs persist longer than others, or can multiply in secondary habitats, they will be overrepresented in the microbial loading analysis compared to their original contribution from fecal material. The recent report of bloom strains of *E. coli* in Australia (78) is an extreme example of growth of indicator bacteria in the environment.

Several studies on the distribution of *Escherichia coli* SPMs in primary versus secondary habitats noted distinct differences in the populations. Of 113 *E. coli* subtypes (determined by MLEE) isolated from bird feces and in the same birds' litter, only 10% of the subtypes were found in both the primary and secondary habitats (106). Another study (33) compared *E. coli* populations from feces of two human couples, each representing a household, and *E. coli* from each household's septic tank using MLEE. This study indicated that *E. coli* populations in a secondary habitat such as a septic tank can differ significantly from those in the primary habitat, from which the organisms originated. Ribotyping of *E. coli* isolated from dog feces, untreated wastewater, and contaminated soil inoculated into water showed that the dominant subtypes in the primary habitat were distinct from those in the secondary habitat, indicating differential survival of *E. coli* subtypes (1). Certain "survivor" strains that could be isolated from uninoculated water and sediments persisted in mesocosms regardless of the inoculum source. *E. coli* populations isolated from swine manure slurry (a pollution source and secondary habitat) were compared to those isolated from soil inoculated with the same slurry (99). Enterobacterial repetitive intergenic consensus-PCR subtyping demonstrated a major shift in community structure in the secondary (manure slurry) habitat compared to the tertiary (soil) habitat, although many subtypes were shared between manure slurry and soil. However, one SPM that was prominent in manure was not recovered from soil, suggesting differential survival of SPMs.

Several studies have examined the persistence of other SIs in secondary habitats. Decay rates of F-specific RNA coliphages in water apparently vary according to type, as type IV phages were shown to be less persistent than types I, II, and III (14). F-specific RNA coliphages also may experience higher inactivation rates in warm waters than in cooler waters (21). *Bacteroides-Prevotella* 16S rRNA gene sequences derived from an equine manure pile were compared to sequences derived from the water immediately downstream of the manure pile, and the sequences were found to cluster according to manure pile versus water habitat, suggesting differential persistence within the *Bacteroides-Prevotella* group (91). Phenotypic analysis (Phene Plate) of *Enterococcus* spp. and fecal coliforms in raw and treated sewage samples showed that populations of comparable diversity and composition could be found in both types of sewage; however, the study was not extended to environmental waters (102).

A study of *Listeria monocytogenes* in facilities making fresh cheese was carried out using ribotyping as the subtyping method (53). One subtype that had previously been identified in human infections was persistent over the 6-month course of the study and was also identified in two plants. This *L. monocytogenes* subtype may be an example of one that is well-adapted to survival outside the host and, therefore, poses a relatively high risk to food safety.

Based on the above phenotypic and genotypic studies, *E. coli* may not be an acceptable candidate as a source tracking organism. Mathematical models of *E. coli* population dynamics in water have shown that in the case of fecal sources that harbor *E. coli* subtypes with variable capacity for survival in secondary habitats, the subtype distribution of strains isolated from water would be unlikely to reflect the distribution in the fecal source (6). For library- and/or culture-independent methods, such as the PCR for *Bacteroides* (7, 8) and *E. coli* toxin genes (57, 58), the primary-habitat-versus-secondary-habitat concern may be somewhat less pressing, but it is still a concern that must be addressed. Current library-independent methods rely almost exclusively on binary data, as an SPM specific for an animal host is either detected in an environmental sample or it is not. However, if the DNA signal is very short-lived in the water compared to indicator organisms and pathogens, the library-independent method could then give false-negative results, indicating that the water is not contaminated when it actually is. Furthermore, while efforts are currently underway to develop quantitative PCR protocols for some SPMs, the accuracy of these methods will also rely, to a certain extent, on issues revolving around primary versus secondary habitats.

Relevance of SI to Regulatory Tools

Indicator bacteria such as coliforms have been used for over a century as indicators of fecal contamination in water (111). In the United States, indicator bacteria (e.g., total coliforms, fecal coliforms, *E. coli*, and enterococci) are the standard by which microbial water quality is measured. Many MST studies, whether they are carried out on bathing beaches, in reservoirs, or on watersheds for TMDL assessments, are often done as responses to monitoring programs that identify water bodies which chronically exceed indicator bacterium standards. Understandably, for assessment of fecal sources, water quality managers prefer an SI that is also a recognized indicator organism. However, as MST methods are tested and validated in the field, a method that utilizes one or more alternative SIs may show greater utility than methods that employ conventional indicator organisms. In such cases, a link between the SI and some combination of public health outcomes, pathogens, and indicators will be required.

An assumption of MST methods that employ alternative SIs is that the results will have some discernible relationship with indicator bacterium levels, pathogens, and public health outcomes. Studies have begun to establish the relationship between conventional indicator bacteria and other SIs. A *Taq* nuclease assay (quantitative PCR) was used to demonstrate a strong correlation between members of the class *Bacteroidetes* and *E. coli* concentrations by the Colilert-18 most-probable-number assay (Idexx Laboratories, Westbrook, Maine) (23).

Some studies have measured the correlation of coliphages with indicator bacteria. Somatic coliphage concentrations in wastewater and seawater in Egypt correlated well with total and fecal coliforms, but not with fecal streptococci (most of the fecal streptococci are enterococci) (26). A study in Italian shellfish-harvesting waters found correlations between coliphages and bacterial indicators (fecal coliforms, *E. coli*, and enterococci), in both overlying waters and sediments (63). In contrast, a study conducted in Boston harbor found no correlation among the following indicators: fecal coliforms, enterococci, somatic coliphages, and F-specific coliphages (79). The failure of an indicator organism and SI to correlate is not a priori a reason to discard the SI, particularly if it is associated with human health risk (see below) (27). The interpretation of the results may, however, prove more complex when the SI is not an indicator organism, particularly in the case of TMDL assessments. With the possible exception of fecal coliforms, *E. coli,* and enterococci, the relationship of SIs to indicator bacteria and each other is far from clear, making this an essential area for further study.

Relevance of SIs to Human Health

The ultimate goal of any MST study is to determine the host species responsible for fecal pollution from among many possible candidates. However, discrimination on a more general scale (e.g., human versus nonhuman) is of practical use in many cases (46, 71, 95). Discrimination at the human/nonhuman level is deemed relevant to human health outcomes due in part to the assumption that human fecal material poses a greater human health risk than other types of fecal material (86, 93). Food-borne outbreaks, particularly those of viral etiology, are frequently the result of improper human hygiene, rather than contamination by animal feces (reviewed in reference 59). Detection of enteric viruses, whose source is exclusively human, has been correlated with gastroenteritis in swimmers in marine waters (40), and a meta-analysis of epidemiological data showed that enteric viruses were strongly associated with gastroenteritis (103). However, "background" levels of indicator bacteria from nonfecal sources are potentially a major source of error in risk assessment studies of recreational waters (55) and often complicate the determination of relationships between indicator bacterium concentrations and human health outcomes. Consequently, SIs that can discriminate human versus nonhuman fecal pollution should be useful, provided they have some association with human health outcomes and/or pathogen loads.

The indicator organism paradigm is based on the assumption that indicator organisms are predictive of human health risk. Much debate and many epidemiological studies have explored this assumption (reviewed in reference 103). The 1986 water quality standards established by the U.S. Environmental Protection Agency for recreational water specify the use of *E. coli* (and not fecal coliforms) measurement for freshwater bodies, and that of enterococci for freshwater and marine water bodies (100). A study on recreational water users in seawater in the United Kingdom found that fecal streptococcus concentrations, but not those of fecal or total coliforms, were correlated with gastroenteritis frequency (56). A meta-analysis of the epidemiological literature on gastroenteritis among recreational water users found that *E. coli* concentrations were significantly associated with gastroenteritis in fresh water, and that *Enterococcus* concentrations were significantly associated with gastroenteritis in marine water (103), supporting the use of these organisms as indicators of human health risk. Coliphages were also predictive of gastroenteritis (103), although fewer studies were available for analysis. The correlation of alternative SIs with human health risk is largely unexplored but will certainly be a growing area as MST methods move from research tools to regulatory applications.

Transferable Methodology

If MST methods are to be employed as regulatory tools, they must be standardized in order to achieve interlaboratory precision. A complete discussion of standardization protocols is beyond the scope of this work, particularly since standardization for different methods requires several different strategies. Concerns that are common to all methods include removal of inhibitors and interfering substances, sampling strategy, sample size, concentration methods, diluents and dilution factors, and data handling. Methods that involve bacterial isolation require standardization of media, incubation conditions and times, counting criteria, and recognition of correct colony morphology. PCR-based methods require extensive standardization, including information concerning the sources of reagents and the type of thermocycler used. Methods based on banding patterns such as rep-PCR must have a standardized protocol for band identification, pattern analyses, and methods used to group isolates (52). Many of the strategies used in the clinical realm may aid in the standardization process, including specialized extraction kits for various matrices and prepackaged reagents for PCR-based methods.

Statistical and other evaluation parameters must also be consistently employed among all laboratories. Moreover, instead of providing simple "percent positive" and "percent negative" reports, analyses should contain positive predictive value and negative predictive value of the tests, which take into account true-positive and false-positive results, or true-negative and false-negative results, respectively (46). Positive predictive value is calculated as the number of positive test results that are actually positive divided by all positive results (all positives = true-positives plus false-positives). The negative predictive value is calculated similarly; i.e., the number of negative test results that are actually negative divided by all negative results (true-negative plus false-negative). It will also be essential to establish the sensitivity of the methods in terms of relevant parameters: e.g., for PCR, the lowest number of cells that can be detected per liter (23); or for library-based methods, the smallest percentage of isolates identified from a given source that can be considered a significant source of contamination (46, 105).

RECOMMENDATIONS FOR THE FUTURE

The prospects for future success of MST methods rest directly in the hands of its practitioners, who must be willing to test assumptions, explore limitations, and develop the broad expertise to choose the most appropriate suite of tools for the specific questions at hand. Library-independent methods are currently experiencing a frequency of use that is similar to that seen in the library-based methods developed several years ago. Earlier MST publications typically described libraries that contained a relatively small number of isolates, and regardless of method or target organism, the correct classification rates reported for these libraries were generally quite high (16, 76, 108). Both in method comparison studies (38, 46, 70, 71, 97) and through general experience, unforeseen limitations for the library-based methods have been identified. Those who continue to use libraries will need to perform stringent quality control, cross-validations, and challenge tests on libraries to demonstrate each method's utility. More recent publications have described primers that can be used to detect source-specific sequences without libraries (84) or, in some cases, without cultivation (7, 22, 27). However, in all of these reports the primers have been tested against a limited number of isolates or samples to check for cross-reactions. The caveats that apply to library-based methods are equally valid for library-independent methods, and it is imperative that assumptions and limitations are rigorously tested.

A nationwide (U.S.) and international network of investigators should be established for the purpose of sharing microbial isolates, DNA, samples, training, and expertise. While development of such a network is discussed at each MST workshop, it is clear that significant funding will be required to achieve this goal. Advancement of MST as a field also requires that scientists, editors, and reviewers are educated about what constitutes a well-designed study. While studies that subtype a small number of isolates using a novel method generally give results that make the method seem feasible, the promise of the early studies is almost never borne out in follow-up studies. The MST field has progressed to the stage at which the practitioners understand what constitutes a robust, well-designed study, and this should be communicated in workshops and review articles. Furthermore, journal editors and reviewers should be willing to publish well-designed studies in which MST methods did not work as envisioned and, thus, report negative results. There is a wealth of data "on the shelf" that could guide future studies, but it often does not appear in the literature.

At a recent (2005) MST workshop in San Antonio, Texas, funded by the Water Environment Research Foundation, it was observed that MST has made great strides for a field that has received little federal funding and no concerted emphasis from any agency. At the same workshop, an MST practitioner stated that it was time for the field to "grow up." Both points of view rang true to the participants and audience, and they serve to capture the current state of MST. Scientists who work in MST should be willing to communicate the challenges associated with MST to other scientists and to managers in the realms of water quality and food safety.

A polyphasic approach to MST may well become the rule, in which extensive exploration of land use, infrastructure, and other watershed characteristics will help determine the particular tools best suited for use in a given water body or watershed. The challenge of source identification in food and water has no simple answers; however, the issues surrounding food and water safety are becoming more pressing as the world's population grows. A commitment to development of the MST field now has the potential to pay great dividends in terms of public health protection in the future.

REFERENCES

1. **Anderson, K. L., J. E. Whitlock, and V. J. Harwood.** 2005. Persistence and differential survival of fecal indicator bacteria in subtropical waters and sediments. *Appl. Environ. Microbiol.* **71**:3041–3048.
2. **Anderson, M. A.** 2003. *Frequency Distributions of Escherichia coli Subtypes in Various Fecal Sources over Time and Geographical Space: Application to Bacterial Source Tracking Methods.* M.S. thesis. University of South Florida, Tampa. http://etd.fcla.edu/SF/SFE0000206/MatthewAnderson.MastersThesis.pdf.
3. **Anderson, M. A., J. E. Whitlock, and V. J. Harwood.** 2003. *Frequency Distributions of Escherichia coli Subtypes in Various Fecal Sources: Application to Bacterial Source Tracking Methods*, abstr. Q-428. Abstr. 103rd Gen. Meet. Am. Soc. Microbiol. 2003. American Society for Microbiology, Washington, D.C.
4. **Aslam, M., G. G. Greer, F. M. Nattress, C. O. Gill, and L. M. McMullen.** 2004. Genotypic analysis of *Escherichia coli* recovered from product and equipment at a beef-packing plant. *J. Appl. Microbiol.* **97**:78–86.
5. **Aslam, M., F. Nattress, G. Greer, C. Yost, C. Gill, and L. McMullen.** 2003. Origin of contamination and genetic diversity of *Escherichia coli* in beef cattle. *Appl. Environ. Microbiol.* **69**:2794–2799.
6. **Barnes, B., and D. M. Gordon.** 2004. Coliform dynamics and the implications for source tracking. *Environ. Microbiol.* **6**:501–509.
7. **Bernhard, A. E., and K. G. Field.** 2000. A PCR assay to discriminate human and ruminant feces on the basis of host differences in *Bacteroides-Prevotella* genes encoding 16S rRNA. *Appl. Environ. Microbiol.* **66**:4571–4574.
8. **Bernhard, A. E., T. Goyard, M. T. Simonich, and K. G. Field.** 2003. Application of a rapid method for identifying fecal pollution sources in a multi-use estuary. *Water Res.* **37**:909–913.
9. **Bettelheim, K. A., A. Kuzevski, R. A. Gilbert, D. O. Krause, and C. S. McSweeney.** 2005. The diversity of *Escherichia coli* serotypes and biotypes in cattle faeces. *J. Appl. Microbiol.* **98**:699–709.

10. **Bjorkman, J., D. Hughes, and D. I. Andersson.** 1998. Virulence of antibiotic-resistant *Salmonella typhimurium*. *Proc. Natl. Acad. Sci. USA* **95**:3949–3953.
11. **Bjorkman, J., I. Nagaev, O. G. Berg, D. Hughes, and D. I. Andersson.** 2000. Effects of environment on compensatory mutations to ameliorate costs of antibiotic resistance. *Science* **287**:1479–1482.
12. **Bonjoch, X., E. Balleste, and A. R. Blanch.** 2004. Multiplex PCR with 16S rRNA gene-targeted primers of *Bifidobacterium* spp. to identify sources of fecal pollution. *Appl. Environ. Microbiol.* **70**:3171–3175.
13. **Borgen, K., G. S. Simonsen, A. Sundsfjord, Y. Wasteson, O. Olsvik, and H. Kruse.** 2000. Continuing high prevalence of VanA-type vancomycin-resistant enterococci on Norwegian poultry farms three years after avoparcin was banned. *J. Appl. Microbiol.* **89**: 478–485.
14. **Brion, G. M., J. S. Meschke, and M. D. Sobsey.** 2002. F-specific RNA coliphages: occurrence, types, and survival in natural waters. *Water Res.* **36**:2419–2425.
15. **Brownell, M. J., M. J. Dontchev, S. D. Shehane, T. M. Scott, J. Lukasik, R. C. Kurz, and V. J. Harwood.** 2005. *Investigation of Microbial Fecal Pollution of a Pristine Florida Beach by BOX-PCR of Escherichia coli and Enterococci and PCR Amplification of Host-Specific Markers*, abstr. Q-406. Abstr. 105th Gen. Meet. Am. Soc. Microbiol. 2005. American Society for Microbiology, Washington, D.C.
16. **Carson, C. A., B. L. Shear, M. R. Ellersieck, and A. Asfaw.** 2001. Identification of fecal *Escherichia coli* from humans and animals by ribotyping. *Appl. Environ. Microbiol.* **67**: 1503–1507.
17. **Carson, C. A., B. L. Shear, M. R. Ellersieck, and J. D. Schnell.** 2003. Comparison of ribotyping and repetitive extragenic palindromic-PCR for identification of fecal *Escherichia coli* from humans and animals. *Appl. Environ. Microbiol.* **69**:1836–1839.
18. **Caugant, D. A., B. R. Levin, and R. K. Selander.** 1984. Distribution of multilocus genotypes of *Escherichia coli* within and between host families. *J. Hyg. (London)* **92**:377–384.
19. **Chern, E. C., Y. L. Tsai, and B. H. Olson.** 2004. Occurrence of genes associated with enterotoxigenic and enterohemorrhagic *Escherichia coli* in agricultural waste lagoons. *Appl. Environ. Microbiol.* **70**:356–362.
20. **Chivukula, V., and V. J. Harwood.** 2004. *Impact of Fecal Pollution on the Microbial Diversity in Natural Waters*, abstr. Q-463. Abstr. Gen. Meet. Am. Soc. Microbiol. 2004. American Society for Microbiology, Washington, D.C.
21. **Cole, D., S. C. Long, and M. D. Sobsey.** 2003. Evaluation of F+ RNA and DNA coliphages as source-specific indicators of fecal contamination in surface waters. *Appl. Environ. Microbiol.* **69**:6507–6514.
22. **Dick, L. K., A. E. Bernhard, T. J. Brodeur, J. W. Santo Domingo, J. M. Simpson, S. P. Walters, and K. G. Field.** 2005. Host distributions of uncultivated fecal *Bacteroidales* bacteria reveal genetic markers for fecal source identification. *Appl. Environ. Microbiol.* **71**:3184–3191.
23. **Dick, L. K., and K. G. Field.** 2004. Rapid estimation of numbers of fecal *Bacteroidetes* by use of a quantitative PCR assay for 16S rRNA genes. *Appl. Environ. Microbiol.* **70**:5695–5697.
24. **Dontchev, M., J. E. Whitlock, and V. J. Harwood.** 2003. *Ribotyping of Escherichia coli and Enterococcus spp. to Determine the Source of Fecal Pollution in Natural Waters*, abstr. Q-426.

Abstr. Gen. Meet. Am. Soc. Microbiol. 2003. American Society for Microbiology, Washington, D.C.

25. **Endley, S., L. Lu, E. Vega, M. E. Hume, and S. D. Pillai.** 2003. Male-specific coliphages as an additional fecal contamination indicator for screening fresh carrots. *J. Food Prot.* **66:**88–93.

26. **Fattouh, F., M. El Senawy, and S. Hassan.** 2004. Recovery of somatic coliphages in wastewater and seawater samples in relation to bacterial indicator organisms and water hydrochemical parameters in Kaiet Bay station, Alexandria. *J. Water Supply Res. Technol. - Aqua* **53:**183–192.

27. **Field, K. G., E. C. Chern, L. K. Dick, J. Fuhrmann, J. Griffith, P. Holden, M. G. LaMontagne, J. Le, B. Olson, and M. T. Simonich.** 2003. A comparative study of culture-independent, library-independent genotypic methods of fecal source tracking. *J. Water Health* **1:**181–194.

28. **Fogarty, L. R., S. K. Haack, M. J. Wolcott, and R. L. Whitman.** 2003. Abundance and characteristics of the recreational water quality indicator bacteria *Escherichia coli* and enterococci in gull faeces. *J. Appl. Microbiol.* **94:**865–878.

29. **Fong, T. T., D. W. Griffin, and E. K. Lipp.** 2005. Molecular assays for targeting human and bovine enteric viruses in coastal waters and their application for library-independent source tracking. *Appl. Environ. Microbiol.* **71:**2070–2078.

30. **Godwin, D., and J. H. Slater.** 1979. The influence of the growth environment on the stability of a drug resistance plasmid in *Escherichia coli* K12. *J. Gen. Microbiol.* **111:**201–210.

31. **Gordon, D. M.** 1997. The genetic structure of *Escherichia coli* populations in feral house mice. *Microbiology* **143:**2039–2046.

32. **Gordon, D. M.** 2001. Geographical structure and host specificity in bacteria and the implications for tracing the source of coliform contamination. *Microbiology* **147:**1079–1085.

33. **Gordon, D. M., S. Bauer, and J. R. Johnson.** 2002. The genetic structure of *Escherichia coli* populations in primary and secondary habitats. *Microbiology* **148:**1513–1522.

34. **Gordon, D. M., and A. Cowling.** 2003. The distribution and genetic structure of *Escherichia coli* in Australian vertebrates: host and geographic effects. *Microbiology* **149:** 3575–3386.

35. **Gordon, D. M., and J. Lee.** 1999. The genetic structure of enteric bacteria from Australian mammals. *Microbiology* **145:**2673–2682.

36. **Griffin, D. W., C. J. Gibson III, E. K. Lipp, K. Riley, J. H. Paul III, and J. B. Rose.** 1999. Detection of viral pathogens by reverse transcriptase PCR and of microbial indicators by standard methods in the canals of the Florida Keys. *Appl. Environ. Microbiol.* **65:**4118–4125.

37. **Griffin, D. W., R. Stokes, J. B. Rose, and J. H. Paul III.** 2000. Bacterial indicator occurrence and the use of an F(+) specific RNA coliphage assay to identify fecal sources in Homosassa Springs, Florida. *Microb. Ecol.* **39:**56–64.

38. **Griffith, J. F., S. B. Weisberg, and C. D. McGee.** 2003. Evaluation of microbial source tracking methods using mixed fecal sources in aqueous test samples. *J. Water Health* **1:** 141–151.

39. **Guan, S., R. Xu, S. Chen, J. Odumeru, and C. Gyles.** 2002. Development of a procedure for discriminating among *Escherichia coli* isolates from animal and human sources. *Appl. Environ. Microbiol.* **68:**2690–2698.

40. **Haile, R. W., J. S. Witte, M. Gold, R. Cressey, C. McGee, R. C. Millikan, A. Glasser, N. Harawa, C. Ervin, P. Harmon, J. Harper, J. Dermand, J. Alamillo, K. Barrett, M. Nides, and G. Wang.** 1999. The health effects of swimming in ocean water contaminated by storm drain runoff. *Epidemiology* **10:**355–363.
41. **Hartel, P. G., J. D. Summer, J. L. Hill, J. V. Collins, J. A. Entry, and W. I. Segars.** 2002. Geographic variability of *Escherichia coli* ribotypes from animals in Idaho and Georgia. *J. Environ. Qual.* **31:**1273–1278.
42. **Hartel, P. G., J. D. Summer, and W. I. Segars.** 2003. Deer diet affects ribotype diversity of *Escherichia coli* for bacterial source tracking. *Water Res.* **37:**3263–3268.
43. **Hartl, D. L., and D. E. Dykhuizen.** 1984. The population genetics of *Escherichia coli*. *Annu. Rev. Genet.* **18:**31–68.
44. **Harwood, V. J., M. Brownell, W. Perusek, and J. E. Whitlock.** 2001. Vancomycin-resistant *Enterococcus* spp. isolated from wastewater and chicken feces in the United States. *Appl. Environ. Microbiol.* **67:**4930–4933.
45. **Harwood, V. J., J. Whitlock, and V. Withington.** 2000. Classification of antibiotic resistance patterns of indicator bacteria by discriminant analysis: use in predicting the source of fecal contamination in subtropical waters. *Appl. Environ. Microbiol.* **66:**3698–3704.
46. **Harwood, V. J., B. Wiggins, C. Hagedorn, R. D. Ellender, J. Gooch, J. Kern, M. Samadpour, A. C. H. Chapman, B. J. Robinson, and B. C. Thompson.** 2003. Phenotypic library-based microbial source tracking methods: efficacy in the California collaborative study. *J. Water Health* **1:**153–166.
47. **He, J. W., and S. Jiang.** 2005. Quantification of enterococci and human adenoviruses in environmental samples by real-time PCR. *Appl. Environ. Microbiol.* **71:**2250–2255.
48. **Heinemann, J. A., R. G. Ankenbauer, and C. F. Amabile-Cuevas.** 2000. Do antibiotics maintain antibiotic resistance? *Drug Discov. Today* **5:**195–204.
49. **Hsu, F. C., Y. S. Shieh, J. van Duin, M. J. Beekwilder, and M. D. Sobsey.** 1995. Genotyping male-specific RNA coliphages by hybridization with oligonucleotide probes. *Appl. Environ. Microbiol.* **61:**3960–3966.
50. **Jenkins, M. B., P. G. Hartel, T. J. Olexa, and J. A. Stuedemann.** 2003. Putative temporal variability of *Escherichia coli* ribotypes from yearling steers. *J. Environ. Qual.* **32:**305–309.
51. **Jiang, S., R. Noble, and W. Chu.** 2001. Human adenoviruses and coliphages in urban runoff-impacted coastal waters of Southern California. *Appl. Environ. Microbiol.* **67:**179–184.
52. **Johnson, L. K., M. B. Brown, E. A. Carruthers, J. A. Ferguson, P. E. Dombek, and M. J. Sadowsky.** 2004. Sample size, library composition, and genotypic diversity among natural populations of *Escherichia coli* from different animals influence accuracy of determining sources of fecal pollution. *Appl. Environ. Microbiol.* **70:**4478–4485.
53. **Kabuki, D. Y., A. Y. Kuaye, M. Wiedmann, and K. J. Boor.** 2004. Molecular subtyping and tracking of *Listeria monocytogenes* in Latin-style fresh-cheese processing plants. *J. Dairy Sci.* **87:**2803–2812.
54. **Katouli, M., A. Lund, P. Wallgren, I. Kuhn, O. Soderlind, and R. Mollby.** 1995. Phenotypic characterization of intestinal *Escherichia coli* of pigs during suckling, post-weaning, and fattening periods. *Appl. Environ. Microbiol.* **61:**778–783.
55. **Kay, D., J. Bartram, A. Pruss, N. Ashbolt, M. D. Wyer, J. M. Fleisher, L. Fewtrell, A. Rogers, and G. Rees.** 2004. Derivation of numerical values for the World Health Organization guidelines for recreational waters. *Water Res.* **38:**1296–1304.

56. **Kay, D., J. M. Fleisher, R. L. Salmon, F. Jones, M. D. Wyer, A. F. Godfree, Z. Zelenauch-Jacquotte, and R. Shore.** 1994. Predicting likelihood of gastroenteritis from sea bathing: results from randomised exposure. *Lancet* **344:**905–909.
57. **Khatib, L. A., Y. L. Tsai, and B. H. Olson.** 2002. A biomarker for the identification of cattle fecal pollution in water using the LTIIa toxin gene from enterotoxigenic *Escherichia coli*. *Appl. Microbiol. Biotechnol.* **59:**97–104.
58. **Khatib, L. A., Y. L. Tsai, and B. H. Olson.** 2003. A biomarker for the identification of swine fecal pollution in water, using the STII toxin gene from enterotoxigenic *Escherichia coli*. *Appl. Microbiol. Biotechnol.* **63:**231–238.
59. **Koopmans, M., and E. Duizer.** 2004. Foodborne viruses: an emerging problem. *Int. J. Food Microbiol.* **90:**23–41.
60. **Kreader, C. A.** 1998. Persistence of PCR-detectable *Bacteroides distasonis* from human feces in river water. *Appl. Environ. Microbiol.* **64:**4103–4105.
61. **Kuhn, I., A. Iversen, L. G. Burman, B. Olsson-Liljequist, A. Franklin, M. Finn, F. Aarestrup, A. M. Seyfarth, A. R. Blanch, X. Vilanova, H. Taylor, J. Caplin, M. A. Moreno, L. Dominguez, I. A. Herrero, and R. Mollby.** 2003. Comparison of enterococcal populations in animals, humans, and the environment—a European study. *Int. J. Food Microbiol.* **88:**133–145.
62. **Leclerc, H., S. Edberg, V. Pierzo, and J. M. Delattre.** 2000. Bacteriophages as indicators of enteric viruses and public health risk in groundwaters. *J. Appl. Microbiol.* **88:**5–21.
63. **Legnani, P., E. Leoni, D. Lev, R. Rossi, G. C. Villa, and P. Bisbini.** 1998. Distribution of indicator bacteria and bacteriophages in shellfish and shellfish-growing waters. *J. Appl. Microbiol.* **85:**790–798.
64. **Lu, L., M. E. Hume, K. L. Sternes, and S. D. Pillai.** 2004. Genetic diversity of *Escherichia coli* isolates in irrigation water and associated sediments: implications for source tracking. *Water Res.* **38:**3899–3908.
65. **Maluquer de Motes, C., P. Clemente-Casares, A. Hundesa, M. Martin, and R. Girones.** 2004. Detection of bovine and porcine adenoviruses for tracing the source of fecal contamination. *Appl. Environ. Microbiol.* **70:**1448–1454.
66. **McLellan, S. L., A. D. Daniels, and A. K. Salmore.** 2001. Clonal populations of thermotolerant *Enterobacteriaceae* in recreational water and their potential interference with fecal *Escherichia coli* counts. *Appl. Environ. Microbiol.* **67:**4934–4938.
67. **McLellan, S. L., A. D. Daniels, and A. K. Salmore.** 2003. Genetic characterization of *Escherichia coli* populations from host sources of fecal pollution by using DNA fingerprinting. *Appl. Environ. Microbiol.* **69:**2587–2594.
68. **McQuaig, S. M., T. M. Scott, J. Lukasik, and S. R. Farrah.** 2005. *Detection of Human Polyomaviruses in Environmental Waters as an Indication of Human Fecal Contamination*, abstr. Q-323. Abstr. 105th Gen. Meet. Am. Soc. Microbiol. 2005. American Society for Microbiology, Washington, D.C.
69. **Milkman, R.** 1973. Electrophoretic variation in *Escherichia coli* from natural sources. *Science* **182:**1024–1026.
70. **Moore, D. F., V. J. Harwood, D. M. Ferguson, J. Lukasik, P. Hannah, M. Getrich, and M. Brownell.** 2005. Evaluation of antibiotic resistance analysis and ribotyping for identification of fecal pollution sources in an urban watershed. *J. Appl. Microbiol.* **99:**618–628.
71. **Myoda, S. P., C. A. Carson, J. J. Fuhrmann, B.-K. Hahm, P. G. Hartel, H. Yampara-Iquise, L. Johnson, R. L. Kuntz, C. H. Nakatsu, M. J. Sadowsky, and M. Samadpour.**

2003. Comparison of genotypic-based microbial source tracking methods requiring a host origin database. *J. Water Health* **1**:167–180.

72. **Nascimento, A. M., C. E. Campos, E. P. Campos, J. L. Azevedo, and E. Chartone-Souza.** 1999. Re-evaluation of antibiotic and mercury resistance in *Escherichia coli* populations isolated in 1978 from Amazonian rubber tree tappers and Indians. *Res. Microbiol.* **150**:407–411.
73. **Noble, R. T., S. M. Allen, A. D. Blackwood, W. Chu, S. C. Jiang, G. L. Lovelace, M. D. Sobsey, J. R. Stewart, and D. A. Wait.** 2003. Use of viral pathogens and indicators to differentiate between human and non-human fecal contamination in a microbial source tracking comparison study. *J. Water Health* **1**:195–207.
74. **Ochman, H., T. S. Whittam, D. A. Caugant, and R. K. Selander.** 1983. Enzyme polymorphism and genetic population structure in *Escherichia coli* and *Shigella*. *J. Gen. Microbiol.* **129**:2715–2726.
75. **Parveen, S., R. L. Murphree, L. Edmiston, C. W. Kaspar, K. M. Portier, and M. L. Tamplin.** 1997. Association of multiple-antibiotic-resistance profiles with point and non-point sources of *Escherichia coli* in Apalachicola Bay. *Appl. Environ. Microbiol.* **63**:2607–2612.
76. **Parveen, S., K. M. Portier, K. Robinson, L. Edmiston, and M. L. Tamplin.** 1999. Discriminant analysis of ribotype profiles of *Escherichia coli* for differentiating human and nonhuman sources of fecal pollution. *Appl. Environ. Microbiol.* **65**:3142–3147.
77. **Pina, C., P. Teixeiro, P. Leite, M. Villa, C. Belloch, and L. Brito.** 2005. PCR-fingerprinting and RAPD approaches for tracing the source of yeast contamination in a carbonated orange juice production chain. *J. Appl. Microbiol.* **98**:1107–1114.
78. **Power, M. L., J. Littlefield-Wyer, D. M. Gordon, D. A. Veal, and M. B. Slade.** 2005. Phenotypic and genotypic characterization of encapsulated *Escherichia coli* isolated from blooms in two Australian lakes. *Environ. Microbiol.* **7**:631–640.
79. **Ricca, D., and J. Cooney.** 1999. Coliphages and indicator bacteria in Boston harbor, Massachusetts. *Environ. Toxicol.* **14**:404–408.
80. **Salyers, A. A., and C. F. Amabile-Cuevas.** 1997. Why are antibiotic resistance genes so resistant to elimination? *Antimicrob. Agents Chemother.* **41**:2321–2325.
81. **Savageau, M. A.** 1983. *Escherichia coli* habitats, cell types, and molecular mechanisms of gene control. *Am. Nat.* **122**:732–744.
82. **Schaper, M., and J. Jofre.** 2000. Comparison of methods for detecting genotypes of F-specific RNA bacteriophages and fingerprinting the origin of faecal pollution in water samples. *J. Virol. Methods* **89**:1–10.
83. **Schaper, M., J. Jofre, M. Uys, and W. O. Grabow.** 2002. Distribution of genotypes of F-specific RNA bacteriophages in human and non-human sources of faecal pollution in South Africa and Spain. *J. Appl. Microbiol.* **92**:657–667.
84. **Scott, T. M., T. M. Jenkins, J. Lukasik, and J. B. Rose.** 2005. Potential use of a host associated molecular marker in *Enterococcus faecium* as an index of human fecal pollution. *Environ. Sci. Technol.* **39**:283–287.
85. **Scott, T. M., S. Parveen, K. M. Portier, J. B. Rose, M. L. Tamplin, S. R. Farrah, A. Koo, and J. Lukasik.** 2003. Geographical variation in ribotype profiles of *Escherichia coli* isolates from humans, swine, poultry, beef, and dairy cattle in Florida. *Appl. Environ. Microbiol.* **69**:1089–1092.

86. **Scott, T. M., J. B. Rose, T. M. Jenkins, S. R. Farrah, and J. Lukasik.** 2002. Microbial source tracking: current methodology and future directions. *Appl. Environ. Microbiol.* **68**:5796–5803.
87. **Selander, R. K., J. M. Musser, D. A. Caugant, M. N. Gilmour, and T. S. Whittam.** 1987. Population genetics of pathogenic bacteria. *Microb. Pathog.* **3**:1–7.
88. **Seurinck, S., T. Defoirdt, W. Verstraete, and S. D. Siciliano.** 2005. Detection and quantification of the human-specific HF183 *Bacteroides* 16S rRNA genetic marker with real-time PCR for assessment of human faecal pollution in freshwater. *Environ. Microbiol.* **7**:249–259.
89. **Seurinck, S., W. Verstraete, and S. D. Siciliano.** 2003. Use of 16S-23S rRNA intergenic spacer region PCR and repetitive extragenic palindromic PCR analyses of *Escherichia coli* isolates to identify nonpoint fecal sources. *Appl. Environ. Microbiol.* **69**:4942–4950.
90. **Simonsen, G. S., H. Haaheim, K. H. Dahl, H. Kruse, A. Lovseth, O. Olsvik, and A. Sundsfjord.** 1998. Transmission of vanA-type vancomycin-resistant enterococci and vanA resistance elements between chicken and humans at avoparcin-exposed farms. *Microb. Drug Resist.* **4**:313–318.
91. **Simpson, J. M., J. W. Santo Domingo, and D. J. Reasoner.** 2004. Assessment of equine fecal contamination: the search for alternative bacterial source-tracking targets. *FEMS Microbiol. Ecol.* **47**:65–75.
92. **Simpson, J. M., J. W. Santo Domingo, and D. J. Reasoner.** 2002. Microbial source tracking: state of the science. *Environ. Sci. Technol.* **36**:5279–5288.
93. **Sinton, L., R. Finlay, and D. Hannah.** 1998. Distinguishing human from animal faecal contamination in water: a review. *N. Zealand J. Mar. Freshwater Res.* **32**:323–348.
94. **Souza, V., M. Rocha, A. Valera, and L. E. Eguiarte.** 1999. Genetic structure of natural populations of *Escherichia coli* in wild hosts on different continents. *Appl. Environ. Microbiol.* **65**:3373–3385.
95. **Stewart, J. R., R. D. Ellender, J. A. Gooch, S. Jiang, S. P. Myoda, and S. B. Weisberg.** 2003. Recommendations for microbial source tracking: lessons from a methods comparison study. *J. Water Health* **1**:225–231.
96. **Stoeckel, D. M., C. M. Kephart, V. J. Harwood, M. A. Anderson, and M. Dontchev.** 2004. *Diversity of Fecal Indicator Bacteria Subtypes: Implications for Construction of Microbial Source Tracking Libraries*, abstr. Q-245. Abstr. Gen. Meet. Am. Soc. Microbiol. 2004. American Society for Microbiology, Washington, D.C.
97. **Stoeckel, D. M., M. V. Mathes, K. E. Hyer, C. Hagedorn, H. Kator, J. Lukasik, T. L. O'Brien, T. W. Fenger, M. Samadpour, K. M. Strickler, and B. A. Wiggins.** 2004. Comparison of seven protocols to identify fecal contamination sources using *Escherichia coli*. *Environ. Sci. Technol.* **38**:6109–6117.
98. **Tartera, C., F. Lucena, and J. Jofre.** 1989. Human origin of *Bacteroides fragilis* bacteriophages present in the environment. *Appl. Environ. Microbiol.* **55**:2696–2701.
99. **Topp, E., M. Welsh, Y.-C. Tien, A. Dang, G. Lazarovits, K. Conn, and H. Zhu.** 2003. Strain-dependent variability in growth and survival of *Escherichia coli* in agricultural soil. *FEMS Microbiol. Ecol.* **44**:303–308.
100. **U.S. Environmental Protection Agency.** 1986. *Bacteriological Ambient Water Quality Criteria for Marine and Fresh Recreational Waters.* EPA-440/5-84/002. U.S. Environmental Protection Agency, Cincinnati, Ohio.

101. **Vali, L., K. A. Wisely, M. C. Pearce, E. J. Turner, H. I. Knight, A. W. Smith, and S. G. Amyes.** 2004. High-level genotypic variation and antibiotic sensitivity among *Escherichia coli* O157 strains isolated from two Scottish beef cattle farms. *Appl. Environ. Microbiol.* **70:**5947–5954.

102. **Vilanova, X., A. Manero, M. Cerda-Cuellar, and A. R. Blanch.** 2004. The composition and persistence of faecal coliforms and enterococcal populations in sewage treatment plants. *J. Appl. Microbiol.* **96:**279–288.

103. **Wade, T. J., N. Pai, J. N. Eisenberg, and J. M. Colford, Jr.** 2003. Do U.S. Environmental Protection Agency water quality guidelines for recreational waters prevent gastrointestinal illness? A systematic review and meta-analysis. *Environ. Health Perspect.* **111:**1102–1109.

104. **Watanabe, T., and Y. Ogata.** 1970. Abortive transduction of resistance factor by bacteriophage P22 in *Salmonella typhimurium*. *J. Bacteriol.* **102:**596–597.

105. **Whitlock, J. E., D. T. Jones, and V. J. Harwood.** 2002. Identification of the sources of fecal coliforms in an urban watershed using antibiotic resistance analysis. *Water Res.* **36:**4273–4282.

106. **Whittam, T. S.** 1989. Clonal dynamics of *Escherichia coli* in its natural habitat. *Antonie Leeuwenhoek* **55:**23–32.

107. **Whittam, T. S., H. Ochman, and R. K. Selander.** 1983. Geographic components of linkage disequilibrium in natural populations of *Escherichia coli*. *Mol. Biol. Evol.* **1:**67–83.

108. **Wiggins, B. A.** 1996. Discriminant analysis of antibiotic resistance patterns in fecal streptococci, a method to differentiate human and animal sources of fecal pollution in natural waters. *Appl. Environ. Microbiol.* **62:**3997–4002.

109. **Wiggins, B. A., P. W. Cash, W. S. Creamer, S. E. Dart, P. P. Garcia, T. M. Gerecke, J. Han, B. L. Henry, K. B. Hoover, E. L. Johnson, K. C. Jones, J. G. McCarthy, J. A. McDonough, S. A. Mercer, M. J. Noto, H. Park, M. S. Phillips, S. M. Purner, B. M. Smith, E. N. Stevens, and A. K. Varner.** 2003. Use of antibiotic resistance analysis for representativeness testing of multiwatershed libraries. *Appl. Environ. Microbiol.* **69:** 3399–3405.

110. **Witte, W.** 1997. Impact of antibiotic use in animal feeding on resistance of bacterial pathogens in humans. *Ciba Found. Symp.* **207:**61–71.

111. **Wolf, H. W.** 1972. The coliform count as a measure of water quality, p. 333–345. *In* R. Mitchell (ed.), *Water Pollution Microbiology*. Wiley Interscience, New York, N.Y.

112. **Woodford, N.** 1998. Glycopeptide-resistant enterococci: a decade of experience. *J. Med. Microbiol.* **47:**849–862.

113. **Workman, S. N., G. E. Mathison, and M. C. Lavoie.** 2005. Pet dogs and chicken meat as reservoirs of *Campylobacter* spp. in Barbados. *J. Clin. Microbiol.* **43:**2642–2650.

Microbial Source Tracking
Edited by Jorge W. Santo Domingo and Michael J. Sadowsky

Molecular Detection and Characterization Tools

3

Suresh D. Pillai and Everardo Vega

The ability to detect microorganisms in foods and water is one of the cornerstones of public health. The desire to "look at" microorganisms fueled the development of early microscopes in the early 17th century. The same need or desire still lingers in the 21st century. However, the questions that we are pursuing today are very different. The past century has seen significant strides in our ability not only to detect specific microorganisms but also to accomplish this task in an efficient, specific, and sensitive manner. Current microbial detection and characterization techniques not only have facilitated the identification of specific microorganisms but also have provided a rather comprehensive understanding of the genetic and metabolic characteristics of the organisms in question. Nucleic acid-based methods (due to their overall sensitivity and specificity compared to other methods) have been the target of extensive investigations over the past many years and will form the basis of discussion in this chapter. This chapter will discuss the technological advances and the current limitations of DNA- and RNA-based methods from a "food perspective." Consequently, the focus of this chapter will be the application and utility of these methods to detect and characterize microorganisms in foods. The discussion will be in the context of whether specific methods can be used to identify the types of microbial contaminants, the extent of contamination, and the possible sources of contamination. In this chapter we suggest a "road map" for the development of detection and characterization tools. The suggested improvements are presented in the context of benchmarking specifications that are required to keep pace with the detection and characterization of current and reemerging microbial pathogens,

SURESH D. PILLAI, Departments of Nutrition and Food Science & Poultry Science, Texas A&M University, College Station, Texas 77843-2472. EVERARDO VEGA, Centers for Disease Control and Prevention, Atlanta, GA 30333.

antibiotic-resistant strains, the improved understanding of the ecology of pathogens in foods, the global nature of food production and distribution, and the movement of peoples around the globe.

Availability of Commercial Molecular Detection Tools: Reality on the Ground

When one attempts to chart the evolution of microbial detection technologies, whether it is in food safety, clinical microbiology, or environmental microbiology, it is evident that major improvements in techniques closely mirror groundbreaking developments in the basic disciplines of microbiology, molecular biology, and immunology. There are a number of reasons for the quick adoption of new detection technologies by research communities. One reason is just the sheer novelty of using a "cutting-edge technology." The ability to adopt or employ a new molecular technology provides the user the means to ask different sets of questions about his/her target organism. A key driving force in the development of new-generation detection and characterization tools is the extraordinary amount of genomic and proteomic information that is now publicly available. There is no doubt that the ability to rapidly detect specific organisms is critically important in clinical medicine, and in food and environmental industries. However, the rate of adoption of these new technologies is vastly different for different end users, such as those in the clinical, food, environmental, and waste industries. Even though the research community may accept a technology, there is no doubt that industry acceptance of the technology may unfortunately never become a reality unless there is regulatory pressure. The criticality of regulatory pressure in catalyzing the commercialization of molecular detection tools is evident when one compares the availability of commercial kits for the food industry to that for the drinking water industry. Since there are no federal regulations requiring the testing of drinking water for pathogens, there is a complete absence of any validated molecular kit for any of the bacterial, viral, or protozoan pathogens that are of concern to the drinking water industry. As a result of current regulations, there is no financial incentive for diagnostic kit manufacturers to invest resources into developing kits for the drinking water industry. Regulatory pressures on the food industry, on the other hand, such as "zero tolerance" for *Listeria monocytogenes* and *Escherichia coli* O157:H7 in ready-to-eat meat products, have spurred or "forced" the food industry to adopt many of these DNA-based technologies. This, in turn, has made it financially attractive for diagnostic companies to invest in the development of molecular technologies specifically for the food industry (42). Currently, the Association of Official Agricultural Chemists (AOAC) has certified DNA hybridization, PCR-, and real-time PCR-based kits already

Table 1 Partial listing of conventional and real-time PCR-based commercial pathogen detection kits for the food industry

Organism	Commercial name	Technology	Manufacturer
E. coli O157	Genevision	Real-time PCR	Warnex
E. coli O157:H7	Genevision	Real-time PCR	Warnex
E. coli O157:H7	BAX	PCR	Qualicon
E. coli O157:H7	Genevision	Real-time PCR	Warnex
E. coli O157:H7	Probelia	PCR	Bio-Rad
Salmonella spp.	Genevision	Real-time PCR	Warnex
Salmonella spp.	Probelia	PCR	Bio-Rad
Salmonella spp.	FoodProof	Real-time PCR	Roche
Salmonella spp.	GENE TRAK	DNA hybridization	Neogen
Salmonella spp.	BAX	PCR	Qualicon
Listeria spp.	GENE TRAK	DNA hybridization	Neogen
Listeria spp.	BAX	PCR	Qualicon
Listeria spp.	Genevision	Real-time PCR	Warnex
L. monocytogenes	Genevision	Real-time PCR	Warnex
L. monocytogenes	FoodProof	Real-time PCR	Roche
L. monocytogenes	Probelia	PCR	Bio-Rad
Campylobacter spp.	Probelia	PCR	Bio-Rad

commercially available for the food industry to detect key food-borne pathogens such as *Listeria monocytogenes* and *E. coli* O157:H7 (Table 1). The detection of specific microbial pathogens is expected to increase worldwide. In 1999, the U.S. food industry performed as many as 144 million microbiological tests (52). These numbers were 23% higher than what was observed in the preceding year. It must be emphasized that, while the number of microbiological tests was indeed significant, only approximately 26% of those done were pathogen-specific tests.

Advent of Molecular Tools

Currently, the number of microbiological tests being performed in the U.S. food industry, and around the developed world, is evenly divided between traditional tests (such as isolation on selective/differential culture media) and rapid screening methods (52). Pathogen detection methods for foods can be broadly divided into quantitative and qualitative methods. The food industry primarily employs qualitative assays. Culture-based quantitative estimation of pathogens is still limited for the most part to most-probable-number-based methods, since a majority of pathogens such as *Salmonella* require an enrichment step prior to detection and enumeration. Quantitative assays in the food industry are used primarily for specialized purposes in evaluating new pathogen intervention technologies, such as electronic pasteurization or validating "kill steps" in developing HACCP protocols (44). A majority of the molecular detection methods, however, only provide qualitative information about the presence or absence of specific pathogens. For example, the Genevision (Warnex, Inc., Canada) and BAX-PCR (Qualicon, Wilmington, Del.) assays require overnight enrichment prior to PCR amplification of the targets. Studies have shown that molecular methods involving DNA, such as DNA hybridization or PCR amplification, are prone to interference from background (sample) matrix components (28, 41, 46, 65). These inhibitory substances, which in many instances have not been completely identified, inhibit the enzymatic reactions to varying degrees. Current PCR-based molecular methods designed for pathogen detection in foods include an enrichment step which serves two purposes: to dilute out the interfering substances in the food matrix, and to provide some assurance that the DNA sequences being detected are viable cells. The inclusion of enrichment steps, however, led to these assays being only qualitative in nature.

Nucleic acid-based detection methods can be either amplification based or probe based. The primary difference between these two categories of methods is whether or not the target sequence is enzymatically amplified. A recurring theme, however, is that in some of the newer methods both the amplification and probe technologies are integrated into the same method (e.g., the TaqMan-PCR assay and Scorpion primer technology) (30, 55). Conventional PCR technologies have undergone numerous and significant improvements, and detection kits based on PCR methods are now commercially available (Table 1). A salient feature of these kits is their ease of use, since all the necessary reagents are often lyophilized into a tablet form within the reaction tube. All that is now required is the addition of a known volume of an adequately enriched sample into the reaction tube and placement in a temperature-cycling instrument. Current versions of the assays and instrumentation include automated gel electrophoresis modules, thereby reducing the potential for human error and laboratory-based contamination issues. There are a number of self-contained, commercial, portable PCR instruments available for field use (5). The heightened awareness of bioterrorism by the federal agencies has helped accelerate the development of these instruments.

There are amplification protocols that do not involve the use of the proprietary PCR-based technology. Kwoh et al. (29) developed one of the first non-PCR amplification protocols, termed the transcription-based amplification system. It involves the sequential use of reverse transcriptase (to form cDNA) followed by RNA polymerase, which catalyzes the synthesis of multiple copies of the original RNA template. Further improvements in the transcription-based amplification system protocol resulted in the development of the self-sustaining sequence replication amplification technology (24). Since heat denaturation is typically not involved in self-sustaining sequence replication, this amplification can be conducted under isothermal conditions, thereby alleviating the need for specialized temperature-cycling instruments. However, these methods require extensive postamplification steps which can be cumbersome and can potentially lead to contamination errors.

Quantitative (Real-Time) PCR

The need to detect single base substitutions within discrete regions of the genome was one of the factors which accelerated the development of quantitative PCR-based methods. Quantitative PCR methods have a variety of terminologies, including TaqMan-PCR, fluorogenic 5′ nuclease assay, quantitative PCR, and "real-time" PCR. The primary difference between these quantitative techniques and conventional PCR methods is that specifically labeled oligonucleotide probes (e.g., TaqMan probes) along with conventional primers are used in the process. More importantly, PCR product formation is detected during the course of the reaction, which is in contrast to the conventional assay, where the reaction products are detected at the end of the protocol using gel electrophoresis. Thus, in real-time PCR, the PCR product is detected in "real time" as opposed to the "end point" detection in the conventional PCR assay. Since the fluorescence measurements take place during the exponential or logarithmic phase of the amplification, it is possible to quantify (in a relative sense) the formation of a specific PCR product. Appropriate standard curves need to be prepared using the same sample matrix and run alongside the experimental samples for absolute quantification. The primers and the TaqMan probes confer additional specificity compared to the conventional PCR assay. Since the assay relies on real-time detection of product formation and involves the detection of fluorescence, it is more sensitive than the conventional end point assay. More importantly, since this assay does not involve any post-PCR handling, there is reduced potential for PCR contamination. Some of the instrumentation available for real-time PCR analyses permits analysis of up to 384 samples, thereby significantly enhancing throughput. Even though the assay currently requires relatively expensive instrumentation, there is a fairly large selection of real-time PCR-related instruments to suit different budgets and needs. In addition to improvements in instrumentation, enhancements and modifications of the basic technology, using specialized primers Amplifluor and Scorpion and light-upon-extension fluorogenic primers, have also occurred (39, 54). These specialized primers and probes can be more expensive than conventional PCR reagents. However, prices have significantly dropped over the last couple of years, partly due to increased demand. Real-time PCR methods based on molecular beacons are already available commercially for detecting food-borne pathogens (Table 1).

DNA Microarrays

There has been considerable progress in the use of DNA microarrays as research tools (13, 15, 23, 60). There are two basic types of microarrays: (i) those that are prepared by spotting DNA probes on glass or membranes, and (ii) those that have oligonucleotides synthesized in situ applied as DNA chips. Detailed discussion of the microarray technology is provided elsewhere in this book. A majority of the current microarray-based studies are aimed at understanding global gene expression patterns of microbial genes. These so-called gene expression assays rely on the immobilization of DNA probes that would be complementary to mRNA targets of interest. The hybridization patterns or signals of the different target genes thereby provide information on gene expression patterns. The primary advantage of microarrays is that thousands of specific sequences can be simultaneously screened. Though the underlying principle is based on Watson-Crick base-pairing rules, thousands of specific DNA sequences can be immobilized or synthesized on extremely small platforms. This provides a significant breakthrough in the number of samples that can be analyzed and the number of target regions that can be screened in the different samples. Hybridization patterns obtained after presenting the microarray to a sample are detected and analyzed using advanced bioinformatics tools. The utility of DNA microarrays for identification was validated by Wang and coworkers, who used microarrays to identify the virus responsible for severe acute respiratory syndrome (SARS) (59). These authors determined that a novel coronavirus was responsible for SARS by hybridizing cell culture-derived viral nucleic acid with 1,000 virus probes on a microarray. Issues such as use of appropriate experimental controls, normalization approaches, and independent verification of microarray data using alternate approaches are extremely critical (7). As expected, approaches for interpreting microarray data are areas of intense current research. There are a number of published studies documenting the use of microarrays for detection, characterization, and gene expression patterns of pathogens (10, 14, 51). The ultimate sensitivity of microarrays for detecting specific organisms in environmental samples is, however, of questionable value in terms of environmental detection. It is well known that significant amounts of mRNA or numbers of target gene copies are needed for reliable microarray results. The ability to obtain large numbers of target copies without amplification (which could introduce bias) is doubtful.

Recent reviews of microarray technologies also call into question the interlaboratory reproducibility and the applicability of microarrays for environmental applications (12).

Challenges Associated with the Application of Genomics- and Proteomics-Based Tools for Molecular Detection and Characterization

Genomics and proteomics have provided a wealth of information for use in the detection and characterization of microorganisms. Paradigm-shifting changes in sequencing technologies have improved the sequencing of organisms, which, in turn, has greatly benefited microbial molecular detection technologies such as gene amplification and probe-based technologies. At the time of writing this chapter, the National Center for Biotechnology Information lists 1,483 completed viral genomes, 230 completed microbial (bacterial) genomes, and an additional 367 bacterial genomes in progress, not including eukaryotes or archaebacteria. Presently, the 16S rRNA gene is the gene of choice for identifying microorganisms because of the publicly available 16S rRNA gene databases. Current environmental metagenomic studies are based primarily on the amplification of the 16S rRNA gene sequences. Analysis of 16S rRNA gene sequences allows testing for diversity of microbial communities as well as the identification of the members of the microbial community. Identification is accomplished by first amplifying the 16S rRNA gene region, then cloning the resulting amplicons, and ultimately sequencing the 16S rRNA gene clones. Though this method is widely used, it is slow, cumbersome, and obviously biased to a large extent because of the preferential amplification at the PCR level and the subjectivity involved with choosing clones to be sequenced. Because of these lingering issues, alternative methods, like fatty acid analysis and community-level physiological profiling, have been developed to gain a better understanding of the microbial community and to identify community members (34).

The identification of the virus responsible for SARS by microarray technology was successful because closely related virus sequences were known and the SARS coronavirus was able to replicate in cell culture, providing an amplification step and a relatively clean background. It must be reiterated that the detection of the SARS virus by microarray technologies in clinical samples (or tissue culture filtrates) is relatively simpler than detecting this virus in environmental samples, where the target would be expected to be in significantly low numbers compared to the background microbial flora and would have to be initially enriched (amplified) prior to detection. DNA microarrays function well when they measure global gene expression within a single population of cells or, as in the SARS coronavirus case, identification of a culturable microorganism within a clinical sample or specimen. The detection of specific target organisms in food or in environmental samples is complicated by the fact that a vast majority of the background organisms are relatively uncharacterized. Recently developed shotgun sequencing approaches, which have been used to identify the metagenomes of diverse habitats, may alleviate this part of the problem in the future (58). Many pathogens, especially viral pathogens, have a low infectious dose. For microarray-based technologies to be of any value, they would have to identify organisms present in very low numbers. Amplification of select genes can be a solution, but the uncharacterized background, amplification inhibitors, and technical difficulty in amplifying all microorganisms of interest simultaneously and without preferential amplification are problems. Another problem is the identification of novel agents. How will novel microorganisms be detected if they are unlike any known sequences? What probes would be used? There is an inherent problem with using known microbial sequences to detect unknown microorganisms. It is almost assured that many novel microorganisms will be missed. This issue is more of a problem when considering the length of the nucleic acid probe. For example, the Affymetrix GeneChip uses a 25-base-pair probe. Even though the Affymetrix technology will soon become a benchmark in clinical investigations, a 25-base-pair probe may not be long enough to discriminate between closely related microorganisms in the natural environment or in foods. Though the use of longer probes by using plasmids or PCR products may provide a solution, it will nevertheless lead to increased base-pair errors. The probes would thus have to undergo rigorous quality assurance/quality control to ensure accuracy and specificity. These are very important issues, especially for novel pathogenic microorganisms. A possible solution to this problem was put forth by Fraser and Dando (20), who suggested looking for specific virulence genes. For example, DNA microarray probes could be designed for pathogenic microorganism virulence genes. Thus, a Shiga toxin probe would be able to detect both *Shigella* spp. and *E. coli* O157:H7. This sacrifices specificity but will allow for the detection of novel agents. Even though this idea does have merit, the virulence factors of many pathogenic microorganisms are not as clear-cut or easily assayed as the Shiga toxin gene. Examples include microorganisms that mimic the host, thereby making an assay for that particular virulence factor unable to discern between the factor and the host, like the *Campylobacter jejuni* lipo-oligosaccharide protein, which mimics human nerve tissue (66). Other virulence factors cause immune system evasion by varying the cell surface epitopes using hypervariable regions, like *Neisseria meningitidis* serotype A and B (66). Also, many virulence factors that help a microorganism survive inside the human body may also be used by nonpathogenic microorganisms to help them survive in the environment, like siderophores (36).

Interest in proteomics for microbial identification and detection has received recent interest because of the more limited number of proteins that are available compared to the total genetic information in an organism. For example, the human genome consists of 20,000 to 25,000 protein-encoding genes, whereas the actual nucleotide count is 3.08 Gbp (26). At any one time about 10% of genes are being actively transcribed (40). Therefore, the number of proteins present is several orders of magnitude less than that of the total pieces of genetic information available for transcription. This narrows down the target of interest for detecting a microorganism. Many major proteins also are expressed in high numbers in or on a microorganism, providing a self-enriching method to detect a microorganism.

Technological advances in proteomic tools, such as matrix-assisted laser desorption ionization–time of flight (MALDI-TOF), have allowed researchers to identify proteins from either purified or raw protein mixtures. The use of MALDI-TOF relies on a database of either gene sequences, for predicted proteins, or protein sequences to identify a protein fingerprint. The bioengineering of antibodies using phage display assays for identifying and selecting antibodies as well as the production of antibodies in *E. coli* has made it possible to rapidly select and produce monoclonal antibodies (8). The growing number of proteins in databases as well as technological advances has given an impetus for using proteomics for identification and detection of microorganisms. Unfortunately, many of the problems associated with DNA microarrays are also applicable to proteomic technologies. Furthermore, antibody-based microarrays encounter much the same problems as those associated with enzyme-linked immunosorbent assays. The complexity, as well as the similarity, of proteins in the environment would make it difficult to identify specific proteins of interest. If the microorganism of interest is commonly found in a habitat, its expressed proteins will be very similar to the background microorganisms (55). If the microorganism is a temporary resident or a passive inhabitant (virus, spore, cyst, or oocyst), then the level of detectable protein may be too low for a positive identification. Optimization and cross-reactivity are problems that would have to be solved for this technology to become effective. This is even more important when considering the complex protein mixtures of many foods and environmental samples. Thus, processing of the actual food or environmental samples would be as important as the microarray assay. Processing difficulties would range from minimal processing for airborne samples to a rigorous processing step required for samples with high organic content, like foods, soils, and certain surface waters.

Other Contemporary Pathogen Detection Technologies

Although there are significant advantages to detecting target organisms using nucleic acid amplification techniques, a number of other detection technologies, based on immunoassays, flow cytometry, terahertz sensors, evanescent wave technologies, MALDI-TOF, immunomagnetics, and neural networks, have also been developed (6, 12, 25, 32, 63). Biosensors and microarrays are thought to find increasing application in identifying the presence of specific pathogens and cellular components such as cell surface receptors, ribosomal RNA molecules, and DNA molecules (4, 9, 12). Biosensors will be used in the future for either real-time detection of specific organisms (if the target numbers are not limiting) or for confirmatory purposes. A number of potential targets for biosensors have been identified. These include the following: cell surface proteins such as porin or siderophore molecules (of which there are approximately 200,000 copies per cell); cell surface polysaccharides, such as lipopolysaccharides (approximately 2,000,000 copies per cell); ribosomal proteins and RNA in rapidly dividing cells (20,000 copies per cell); nonribosomal RNA molecules (100 to 1,000 molecules per cell); and nonribosomal proteins (3,000 molecules per cell). It can be expected that the current nucleic acid-based molecular methods, such as probe and gene amplifications, will get integrated into biosensor technologies in the future. These new developments, which will occur at the interfacial fields of traditional scientific disciplines such as microbiology, biochemistry, and electrical engineering, highlight the growing importance of multidisciplinary collaborations (42).

Tracking the origins of food-borne pathogens and identifying the pathways by which they enter the food chain is critically important in preventing food-borne illnesses. Characterization of bacterial strains helps in epidemiological studies. There are several nongenetic methods or phenotypic methods, such as serotyping, antimicrobial susceptibility testing, phage typing, and multilocus enzyme electrophoresis. The genetic or molecular pathogen characterization techniques that are commonly used to characterize food-borne pathogens are plasmid typing (38), pulsed-field gel electrophoresis (PFGE) (53), restriction fragment length polymorphism analysis (22), amplified fragment length polymorphism (AFLP) analysis (16), random amplified polymorphic DNA PCR (61), arbitrarily primed PCR (64), repetitive element PCR (21), and multilocus sequence typing (17). The Centers for Disease Control and Prevention (CDC) use standardized PFGE protocols as part of the PulseNet System for tracking food-borne pathogens. Foley and Walker (19) recently provided a qualitative comparison of the different tools for molecular differentiation of food-borne pathogens (Table 2).

Table 2 Characteristics of contemporary bacterial differentiation methods[a]

Method[b]	Discrimination	Reproducibility	Difficulty	Relative cost	Relative time
Serotyping	Low	High	High	Moderate	Moderate
Antibiotic resistance profiling	Low	High	Low	Low	Low
MLEE	Moderate	High	High	Moderate	High
Plasmid profiling	Low	High	Low	Low	Low
PFGE	High	High	Moderate	High	High
Ribotyping	Moderate	High	Moderate	High	Moderate
Multilocus sequence typing	High	High	High	High	High
AFLP	High	High	Moderate	High	Moderate
RAPD PCR	Moderate	Moderate	Moderate	Moderate	Moderate
Rep-PCR	Moderate	Moderate	Moderate	Moderate	Moderate

[a]Modified from reference 19.
[b]MLEE, multilocus enzyme electrophoresis; RAPD, random amplified polymorphic DNA; Rep, repetitive element PCR.

Critical Issues Surrounding Molecular Methods for Detecting Organisms in Foods and the Environment

Detection Sensitivity

Organisms, whether they are pathogens in foods, drinking water, or surface water, are generally present in low numbers. To detect these organisms, their population sizes are normally increased by enrichment procedures which are performed in the laboratory. The presence of the target organisms in low numbers is extremely significant in terms of the application of molecular techniques (assay sensitivity), as shown in Fig. 1. Even though molecular methods are often touted as being highly sensitive (detection limit of 1 to 10 gene targets), they are generally not of value if they are directly employed for detecting organisms in food or environmental samples. These methods can thus only perform if there are significant levels of a contaminant in a sample, or if multiple small aliquots of the same sample are analyzed. As would be expected, current protocols for use with molecular methods are significantly hampered by this sample volume limitation. Sample processing and sample volumes used for detection can be considered to be the two most critical aspects of molecular assays for detecting pathogens in food or environmental samples. This is because most molecular assays, such as PCR and real-time PCR assays, use sample volumes ranging from 2 to 10 μl/assay. Real-time PCR assays, microarrays, and other technologies, such as biosensors, which are thought to be "breakthrough technologies" for detecting pathogens are at present significantly limited due to their "maximum sample volume" limitations. For example, if we assume a virulence gene expression (microarray-based) biosensor

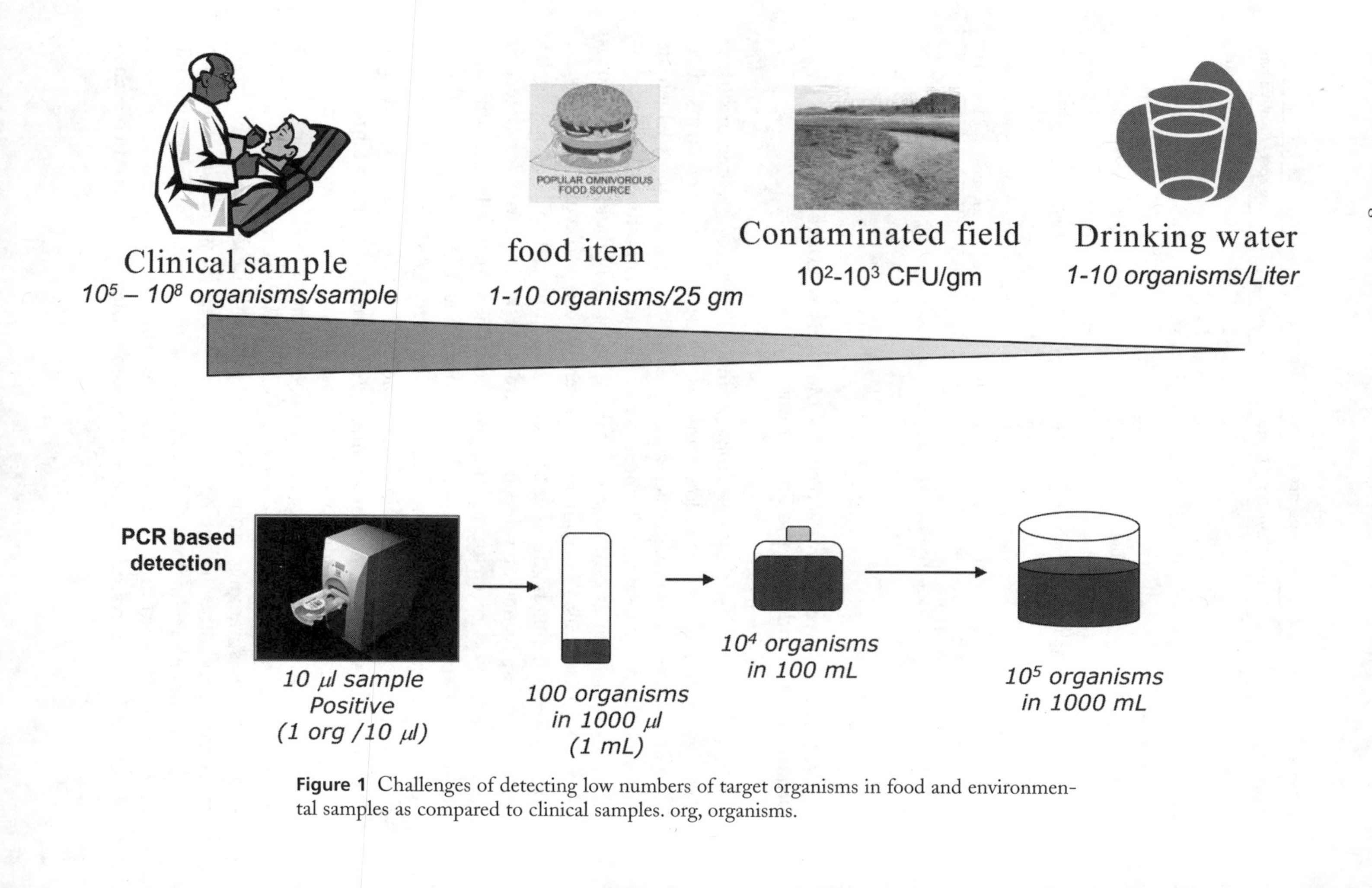

Figure 1 Challenges of detecting low numbers of target organisms in food and environmental samples as compared to clinical samples. org, organisms.

(which employs a 50-μl sample volume for analysis) has been developed which can detect one cell of a target organism and the biosensor/microarray is to be used to detect that organism in 500 g of soil or 500 gallons of processed orange juice, it requires that either 500 g of soil or 500 gallons of juice be concentrated down to 50 μl. This 500-g or 500-gallon volume should contain a total of 3.6×10^7 cells of the target organism! Only under these circumstances would the biosensor detect the organism (if the 50-μl aliquot of the sample is used for the biosensor analysis). Target organism concentrations below this threshold would be below the detection sensitivity of the biosensor and result in a false-negative result. As another example, let us assume that 100 g of a contaminated soil sample contains 10 target organisms of specific bioremediation potential. If this sample was homogenized in 200 ml of buffer, the buffer would theoretically contain only 0.05 organisms per 1 ml. If this sample was tested by any assay (conventional or molecular), it would invariably result in a "nondetect." In other words, only assays (molecular or conventional) that employ 20 ml or larger will be at or greater than the one-organism detection limit and thus capable of detecting the target organism. Thus, for biosensors or microarrays, or any other molecular method, to be of any practical value, the sample has to be adequately concentrated so that the contaminants in the concentrated samples are within the sensitivity thresholds of the molecular assay. It is the extraction efficiency of the sample processing protocol, and not really the detection sensitivity of the assay, that is the critical bottleneck. In order to avoid significant confusion regarding the sensitivity thresholds, the detection sensitivity of the assay and the overall sensitivity of the method have to be clearly specified and understood. Very often the sensitivity threshold of an assay is erroneously calibrated by determining the smallest number of organisms that can be detected in the sample that is to be analyzed. The ultimate detection sensitivity of an assay should be based on the minimum number of organisms that need to be present in the sample prior to sample processing. Thus, it is evident how critical the detection sensitivity is, especially when attempting to detect pathogens in foods or identify unique genotypes in the environment. The "Achilles heel" in the use of molecular technologies for pathogen detection is the inability to obtain suitable samples for analysis. The current lack of effective microbial capture and concentration methods to recover small numbers of potential targets from food, water, and soil samples into a suitable sample for analysis is the key hurdle.

Sample Processing

The generalized protocol for the initial steps in processing food samples is that a defined amount of sample is homogenized (or "stomached") in approximately 250 ml of a suitable buffer, and an aliquot of the homogenized sample is enriched using specialized selective media or directly used in the molecular assay. All commercial PCR and real-time PCR assays require at least an overnight enrichment step. The basic concept of homogenizing or "stomaching" a large portion of the sample is advantageous because it increases the likelihood of detecting pathogens that may be at low concentrations or irregularly distributed on the sample. The use of large sample volumes for the initial homogenization also increases the likelihood that larger amounts of the potential inhibitory compounds will get copurified with the target organisms. Thereby, concentration or purification of samples may increase unwanted "components" that may cause problems in molecular or infectivity assays further downstream. These inhibitory components can also exert variable effects on the enzymatic reactions. Thus, in some instances, using large volumes of the original homogenate for plating or molecular analysis can be counterproductive. This is primarily because the detection sensitivity of the overall assay is proportionally reduced, due to an increase of inhibitory compounds. These inhibitors need to be removed to prevent false-negative results. Commercial products such as the Soil-DNA kit (MoBio, Solano, Calif.) and PrepMan (Applied Biosystems, Foster City, Calif.) are available for "cleaning" or "purifying" samples. These kits are somewhat effective in reducing or removing inhibitory components (47). Instrumentation for automated extraction of DNA from low-biomass samples is currently available (IGene Diagnostics, Menlo Park, Calif.). It is important to realize that whenever such sample purification kits are used there is a very high likelihood for potential target nucleic acids to be reduced also. This is not of particular concern in clinical samples which contain a large number of potential targets. However, it can be a significant problem when working with food and environmental samples (43, 45).

Though the purpose of large volume concentrations and enrichments is to help enhance detection sensitivity, the methods to detect specific groups of organisms can be vastly different. For example, a bacterial enrichment or concentration could take from as little as a few minutes up to 72 hours depending on the method used. A viral or protozoan assay could take 24 to 48 hours to sample and process and an additional 1 to 4 weeks to assay for infectivity or a few hours to do a molecular analysis. The method will also be dependent on the sample and the organic load of the sample.

Due to the physiological differences among and between organisms (bacteria versus viruses versus protozoa), it is difficult to envision a single protocol to concentrate all microorganisms of interest. The fact that a protocol to concentrate all or many microorganisms has yet to be developed is a problem that has to be solved before the first microbial biosensor technologies become truly useful. Recent studies have highlighted the differences between different groups of viruses in the way they attach to lettuce surfaces (57). Though the general mechanism by which enteric viruses attach to sediments is relatively well known, our studies suggest that the current model for virus attachment to sediments does not adequately explain the attachment of viruses to lettuce. These findings have major implications in the recovery of viruses from lettuce per current sample processing protocols.

Given the variety of food and environmental samples (soils, sludges, water, waste-amended soils, etc.), and the variability in target organism loads, additional research in sample processing is critically needed. The possible research areas include efficient sample capture, concentration, and purification methods. The ability to capture multiple target organisms without concentrating potential inhibitory components will be a key benchmark of such technologies in the future.

Regulatory Acceptance of Molecular Methods

Regulatory acceptance of molecular methods for microbial identification and characterization is a substantial hurdle that must be overcome. As mentioned earlier in the chapter, although a number of molecular methods have been validated for federal reporting purposes in the food industry, there are no molecular methods that have been approved for use in the drinking water industry. One key hurdle with the acceptance of molecular methods is the inability to differentiate between live and dead microorganisms. Molecular methods, such as direct PCR assays, have often been cited as not being relevant for food safety applications since they do not provide any indication of whether the pathogen is viable. There is validity to both sides of the argument as to whether or not direct molecular methods are reliable indicators of the presence of viable, infectious organisms. Commercial assays developed for the food industry have overcome this hurdle by incorporating an initial enrichment step, thereby confirming the detection of viable organisms. Current AOAC-approved PCR-based methods, such as BAX-PCR assays, solve the issue of viability by incorporating an enrichment step prior to gene amplification. The AOAC Research Institute has certified more than 40 commercial pathogen and toxin test kits since 1991, when the Performance-Tested Methods program was created (56). The Performance-Tested Methods program was initiated to provide an independent third-party verification of a detection kit's capabilities (2). For organisms that must maintain an internal energy potential, internalization of dyes or binding of nucleic acid has been used to differentiate between live and dead microorganisms. The dye exclusion methods (which in the truest sense are not molecular assays) for cysts and oocysts of *Giardia* spp. and *Cryptosporidium* spp. have been approved by the United States Environmental Protection Agency (U.S. EPA) for monitoring purposes (18). In addition to technical challenges, there are also fundamental questions related to pathogenicity, virulence, and viability. For pathogen-monitoring purposes, questions such as, "What level of identification is acceptable?" and "Is genus level identification acceptable?" still linger. Even though identifying all *Salmonella* spp. is important, it may not be important to identify all *Bacillus* spp. strains. There are 2,463 *Salmonella* serotypes. Would all serotypes need to be detected or only the most common *Salmonella* serotypes? The answers to these questions would be dependent on the type of assay desired, type of foods, and whether the assay is for routine pathogen surveillance or for end-product testing. The applicability of a particular molecular method for detecting pathogens in food samples, however, also depends on the question being asked. If the question is whether or not the sample harbored specific pathogens (irrespective of their current viability) at some point of time, then it is difficult to argue against the use of molecular methods. Even though positive PCR results signify the presence of a specific microbial contamination, they do not, however, indicate any particular measure of health risks. Conversely, negative PCR results do not suggest that there is a lack of contamination, nor do they provide any indication of the health risks. A negative result is very difficult to interpret without a clear delineation of the detection sensitivity of the assay (33). Detection limit is, in turn, influenced by factors such as the efficiency of target recovery, the presence of inhibitory compounds, and the volume of sample that is analyzed.

Loge et al. (33) have recently proposed a risk-based framework for providing biological relevance to PCR assays. The framework not only provides a means of assessing the health risks associated with PCR results but also provides a tool to identify the impacts of future improvements in PCR-based assays. The framework is based on a method for quantifying the detection limits of any PCR-based assay. The detection limit can be quantified using the relationship

$$\text{Detection limit} = \frac{RI}{V_f} \cdot \frac{S}{\%V_p}$$

where $R = 1/\text{Re}$ (unitless), Re is the recovery of organisms from the sample, I is the inverse of the dilution factor necessary to remove compounds that inhibit PCR (unitless), V_f is the volume of the sample, S is the sensitivity of the PCR assay (CFU), and $\%V_p$ is the fraction of the concentrated sample analyzed by PCR.

For pathogens currently in the zero-tolerance status, such as *E. coli* O157:H7 and *Listeria monocytogenes*, the use of direct PCR results to institute effective disinfection programs or voluntary recalls can be justified. Given the number of unknowns surrounding pathogen/target organism distribution within food and environmental samples, sampling, and sample processing, it is risky to assume that PCR results are of insignificant value since positive results can originate from nonviable cells. Opponents of molecular methods often also use the argument that there is a lack of significant correlation between the results of molecular assays and conventional assays when samples are analyzed. This lack of correlation is not surprising, since the assay results depend on the sample volumes being employed, the distribution of organisms within the original sample (thereby dictating the presence or absence of the organisms in sample aliquots), and the detection sensitivity of the different assays. It would be erroneous to explain that increased detection of target organisms by molecular methods is only due to detection of free nucleic acids or nonviable cells.

From a regulatory standpoint, there is a growing demand that molecular methods be given serious consideration. This is partly because U.S. regulatory agencies are attempting to base all regulatory decisions on a risk-based framework. For some human viral pathogens, such as noroviruses (Norwalk-like viruses), which are responsible for over 66% of all food-borne human illnesses, culture-based methods are currently unavailable (35). Given the public health risk importance of some specific pathogens such as noroviruses in terms of food-borne illnesses, it would not be surprising that regulatory agencies accept the use of molecular methods for such pathogens in the near future. Regulatory approval of molecular methods implies that strict quality assurance (QA)/quality control (QC) controls have to be developed for the various techniques, as well as for the calibration of method performance and interlaboratory validation. Currently, for PCR-based methods QA/QC procedures include the integration of internal spiked sample controls and the sequencing of PCR products.

A number of studies have shown that changing demographics, inherent changes in the pathogens themselves, changes in food preparation, and changes in distribution and consumption have the potential to adversely affect human health (1, 50). Quantitative risk assessment, which will ultimately help in reducing these risks, is a possible framework for designing programs to reduce food-borne disease (31). Hazard identification and exposure assessment are key components of quantitative microbial risk assessment; molecular methods can be used in both.

Molecular methods will find increasing applications in risk assessment in the future. Regulatory agencies such as the U.S. Department of Agriculture (USDA) and the Food and Drug Administration (FDA) would have to deal with the issue of emerging and reemerging infectious organisms and agents such as *Cyclospora*, enteric viruses, and prions in foods. These agencies have the responsibility to identify the critical contaminants and adequately assess the potential risks involved. Regulatory agencies would have to ultimately prioritize these infectious organisms for regulatory purposes. At present, there is a tendency towards the establishment of a zero-tolerance goal for certain microbial contaminants. Indicator organisms such as *E. coli* are also used to show the possible presence of fecal contamination, thereby indirectly indicating the possible presence of pathogens. However, a number of studies have shown that using indicator organisms can be fraught with limitations, since fecal coliforms and *E. coli* do not exhibit the same resistance to disinfectants and decontamination approaches as do protozoa and viruses.

Identifying the virulence function activity relationships of food-borne pathogens is possible with molecular methods. The National Research Council (NRC) in their study on drinking-water contaminants has proposed the use of virulence function activity relationships as a method to predict the ultimate virulence of microbial contaminants (37). The concept of using a multitude of characteristics (to prioritize the contaminants in terms of virulence), rather than relying on a few characteristics, has significant advantages. Since the ultimate virulence of a pathogen depends not only on the organism's virulence attributes but also on the host susceptibility, as well as the ecology of the organism in the substrate(s) that the host is exposed to, making use of multiple "descriptors" provides a science-based risk categorization process. We term the process of understanding the molecular ecology and characteristics of pathogens in matrices that are relevant to human health "substrate-activity-function-virulence relationships" (Fig. 2). The NRC has recommended that identifying the critical contaminants (microbial and chemical) follow a stepwise process. The NRC recommends that the process of selecting the "Critical Contaminant List" (CCL) from a "Preliminary CCL" (PCCL) be based on neural network analysis of all available genomic, proteomic, and epidemiological databases. Current culture-based methods are seriously limited in their ability to provide the type of information needed for such an endeavor. Molecular methods for detection and genetic/molecular characterization can be expected to fill this niche.

Personnel and Laboratory Infrastructure Needs for Molecular Methods

There is little doubt that many of the current molecular methods of pathogen detection have significant advantages over conventional microbiology methods. However, the reality is that over 90% of the current microbiology tests used in the food industry (where commercial molecular assays are available) still rely on conventional agar-plating techniques. Some of the concerns that the food industry has about accepting emerging molecular methods are the cost, the level of technical expertise required (to perform the tests and interpret the test results), and the laboratory infrastructure required. Laboratory-

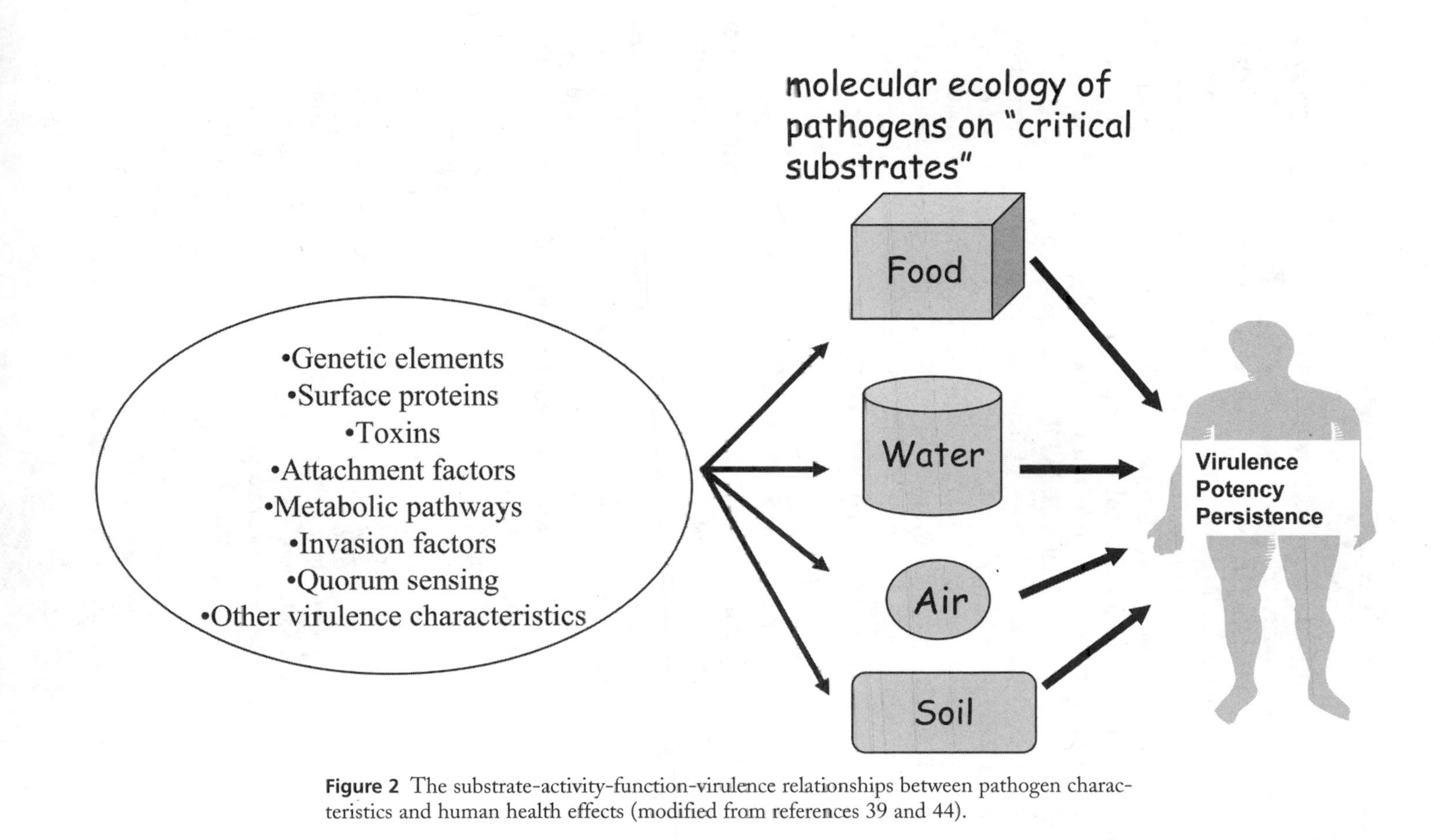

Figure 2 The substrate-activity-function-virulence relationships between pathogen characteristics and human health effects (modified from references 39 and 44).

based contamination very often can invalidate PCR results (3). Accidental human-induced cross-contamination of reagents and samples is still a very serious problem facing many laboratories that employ molecular methods, especially PCR-based techniques. Specific training of laboratory personnel and changes in the sample flowthrough patterns in the laboratory as well as the actual physical layout of the laboratory can be instrumental in reducing the potential for cross contamination. As the detection and characterization methods become more specialized and sophisticated, the equipment needs and the expertise levels of a molecular/microbiological QA/QC laboratory will undoubtedly increase. Molecular methods will not totally replace conventional microbiology methods. While molecular methods will allow for faster, more sensitive, and more in-depth detection and characterization capabilities, they will carry with them an increased need for internal quality control and personnel who are adequately trained to ensure accuracy and proper interpretation of their results. In the final analysis, however, it will be regulatory and liability pressures which ultimately dictate whether or not, and to what extent, a specific industry embraces molecular methods.

FUTURE OUTLOOK AND RESEARCH STRATEGIES

The scientific community, the regulatory community, and the general public need to realize that it is impossible and incorrect to cubbyhole food safety separately from environmental quality and public health. As shown in Fig. 3, the organisms that are of concern to food safety find their origins in the environment, in farm animals, and in humans. Organisms are transmitted between humans, animals, the environment, and foods through air, water, soil, and equipment routes. Multiple disciplines and factors ultimately impact the safety and quality of foods. Although the transmission of hepatitis A virus (HAV) infection may be decreased by vaccinating food service workers, there are other routes for hepatitis A transmission that are extremely difficult to control. Safety in food service operations can depend highly upon the incoming food ingredients. Enteric viruses such as HAV can enter the food chain through contaminated irrigation water or wash water. In 1988, 202 cases of HAV were reported in one outbreak, and upon investigation, the outbreak was traced back to commercially distributed lettuce. Contamination of the lettuce was found to have occurred prior to local distribution (48). In 2003, approximately 555 people were infected with HAV at one restaurant through contaminated green onions. HAV was traced back to contamination in the distribution system or during growing, harvesting, packing, or cooling (11). Pathogens such as *L. monocytogenes* find ecological niches (to survive) within food processing facilities in drains and on food-handling equipment. Thus,

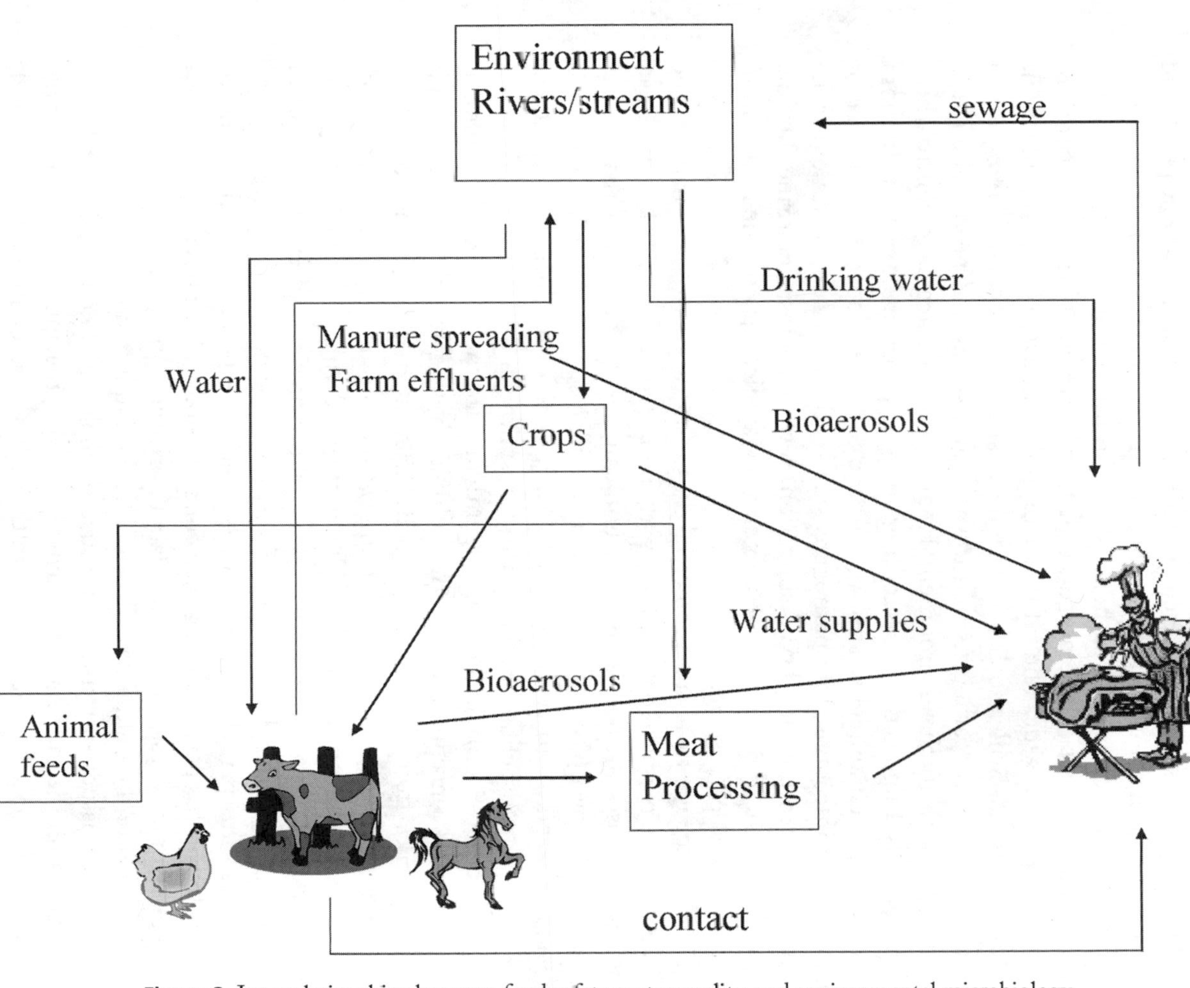

Figure 3 Interrelationships between food safety, water quality, and environmental microbiology.

the design, the material, and the ability to clean and disinfect the equipment all have significant impacts on our ability to control this pathogen.

In the future, improved analytical methods will lead to the identification of the environmental reservoirs of a number of pathogens that are of relevance to food safety. Improved surveillance and tracking studies will lead to the identification of many of these pathogens that could be classified as emerging or reemerging pathogens. Studies on the biotic and abiotic factors that influence the emergence of these pathogens will gain increased attention worldwide. The influence of global climate change on infectious disease trends will also be another intensively studied area. The fate of newly identified pathogens in air, water, and soil would need to be elucidated to identify the environmental management strategies to control these pathogens. Improved waste handling and waste treatment systems for both municipal and animal wastes would have to be developed to minimize or eliminate infections from pathogens such as SARS. Water treatment technologies incorporating UV disinfection, reverse osmosis, chlorine dioxide, and ozone need to be evaluated for specific applications on the farm and in postharvest processing.

Even though there are commercial diagnostic assays to detect many of the key food-borne bacterial pathogens, there are no commercial kits to detect enteric viruses in food. This is an unfortunate situation given that viruses account for over 65% of known food-borne illnesses in humans. Infected food handlers and a variety of food items such as fresh produce, frosted bakery products, and shellfish are known to transmit viral infections to the human population (49, 62). Methods are urgently needed to process food samples, to extract and concentrate viruses, and, ultimately, to detect the viruses without interference from the sample matrices. Current methods to detect enteric viruses using tissue culture are extremely time consuming and labor intensive and are consequently expensive. Qualitative molecular methods that provide only a cursory positive/negative result for the presence of enteric viruses in food may have only limited applicability given the potential to detect noninfectious virus particles. There have been a few reports on the detection of infectious virus particles using molecular methods (27). These methods need to be rigorously tested in multiple laboratories and on different types of samples to prove their ultimate utility for the food industry.

Critical issues surrounding the use of molecular methods for pathogen detection in the food industry include choosing the appropriate sample volumes to be tested, sample concentrations, and purification procedures, and, ultimately, regulatory acceptance of the molecular methods. Additionally, the proper uses of indicator microorganisms, microbial risk assessment studies, and reemerging organisms are issues that regulatory agencies will need to address before implementation of new technologies. Although spectacular breakthroughs have been achieved in terms of development of biosensors for pathogen detection, much work still remains to be done with respect to the first step in detecting pathogens, namely, sample processing. The processing has to be extremely efficient at recovering small numbers of organisms in and on food samples but also has to be amenable to downstream applications such as molecular assays (i.e., PCR, quantitative PCR, or biosensors) without any interference from the sample matrix. PCR is ideally suited for the food industry. It may provide more effective product quality monitoring and prevent costly recalls. Even though molecular methods will not totally replace conventional microbiological assays, molecular methods in the future can allow for faster, more sensitive, and better characterization capabilities. This will entail an increased awareness of QA/QC as it pertains to molecular methods and a trained workforce that can keep up with the rapidly changing technologies. The CDC have supported the nationwide PulseNet, which relies on PFGE fingerprinting and is considered the gold standard for epidemiological tracking. In the future, methods to differentiate bacterial strains will be developed that are rapid, reproducible, and, hopefully, automated. The current 16S rRNA gene-based fingerprinting method (termed riboprinting), though automated and sophisticated, is still far from being as discriminatory as PFGE. It is critical to improve and devise new discriminatory tools. Methods based on multilocus variable number tandem repeat analysis show promise. It is new methodologies such as these that will help us to move forward into the future with more discriminatory results and with a more rapid response time.

REFERENCES

1. **Altekruse, S. F., M. L. Cohen, and D. L. Swerdlow.** 1997. Emerging foodborne diseases. *Emerg. Infect. Dis.* **3**:285–293.
2. **AOAC International.** 2003. *Performance Tested Methods*. [Online.] http://www.aoac.org/testkits/testedmethods.html.
3. **Baselski, V. S.** 1996. The role of molecular diagnostics in the clinical microbiology laboratory. *Clin. Lab. Med.* **16**:49–60.
4. **Bavykin, S. G., J. P. Akowski, V. M. Zakhariev, V. E. Barsky, A. N. Perov, and A. D. Mirzabekov.** 2001. Portable system for microbial sample preparation and oligonucleotide microarray analysis. *Appl. Environ. Microbiol.* **67**:922–928.
5. **Belgrader, P., S. Young, B. Yuan, M. Primeau, L. A. Christel, F. Pourahmadi, and M. A. Northrup.** 2001. A battery-powered notebook thermal cycler for rapid multiplex real-time PCR analysis. *Anal. Chem.* **73**:286–289.
6. **Bernardo, K., N. Pakulat, M. Macht, O. Krut, H. Seifert, S. Fleer, F. Hunger, and M. Kronke.** 2002. Identification and discrimination of *Staphylococcus aureus* strains using matrix-assisted laser desorption/ionization–time of flight mass spectrometry. *Proteomics* **2**:747–753.

7. **Bowtell, D., and J. Sambrook (ed.).** 2003. *DNA Microarrays: a Molecular Cloning Manual.* Cold Spring Harbor Laboratory Press, Cold Spring Harbor, N.Y.
8. **Brekke, O. H., and I. Sandlie.** 2003. Therapeutic antibodies for human diseases at the dawn of the twenty-first century. *Nat. Rev. Drug Discov.* **2**:52–62.
9. **Call, D. R., F. J. Brockman, and D. P. Chandler.** 2001. Detecting and genotyping Escherichia coli O157:H7 using multiplexed PCR and nucleic acid microarrays. *Int. J. Food Microbiol.* **67**:71–80.
10. **Call, D. R., M. K. Borucki, and T. E. Besser.** 2003. Mixed genome microarrays reveal multiple serotype and lineage-specific differences among strains of *Listeria monocytogenes. J. Clin. Microbiol.* **41**:632–639.
11. **Centers for Disease Control and Prevention.** 2003. Hepatitis A outbreak associated with green onions at a restaurant—Monaca, Pennsylvania, 2003. *Morb. Mortal. Wkly. Rep.* **52**: 1155–1157.
12. **Chandler, D. P., and A. E. Jarrell.** 2003. Enhanced nucleic acid capture and flow cytometry detection with peptide nucleic acid probes and tunable-surface microparticles. *Anal. Biochem.* **312**:182–190.
13. **Chandler, D. P., and A. E. Jarrell.** 2005. Taking arrays from the lab to the field: trying to make sense of the unknown. *BioTechniques* **38**:591–600.
14. **Chizhikov, V., A. Rasooly, K. Chumakov, and D. D. Levy.** 2001. Microarray analysis of microbial virulence factors. *Appl. Environ. Microbiol.* **67**:3258–3263.
15. **Cho, J. C., and J. M. Tiedje.** 2002. Quantitative detection of microbial genes by using DNA microarrays. *Appl. Environ. Microbiol.* **68**:1425–1430.
16. **Desai, M., E. J. Threlfall, and J. Stanley.** 2001. Fluorescent amplified-fragment length polymorphism subtyping of the *Salmonella enterica* serovar Enteriditis phage type 4 clone complex. *J. Clin. Microbiol.* **39**:201–206.
17. **Enright, M. C., and B. G. Spratt.** 1999. Multilocus sequence typing. *Trends Microbiol.* **7**:482–487.
18. **Environmental Protection Agency.** 2001. *Method 1623: Cryptosporidium and Giardia in Water by Filtration/IMS/FA.* EPA 821-R-01-025. [Online.] http://www.epa.gov/waterscience/methods/1623.pdf.
19. **Foley, S. L., and R. D. Walker.** 2004. Methods for differentiation among bacterial foodborne pathogens, p. 303–316. *In* R. C. Beier, S. D. Pillai, T. D. Phillips, and R. L. Ziprin (ed.), *Preharvest and Postharvest Food Safety: Contemporary Issues and Future Directions.* IFT Press Series. Blackwell Publishing, Ames, Iowa.
20. **Fraser, C. M., and M. R. Dando.** 2001. Genomics and future biological weapons: the need for preventive action by the biomedical community. *Nat. Genet.* **29**:253–256.
21. **Garaizar, J., N. Lopez-Molina, I. Laconcha, D. L. Baggesen, A. Rementeria, A. Vivanco, A. Audicana, and I. Perales.** 2000. Suitability of PCR fingerprinting, infrequent-restriction-site PCR, and pulsed-field gel electrophoresis, combined with computerized gel analysis, in library typing of *Salmonella enterica* serovar Enteriditis. *Appl. Environ. Microbiol.* **66**:5273–5281.
22. **Gendel, S. M., and J. Ulaszek.** 2000. Ribotype analysis of strain distribution in Listeria monocytogenes. *J. Food Prot.* **63**:179–185.
23. **Gilles, P. N., D. J. Wu, C. B. Forster, P. J. Dillon, and S. J. Chanock.** 1999. Single nucleotide polymorphic discrimination by an electronic dot blot assay on semiconductor microchips. *Nat. Biotechnol.* **17**:365–370.

24. **Guatelli, J. C., T. R. Gingeras, and D. D. Richman.** 1989. Nucleic acid amplification in vitro: detection of sequences with low copy numbers and application to diagnosis of human immunodeficiency virus type 1 infection. *Clin. Microbiol. Rev.* **2**:217–226.

25. **Hutchinson, A. M.** 1995. Evanescent wave biosensors. Real-time analysis of biomolecular interactions. *J. Mol. Microbiol. Biotechnol.* **3**:47–54.

26. **International Human Genome Sequencing Consortium.** 2004. Finishing the euchromatic sequence of the human genome. *Nature* **431**:931–945.

27. **Ko, G., T. L. Cromeans, and M. D. Sobsey.** 2003. Detection of infectious adenovirus in cell culture by mRNA reverse transcription-PCR. *Appl. Environ. Microbiol.* **69**:7377–7384.

28. **Kreader, C. A.** 1996. Relief of amplification inhibition in PCR by bovine serum albumin or T4 gene 32 protein. *Appl. Environ. Microbiol.* **62**:1102–1106.

29. **Kwoh, D. Y., G. R. Davis, K. M. Whitfield, H. I. Chappelle, L. J. DiMichele, and T. R. Gingeras.** 1989. Transcription-based amplification and detection of amplified human immunodeficiency virus type 1 with a bead based sandwich hybridization format. *Proc. Natl. Acad. Sci. USA* **86**:1173–1177.

30. **Lakowicz, J. R.** 1999. *Principles of Fluorescent Spectroscopy*, 2nd ed., p. 698. Plenum Press, New York, N.Y.

31. **Lammerding, A. M., and G. M. Paoli.** 1997. Quantitative risk assessment: an emerging tool for emerging foodborne pathogens. *Emerg. Infect. Dis.* **3**:483–487.

32. **Liu, Y., J. Ye, and Y. Li.** 2003. Rapid detection of Escherichia coli O157:H7 inoculated in ground beef, chicken carcass, and lettuce samples with an immunomagnetic chemiluminescence fiber-optic biosensor. *J. Food Prot.* **66**:512–517.

33. **Loge, F. J., D. E. Thompson, and D. R. Call.** 2002. PCR detection of specific pathogens: a risk based analysis. *Environ. Sci. Technol.* **36**:2754–2759.

34. **Matos, A., J. L. Garland, and F. W. Fett.** 2002. Composition and physiological profiling of sprout-associated microbial communities. *J. Food Prot.* **65**:1903–1908.

35. **Mead, P. S., L. Slutsker, V. Dietz, L. F. McCaig, J. S. Brese, C. Shapiro, P. M. Griffin, and R. V. Tauxe.** 1999. Food-related illness and death in the United States. *Emerg. Infect. Dis.* **5**:607–625.

36. **Molina, M. A., J. L. Ramos, and M. Espinosa-Urgel.** 2003. Plant-associated biofilms. *Rev. Environ. Sci. Biotechnol.* **2**:99–108.

37. **National Research Council.** 2001. *Classifying Drinking Water Contaminants for Regulatory Consideration.* National Academy Press, Washington, D.C.

38. **Nauerby, B., K. Pedersen, H. H. Dietz, and M. Madsen.** 2000. Comparison of Danish isolates of *Salmonella enterica* serovar Enteritidis PT9a and PT11 from hedgehogs (*Erinaceus europaeus*) and humans by plasmid profiling and pulsed-field gel electrophoresis. *J. Clin. Microbiol.* **38**:3631–3635.

39. **Nazarenko, I., B. Lowe, M. Darfler, P. Ikonomi, D. Schuster, and A. Rashtchian.** 2002. Multiplex quantitative PCR using self-quenched primers labeled with a single fluorophore. *Nucleic Acids Res.* **30**:37–43.

40. **Pandey, A., and M. Mann.** 2000. Proteomics to study genes and genomes. *Nature* **405**:837–846.

41. **Pena, J., S. C. Ricke, C. L. Shermer, T. Gibbs, and S. D. Pillai.** 1999. A gene amplification-hybridization sensor based methodology to rapidly screen aerosol samples for specific bacterial gene sequences. *J. Environ. Sci. Health A* **34**:529–556.

42. **Pillai, S.D.** 2004. Molecular methods for microbial detection, p. 455. *In* R. C. Beier, S. D. Pillai, T. D. Phillips, and R. L. Ziprin (ed.), *Pre-harvest and Post-Harvest Food Safety: Contemporary Issues and Future Directions.* Institute of Food Technologists/Iowa State Press, Ames, Iowa.
43. **Pillai, S. D.** Bacteriophages as indicators. *In* S. Goyal (ed.), *Food Virology*, in press. Kluwer Publishers, Norwell, Mass.
44. **Pillai, S. D., and J. Totten.** 2003. Molecular methods for microbial detection and characterization, p. 255–264. *In* A. Pandey (ed.), *Concise Encyclopedia of Bioresource Technology.* The Harworth Press, New York, N.Y.
45. **Pillai, S. D., and S. C. Ricke.** 1995. Strategies to accelerate the applicability of gene amplification protocols for pathogen detection in meat and meat products. *Crit. Rev. Microbiol.* **21**:239–261.
46. **Pillai, S. D., K. L. Josephson, R. L. Bailey, C. P. Gerba, and I. L. Pepper.** 1991. Rapid method for processing soil samples for polymerase chain reaction amplification of specific gene sequences. *Appl. Environ. Microbiol.* **57**:2283–2286.
47. **Roe, M. T.** 2002. *Prevalence of Class 1 and Class 2 Integrons in Poultry Processing and a Natural Water Ecosystem.* M.S. thesis. Texas A&M University, College Station, Tex.
48. **Rosenblum, L. S., I. R. Mirkin, D. T. Allen, S. Safford, and S. C. Hadler.** 1990. A multifocal outbreak of Hepatitis A traced back to commercially distributed lettuce. *Am. J. Public Health* **80**:1075–1079.
49. **Sair, A. I., D. H. D'Souza, and L. A. Jaykus.** 2002. Human enteric viruses as causes of foodborne disease. *Comp. Rev. Food Sci. Food Safety* **1**:73–89.
50. **Smith, J. L., and P. M. Fratamico.** 1995. Factors involved in the emergence and persistence of food-borne diseases. *J. Food Prot.* **58**:696–708.
51. **Stintzi, A.** 2003. Gene expression profile of *Campylobacter jejuni* in response to growth temperature variation. *J. Bacteriol.* **185**:2009–2016.
52. **Strategic Consulting.** 2000. *Pathogen-Testing in the U.S. Food Industry.* Strategic Consulting, Woodstock, Vt.
53. **Swaminathan, B., T. J. Barrett, S. B. Hunter, and R. V. Tauxe.** 2001. PulseNet: the molecular subtyping network for foodborne bacterial disease surveillance, United States. *Emerg. Infect. Dis.* **7**:382–389.
54. **Thelwell, N., S. Millington, A. Solinas, J. Booth, and T. Brown.** 2000. Mode of action and application of Scorpion primers to mutation detection. *Nucleic Acids Res.* **28**:3752–3761.
55. **Tringe, S. G., C. Von Mering, A. Kobayashi, A. A. Salamov, K. Chen, H. W. Chang, M. Podar, J. M. Short, E. J. Mathur, J. C. Detter, P. Bork, P. Hugenholtz, and E. M. Rubin.** 2005. Comparative metagenomics of microbial communities. *Science* **308**:554–557.
56. **Vasavada, P. C.** 2001. Getting really rapid test results. *Food Safety* **7**:28–38.
57. **Vega, E., J. Smith, J. Garland, A. Matos, and S. D. Pillai.** Variability of virus attachment patterns to butterhead lettuce. *J. Food Prot.*, in press.
58. **Venter, J. C., K. Remington, J. F. Heidelberg, A. L. Halpern, D. Rusch, J. A. Eisen, D. Wu, I. Paulsen, K. E. Nelson, W. Nelson, D. E. Fouts, S. Levy, A. H. Knap, M. W. Lomas, K. Nealson, O. White, J. Peterson, J. Hoffman, R. Parsons, H. Baden-Tillson, C. Pfannkoch, Y. H. Rogers, and H. O. Smith.** 2004. Environmental genome shotgun sequencing of the Sargasso Sea. *Science* **304**:66–74.

59. **Wang, D., A. Urisman, Y. Liu, M. Springer, T. G. Ksiazek, D. D. Erdman, E. R. Mardis, M. Hickenbotham, V. Magrini, J. Eldred, J. P. Latreille, R. K. Wilson, D. Ganem, and J. L. DeRisil.** 2003. Viral discovery and sequence recovery using DNA microarrays. *PLoS Biol.* **1**:257–260.

60. **Wei, Y., J. M. Lee, C. Richmond, F. R. Blattner, J. Rafalski, and R. A. LaRossa.** 2001. High density microarray-mediated gene expression profiling of *Escherichia coli*. *J. Bacteriol.* **183**:545–556.

61. **Welsh, J., and M. McClelland.** 1990. Fingerprinting genomes using PCR with arbitrary primers. *Nucleic Acids Res.* **18**:7213–7218.

62. **Widdowson, M. A., A. Sulka, S. N. Bulens, R. S. Beard, S. S. Chaves, R. Hammond, E. D. Salehi, E. Swanson, J. Totaro, R. Woron, P. S. Mead, J. S. Bresee, S. S. Monroe, and R. I. Glass.** 2005. Norovirus and foodborne disease, United States, 1991–2000. *Emerg. Infect. Dis.* **11**:95–102.

63. **Widmer, K. W., K. H. Oshima, and S. D. Pillai.** 2002. Identification of *Cryptosporidium parvum* oocysts by an artificial neural network approach. *Appl. Environ. Microbiol.* **68**:1115–1121.

64. **Williams, J. G., A. R. Kubelik, K. J. Livak, J. A. Rafalski, and S. V. Tingey.** 1990. DNA polymorphisms amplified by arbitrary primers are useful as genetic markers. *Nucleic Acids Res.* **18**:6531–6535.

65. **Wilson, I. G.** 1997. Inhibition and facilitation of nucleic acid amplification. *Appl. Environ. Microbiol.* **63**:3741–3751.

66. **Wren, B. W.** 2000. Microbial genome analysis: insights into virulence, host adaptation and evolution. *Nat. Rev. Genet.* **1**:30–39.

Microbial Source Tracking
Edited by Jorge W. Santo Domingo and Michael J. Sadowsky

Molecular Subtyping, Source Tracking, and Food Safety

4

Thomas S. Whittam and Teresa M. Bergholz

Food-borne pathogens have been studied intensively, often prompted by conspicuous epidemics of disease. As a consequence of these investigations, detailed systems for molecular typing and extensive databases of strains and genotypic characteristics have accumulated. In addition to its great value in the epidemiological investigation of outbreaks, application of standardized subtyping methods is being extended to fundamental studies of the population biology and evolution of pathogens. Our purpose in this chapter is to highlight the use of molecular subtyping of microbes in application to understanding the transmission of food-borne illness and food safety.

Bacteria, viruses, and parasites typically enter the food supply through contamination of raw food ingredients or postprocessing contamination of products. The primary sources of microbes in raw food commodities are soil and water, the intestinal tracts of animals, and animal hides. During and after processing, foods can become contaminated via food contact surfaces, food handlers, and air and dust. Understanding the routes and sources of microbial contamination of food is essential to the control of these organisms in the food supply.

Our focus here is on the application of molecular subtyping methods for identifying and tracing sources of microbial pathogens in food and food-borne human diseases. It should be noted, however, that in addition to the food-borne pathogens, harmless or nonpathogenic microorganisms can also contaminate the food supply and contribute to spoilage. These microbes are generally part of the natural microbiota of raw foods and food products, which serves as an ideal growth medium for many organisms with nearly

Thomas S. Whittam and Teresa M. Bergholz, National Food Safety and Toxicology Center, 165 Food Safety and Toxicology Building, Michigan State University, East Lansing, MI 48824.

neutral pH, high water activity, and richness in nutrients. If left unchecked, growth of these harmless microorganisms on foods results in off odors and flavors and, ultimately, in an inedible food product. Many of the methods developed to inhibit the growth of nonpathogenic microbes in foods also restrict the growth of food-borne pathogens.

Food-Borne Pathogens and Public Health

Scientists from the Centers for Disease Control and Prevention (CDC) have estimated that food-borne disease accounts for 76 million illnesses and 5,000 deaths each year in the United States (84). The spectrum of food-borne pathogens includes bacteria, viruses, and parasites (Table 1). Some pathogens, such as *Shigella* or norovirus, naturally circulate only in humans, while others primarily reside in specific animals or environments. Some pathogens are virtually always transmitted to humans through food products, whereas others can be transmitted by other routes, including food. Together, these pathogens cause an array of acute illnesses, mostly involving vomiting or diarrhea. In addition, food-borne infections can trigger chronic sequelae, including, for example, renal disease resulting from damage caused by a toxin produced by *Escherichia coli* and autoimmune neurological disease resulting from molecular mimicry between *Campylobacter jejuni* lipo-oligosaccharides and human gangliosides (73, 133). The total economic burden of food-borne diseases on the U.S. population is considerable, and has been estimated to cost billions of dollars in sick time, hospitalizations, and product loss (116).

To understand the epidemiology of food-borne disease in the United States, the CDC, the U.S. Department of Agriculture (USDA), and the Food and Drug Administration (FDA) established the Foodborne Diseases Active Surveillance Network (FoodNet). FoodNet is a national surveillance system for notifiable diseases that includes sites in 10 states and covers approximately 15% of the U.S. population. Information collected from these sites is used to estimate the number of cases of illness due to various food-borne pathogens, to monitor trends in food-borne diseases, and to determine the proportion of diseases attributable to specific foods (Fig. 1).

One of the most common food-borne diseases caused by a bacterial pathogen is salmonellosis, gastroenteritis resulting from infection by one of the many serotypes of *Salmonella enterica*. The number of cases based on 2004 FoodNet data (23) is 147 per million persons, with 40% of those cases occurring in children under 15 years old. *Campylobacter* infections, most often caused by *C. jejuni* and occasionally by *Campylobacter coli*, are the second most common bacterial food-borne infection, with the number of reported cases at 129 per million persons. It has been estimated that greater than 2 million cases of illness are due to *C. jejuni* infections, with approximately 2,000 deaths

Table 1 Bacterial, viral, and parasitic agents that contribute to the burden of food-borne illness in the United States

Agent	Estimated total illness[a]	Food vehicle(s)	Environmental source(s) and comments
Bacterial pathogens			
Campylobacter jejuni and *C. coli*	2,453,926	Raw and undercooked meat and poultry, raw milk, and untreated water	Reservoir in poultry, pigs, and cattle; most common bacterial cause of diarrhea in the United States
Salmonella enterica, nontyphoidal	1,412,498	Raw and undercooked eggs, undercooked poultry and meat, dairy products, seafood, fruits, and vegetables	Bovine and poultry; most common cause of death from food-borne illness
Shigella sonnei	448,240	Salads, milk and dairy products, and unclean water	No animal reservoir, circulates among humans; highly infectious
Escherichia coli O157:H7 and other Shiga toxin-producing *E. coli* strains	110,220	Meat, especially undercooked or raw hamburger, produce, and raw milk	Reservoir in ruminants; produces deadly toxins that can result in hemolytic uremic syndrome
Listeria monocytogenes	2,518	Dairy products including soft cheeses, meat, poultry, seafood, and produce; a problem in ready-to-eat foods	Reservoir in soil and water; grows at refrigeration temperatures
Vibrio vulnificus	94	Raw or undercooked shellfish	Causes gastroenteritis or primary septicemia
Bacterial toxins			
Clostridium perfringens and *C. botulinum*	248,578	Home-prepared foods and herbal oils; honey, particularly when given to children less than 12 mos. of age	Toxin causes botulism, a life-threatening illness that can prevent the breathing muscles from moving air in and out of the lungs
Staphylococcus aureus	185,060	Ingesting cooked foods high in protein (e.g., cooked ham, salads, bakery products, dairy products) typically contaminated via food handlers	Staphylococcal enterotoxins cause vomiting in short time

(continued)

Table 1 Bacterial, viral, and parasitic agents that contribute to the burden of food-borne illness in the United States *(continued)*

Agent	Estimated total illness[a]	Food vehicle(s)	Environmental source(s) and comments
Viruses			
Norovirus	23,000,000	Any food can be contaminated with norovirus if handled by someone who is infected with this virus	Leading cause of diarrhea in the United States; approximately 40% of cases are food-borne
Rotaviruses	3,900,000	Transferred to ready-to-eat foods via infected food handlers	Very low rate of food-borne transmission
Hepatitis A	83,391	Transmitted to foods via infected food handlers	Peak time for transmission is approximately 2 wks. before person shows symptoms
Parasites			
Giardia lamblia	2,000,000	Fresh produce and infected food handlers	Recreational water is the major source of transmission; roughly 10% of cases are assumed to be food-borne
Cryptosporidium parvum	300,000	Fresh produce, typically contaminated via irrigation or wash water	Spread in contaminated water or person-to-person; roughly 10% of cases are assumed to be food-borne
Toxoplasma gondii	225,000	Meat, primarily pork	Causes toxoplasmosis, a severe disease that can produce central nervous system disorders, particularly mental retardation and visual impairment in children
Cyclospora	16,250	Fresh produce, typically contaminated via irrigation or wash water	Causes diarrhea that can last up to 7 wks; little is known about the life cycle of the parasite

[a]Annual number of cases for total U.S. population (84).

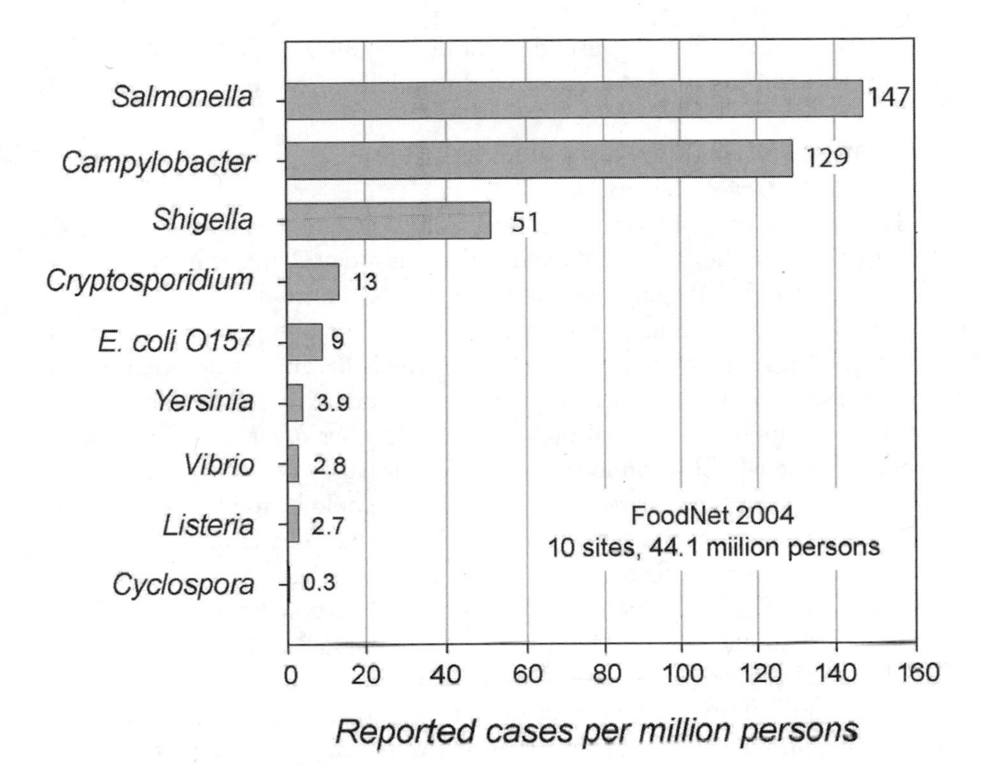

Figure 1 Reported cases of food-borne illness in 2004 for 10 FoodNet sites (California, Colorado, Connecticut, Georgia, Maryland, Minnesota, New Mexico, New York, Oregon, and Tennessee).

each year in the United States (84). Shigellosis is the third most common bacterial disease with most (>75%) cases attributed to a single serotype, *Shigella sonnei* (Table 1). Illness attributed to other food-borne agents is not as prevalent but is of great importance because of the severity of disease and the nature of the susceptible populations. For example, a single serotype of Shiga toxin-producing *E. coli*, O157:H7, accounts for nine cases of disease per million people. Similarly, a total of 696 cases of listeriosis, resulting from infection with *Listeria monocytogenes*, were reported for 2003, with the highest incidence occurring in infants less than 1 year of age (24).

Data collected through the FoodNet system as well as other surveillance systems are beginning to provide more precise estimates of the burden of food-borne disease nationwide (Table 1). Surveys of people within FoodNet sites estimate that there are approximately 1.4 self-reported episodes of diarrhea per person per year (54). Surprisingly, in the total estimate of food-borne disease in the United States, known pathogenic agents account for only 19% of cases. This suggests that more than 80% of the food-borne illness in this country cannot be attributed to a known cause.

Population Biology of Food-Borne Pathogens

Population Genetic Diversity

The infusion of methods and concepts from population genetics into microbiology began in the 1980s (106) and continues today. A major impact on the view of bacterial pathogens associated with food-borne disease was accomplished by the application of multilocus enzyme electrophoresis (MLEE) to a variety of bacterial species. This method, which became established as the principal method for empirical population genetics in the 1960s and 1970s, was broadly applied to natural populations of higher organisms. The attractiveness of the MLEE technique was that it uncovered polymorphisms in the primary structure of conserved proteins and thus could be used to index allelic variation at genes encoding housekeeping functions. Therefore, evolutionary geneticists had for the first time a general tool for assessing the amount and organization of genetic variation in animal and plant populations in nature.

The early applications of MLEE to bacteria associated with food-borne infectious diseases revealed several general findings:

Most clinically recognized species harbor substantial allelic variation in conserved protein-coding genes. This variation comprises the fraction of allelic variation that results in amino acid replacements that influence electrophoretic migration rates of the corresponding proteins. The fraction of DNA sequence variation that is electrophoretically detectable by conventional means has been estimated to be approximately 9%, based on comparisons of *E. coli* and *Salmonella* gene sequences (122). The fact that multiple alleles could be detected at many loci indicates that the results of MLEE analysis would provide a rich system for resolving multilocus genotypes (i.e., electrophoretic types [ETs]) of diverse collections of isolates.

Certain multilocus enzyme genotypes are highly overrepresented in samples of isolates from clinical cases or other sources. These genotypes represent bacterial clones that have achieved high prevalence and often account for many cases of disease. For example, analysis of variation for 16 enzyme loci in 175 strains of *L. monocytogenes* uncovered 45 distinct ETs. One clone was associated with epidemics in the 1980s attributed to contamination of soft cheese in western Switzerland and in Los Angeles, California (93). Such overrepresentation of specific ETs as well as strong linkage disequilibrium between loci is indicative of a clonal population structure. Clonal population structures were also observed in similar studies of other food-borne pathogens including the nontyphoidal serovars of *Salmonella* (11, 105) and Shiga toxin-producing *E. coli* O157 (123, 124).

Host Adaptation and Specificity

The question of how bacterial pathogens evolve and adapt to new hosts is crucial to understanding the fundamental basis for the origin of infectious diseases as well as the emergence of new pathogens. What is the nature and molecular basis of species barriers? What restricts certain pathogens to infecting only humans, whereas others infect humans and animals? What factors determine host specificity and how is host specificity linked to virulence? In general, there is very limited knowledge about the genes and mechanisms involved in host adaptation, although host specificity varies among human pathogens. For example, among the antigenic variants of *Salmonella*, some serotypes lack host specificity, such as *S. enterica* serotypes Typhimurium and Enteritidis, and can cause disease in both humans and livestock. These pathogens tend to cause less severe disease and lower mortality rates than host-specific serovars such as *S. enterica* serotype Typhi, which infects only humans, is more virulent, and contributes to a greater mortality rate (10). A parallel situation occurs between pathogenic *E. coli* strains and closely related *Shigella* strains (94). Some diarrhea-causing *E. coli* strains, such as serotype O26:H11, can elicit disease in humans and a variety of ruminants, whereas *Shigella* strains can cause a more severe disease (i.e., bacillary dysentery) only in humans and closely related primates. Such specialization on certain host species has practical implications; progress in elucidating mechanisms of pathogenesis is often hampered by lack of realistic animal models of disease for several highly host-adapted pathogens.

Knowledge of the molecules involved in host specificity has led to several important breakthroughs. For example, listeriosis is a severe food-borne disease caused by infection with *L. monocytogenes*. In human disease, these bacteria invade epithelial cells and breach the intestinal barrier; however, mice and rats are not susceptible to infection. The species specificity depends in part on the interaction between a bacterial surface protein (internalin) and a receptor (E-cadherin) expressed on human epithelial cells. A single amino acid position in E-cadherin appears to be critical: human E-cadherin contains a proline residue at position 16, whereas mouse and rat E-cadherin contain a glutamic acid that prevents the interaction with internalin. By expressing human E-cadherin in transgenic mice, researchers have engineered a relevant mouse model for the human infection (69). This model is now used to elucidate key steps in pathogenesis and is being refined to provide a more accurate picture of human listeriosis (39).

Molecular Methods of Subtyping Food-Borne Pathogens

The most appropriate molecular subtyping method for a given application depends on both the question that is being addressed and the level of variation in the bacterium or pathogen that is being studied (Table 2). In some cases, isolates are so similar genetically that only the most sensitive methods can detect differences based on the most rapidly changing genetic characteristics of the microorganisms. In other cases, the most appropriate methods target more slowly changing genes where differences did not accumulate so rapidly. The choice of method also depends on the level of analysis. For molecular epidemiology, often the question is simply one of discrimination: are isolates the same or different? Isolates that are identical or indistinguishable based on a sensitive method can often be epidemiologically linked and used to identify clusters of cases with a common source of contamination. In other cases, strains may be phenotypically different based on the type or presence of specific antigens, and the primary question is: how different are these strains genetically? In such cases, subtyping methods providing data that can be interpreted in terms of the nature of genetic changes (for example, the number of point mutations) are more appropriate.

Restriction Fragment Length Polymorphism (RFLP) Ribotyping

Ribosomal DNA (rRNA gene) fingerprinting (ribotyping) is a technique in which genomic DNA is digested by a restriction endonuclease and electrophoresed on an agarose gel, and then probes homologous to rRNA gene sequences are hybridized to the samples through Southern hybridization. Ribotyping reduces the complexity of genomic DNA restriction profiles by focusing on only those genomic DNA fragments that contain nucleotide sequences homologous to the rRNA or rRNA gene probe. Although the genes encoding 16S and 23S rRNA contain hypervariable regions, there is a high degree of sequence conservation in those genes that allows rRNA from *E. coli* to be used as a universal probe for all bacteria. Ribotyping is useful for discriminating subtypes among specific foodborne pathogens. For example, ribotyping was shown to differentiate subtypes of *L. monocytogenes* with a high degree of reproducibility between laboratories (112) and produced comparable results to MLEE (7). However, ribotyping could not distinguish between *L. monocytogenes* strains of serotypes 1/2b and 4b, and thus, alternative subtyping methods may be more applicable (45). This also was true for the study of *E. coli* O157:H7 isolates in which pulsed-field gel electrophoresis (PFGE) and phage typing were found to be more discriminatory (80).

A modified ribotyping system, using a mixture of PstI and SphI for restriction digestion of DNA (PS ribotyping), has proven useful for subtyping *S. enterica* serotype Enteritidis strains. This method was shown to be more discriminatory than phage typing and PFGE combined and is useful for investigating outbreaks involving isolates with identical phage types (28).

Table 2 Summary of potential advantages and disadvantages of some of the alternative subtyping methods for food-borne pathogens

Typing method	Advantages	Limitations
Ribotyping	Fingerprint differences result from many classes of genomic change (e.g. mutations, gene gains and losses); automated fingerprinting systems available; public databases available for some pathogens	Must be optimized for different organisms; data portable as images; analysis limited to pattern matching
PFGE profiling	Fingerprint differences result from many classes of genomic change (e.g. mutations, gene gains and losses); differences accumulate fast enough for outbreak investigation; no prior genomic knowledge required	Gel-based system is labor intensive; genetic basis of fingerprint changes usually not known; data portable as images; analysis limited to pattern matching
AFLP	High sensitivity to genetic changes; fragment differences result from many classes of genomic differences that accumulate fast enough for outbreak investigation; no prior genomic knowledge required	Genetic basis of fingerprint changes not immediately known; data portable as images; analysis limited to pattern matching; high expense per isolate
MLST	Determines exact nucleotide differences for conserved loci, mostly one class of genetic change (synonymous mutations); data amenable to population genetic and phylogenetic analysis; data portable and internet accessible	Requires prior knowledge of gene sequences; insufficient rate of change for many applications in molecular epidemiology; does not detect genomic alterations outside of loci of interest; high expense per isolate
Microarray	Detects differences in gene content; simultaneously monitors all genes in a genome; hybridization conditions can vary sensitivity	Cannot detect new genes; limited map information; requires known reference genome; high expense per isolate
RAPD	Rapid PCR-based method detects differences in genome content between strains based on randomly amplified DNA fragments; no prior genomic knowledge is required	Genetic basis of RAPD fingerprints not immediately known; may be difficult to reproduce between laboratories

PFGE Profiling

PFGE profiling is a DNA fingerprinting method which is based on the restriction digestion of purified genomic DNA. It is currently considered the gold-standard method for subtyping food-borne pathogens. Briefly, bacteria are grown in broth or on solid medium and are combined with molten agarose. The resulting agarose plugs, containing whole bacteria, are then subjected to detergent-enzyme lysis and whole-genome digestion using a rare cutting restriction enzyme. The enzymatic digestion results in large DNA fragments (10 to 800 kb in length), which are electrophoresed under alternating electric currents, thereby producing a banding pattern or DNA fingerprint.

PFGE forms the basis for PulseNet, a national molecular subtyping network that was established in 1996 by the CDC and is now utilized by various state public health laboratories and food safety laboratories at the FDA and USDA (111). PulseNet has developed standardized methods for PFGE to ensure uniform interpretation of banding patterns. In most cases, gels are scanned and converted to digital images, which can then be analyzed using specialized computer software.

For many food-borne pathogens, PFGE results in superior resolution of subtypes compared to other typing methods. For instance, PFGE is useful for discriminating between isolates of *E. coli* O157:H7 (80) and was shown to identify outbreaks of O157:H7 that were not detectable using traditional epidemiological methods (Fig. 2) (12). In addition, PFGE combined with plasmid profiling was found to be superior to ribotyping in distinguishing outbreak strains of *Shigella sonnei* (70) and was critical for understanding the scope of a *C. jejuni* outbreak in Kansas (40).

The major advantage of PFGE profiling is that the technique can be widely applied to different bacterial species and is relatively simple and straightforward to perform in the laboratory. The procedures are easily standardized, quality control can be readily tested, and the reagents required are relatively inexpensive. The main strength of this technique is that it detects a variety of genetic events that alter bacterial genomes and generate distinctive PFGE fragment profiles. In addition to point mutations, these genetic events include acquisition of mobile genetic elements, such as lysogenic bacteriophages (67), changes in the copy numbers of insertion sequences, and other chromosomal insertions, transpositions, and deletion events. On the whole, the combined effect of these events results in rapid fingerprint divergence and can lead to differences accumulated in epidemiological real time. As a result, PFGE is an extremely sensitive method that can detect differences between even very closely related strains, so that isolates with indistinguishable PFGE profiles can be classified as epidemiologically linked with a high degree of confidence. Thus, during outbreak investigations, cases and

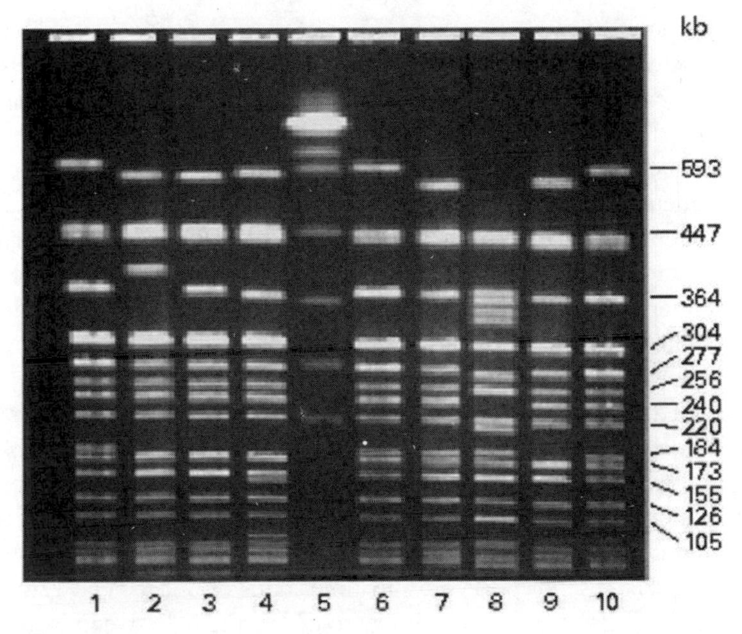

Figure 2 PFGE patterns of *E. coli* O157:H7. Lanes 1, 6, and 10 represent isolate G5244, which is a standard strain used to characterize molecular size. Lanes 3 and 9 represent isolates from cases involved in a single cluster. Lanes 2, 4, 7, and 8 represent isolates from other sporadic cases of disease. Lane 5 is an additional molecular size standard (12).

sporadic cases can be separated by molecular data and the extent of an outbreak and linkage to contaminated food or water sources can be inferred.

The broad applicability of PFGE, however, is limited because in many cases bacteria can change rapidly and develop different PFGE profiles, thereby contributing to a loss of information about the history of divergence. This can occur on the order of hundreds of generations. A second disadvantage of this technique is that it requires dedicated hands-on labor in the laboratory and is not readily amenable to automation, in contrast to subtyping methods based on nucleotide sequencing and restriction fragment analysis techniques (e.g., amplified fragment length polymorphism [AFLP] or variable number of tandem repeats [VNTR]).

AFLP

AFLP analysis, like PFGE profiling, also allows examination of whole-genome polymorphisms, but by a different approach. Briefly, target DNA is digested with two different restriction enzymes, and adaptor oligonucleotides are ligated to the sticky-ended DNA fragments. A PCR method is then used to amplify a subset of these fragments, which are then separated and detected by an automated sequencer system. The restriction fragments analyzed are small, and even a single base mutation can be detected. The use of different sets of restriction enzymes or different primer pair combinations can generate large numbers of different AFLP fingerprints without prior knowledge of the genomic sequence. Fluorescent AFLP (FAFLP), which uses fluorescent dye-labeled primers, is reproducible and capable of standardization.

This technique has the high sensitivity of PFGE profiling and can be adapted to automatic analysis for higher throughput using fluorescent dye-labeled primers (FAFLP) and an automated sequencer (57). An additional benefit of this technique is that the genetic basis of fragment differences can be readily investigated. Specific AFLP-produced bands can be excised from gels, amplified, cloned, and further characterized by restriction analysis, subsequent sequencing, or hybridization (57). AFLP systems for high-resolution genotyping of food-borne pathogens have been applied to outbreak investigation and source tracking of *C. jejuni* (35), pathogenic *E. coli* (49), *Salmonella* (32, 98), and *Clostridium perfringens* (83). A comparison of the molecular subtyping of *S. enterica* serotype Typhimurium strains of phage type 126 demonstrated that FAFLP has a greater ability to discriminate between isolates than PFGE (98). A similar level of high resolution and discrimination of genotypes was seen for *E. coli* O157:H7 (49).

Multilocus Sequence Typing (MLST)

MLST is based on similar principles as is multilocus enzyme electrophoresis (78) and is now being applied and developed for a variety of human pathogenic bacteria. The fundamental idea is that the method uncovers genetic variation in multiple conserved genes, and this genetic variation is used to classify strains, identify clonal groups, and elucidate the history of divergence of the chromosomal background. Briefly, the method relies on determining the nucleotide sequence of regions of approximately 500 base pairs from multiple genetic loci distributed around the genome of a particular bacterial species. The genes to be sequenced are amplified by specific PCR primers and the sequencing is generally performed using an automated sequencer. The principal advantage of this technique is that most of the genetic variation uncovered in conserved structural genes is selectively neutral or nearly so. Because protein-encoding genes with housekeeping functions are often the targets, most mutational differences between alleles are in the third positions and other synonymous sites (i.e., those that do not lead to an amino acid replacement) in codons. Neutral mutations accumulate at a nearly uniform rate in populations and there is a well-developed theoretical and statistical methodology for analyzing this type of nucleotide sequence information. In addition, there are new analytical tools for special application to bacterial pathogens, such as BURST analysis (37), which can be applied to populations that experience high rates of recombination (109). The procedures for MLST are readily standardized; the data are portable between laboratories and expressed as nucleotide strings. In addition, the laboratory methods are easily automated to make use of robotics and automated sequencers so that rapid, high-throughput analysis is possible (78).

The primary disadvantage associated with MLST is that it has limited applicability in outbreak investigations or for organisms with low levels of genetic variation, such as species with low levels of variation (e.g., *Mycobacterium tuberculosis*) or various isolates of a pathogenic clone. For example, a recent comparison of more than 75 *E. coli* O157:H7 strains revealed only two single point mutations in the sequencing of seven conserved genes (88). In this case, the variation detected accumulates so slowly that there is an insufficient rate of change for application in molecular epidemiology, outbreak investigation, and source tracking. However, this is a strength when MLST is used in concert with other techniques in which the multilocus genotypic data provide a testable framework against which other types of genetic and phenotypic data can be measured. A second disadvantage is that the PCR primers are specific for sequences within a species or closely related groups of strains, but new primers generally have to be designed, tested, and optimized for each species of organism. The third disadvantage is that the expense per locus or strain is often high because of the costs associated with DNA polymerase, sequencing reaction components, and equipment operation and maintenance costs. Consequently, many epidemiological studies involving hundreds or thousands of isolates can be cost prohibitive.

An MLST-based typing scheme has been developed for *C. jejuni* (34) and tested for applicability in outbreak investigations (100). MLST alone did not resolve strains between outbreaks as well as PFGE, but combining MLST with sequencing of the highly variable *flaA* gene, which encodes the major subunit of the flagellum, led to a level of discrimination equivalent to that of PFGE for outbreak investigations. MLST also has been evaluated for differentiating *L. monocytogenes* strains. In this analysis, it was found that virulence genes, in addition to housekeeping genes, were necessary to allow maximum subtype discrimination (19). Therefore, in order to provide better discriminatory power for studying *L. monocytogenes*, MLST can be performed only on virulence genes and virulence-associated genes (MVLST), which is comparable to PFGE and provides greater discrimination than ribotyping (134). MVLST, however, is not useful for some bacterial species, such as *E. coli* O157:H7 (41). In this case, MVLST as well as MLST of housekeeping genes (88) was found to be poorly discriminatory between isolates compared to PFGE.

Microarrays

There are a variety of types, methods, and uses of DNA microarrays, which can be applied to questions of food-borne pathogen detection, molecular pathogenesis, and pathogen evolution and emergence. The two major types of microarrays are (i) printed arrays, in which PCR products represent the majority of genes (open reading frames) occurring within the genome and (ii) oligonucleotide arrays, in which short probes are either deposited or synthesized in situ in an array format to detect specific segments of DNA. Whole-genome microarrays have been used to detect and measure differences in gene content between strains when the array is based on a reference genomic sequence (101). Microarrays also have been adapted to detect a variety of bacterial species and microbial pathogens in clinical and environmental samples (27, 129).

The main strength of the microarray approach is that thousands of specific gene hybridizations can be tested simultaneously under the same conditions, so that, for example, an entire genome can be examined at once in a single experiment. The main disadvantage, however, is that the variation detected from strain to strain cannot be ascribed to particular locations in the genome or mapped to specific positions. In addition, differences in gene copy number or cross-hybridization to duplicated genes can confound the interpretation.

In application to food-borne pathogens, there are two different approaches for utilizing genomic array data. The first approach examines differences in gene content; that is, the presence or absence of genes (125). This application can determine whether certain genes or gene combinations are diagnostic for specific subtypes or species (118). For example, Call and colleagues (21) developed and utilized a mixed-genome microarray to both fingerprint isolates of *L. monocytogenes* and identify genes that were unique to different phylogenetic lineages (20). The second approach involves using microarrays to investigate differences in gene expression to assess which genes are turned off and on at the level of transcription in response to environmental conditions or during the course of infection.

RFLP-PCR–Based Assays of Single Genes

If knowledge of given gene sequences is available, then pathogens can be typed based on allelic variation within these specific genes. In order to do this, PCR primers are designed based on the known sequence, and gene amplification is performed using PCR followed by digestion of the PCR product with restriction endonucleases and agarose gel electrophoresis. Different banding patterns are produced, which are based on nucleotide changes at the restriction sites for specific enzymes. Genes selected for this typing method are typically genes encoding virulence factors or antigenic cell surface structures, such as flagella. For instance, genes encoding virulence factors listeriolysin O and internalin and actin polymerization are routinely utilized to classify strains of *L. monocytogenes* (102, 127). In most cases, however, the PCR-RFLP subtyping method is most discriminatory when utilized along with another subtyping method, such as ribotyping.

RFLP of the *flaAB* genes encoding the flagellin of *C. jejuni*, for example, has been successfully employed to subtype strains (6). Further studies indicated that typing *flaA* and *flaB* separately improved discrimination of *C. jejuni* subtypes (92). In addition, RFLP of *flaA* was successfully utilized to identify recurring subtypes of *C. jejuni* isolated from retail chicken (77), as well as to type 233 strains of *C. jejuni* isolated from a variety of human and animal hosts (90). In both cases, RFLP provided superior discrimination compared to conventional serotyping.

PCR-Based Assays for Genetic Fingerprinting

Bacterial pathogens also can be typed based on genome-wide polymorphisms. Similar to PFGE, random amplification of polymorphic DNA (RAPD) can detect many different types of genetic events that alter bacterial genomes. Single arbitrarily chosen PCR primers are selected to amplify DNA from a set of strains (121, 128), which can be distinguished based on differences in banding patterns following agarose gel electrophoresis. The main advantages of the RAPD system are that no prior knowledge of the genome sequence is necessary and that it is also less time consuming than other DNA fingerprinting methods, such as PFGE. The primary disadvantage involves the reproducibility of fingerprints, which can be problematic due to differences in DNA preparations between laboratories. However, improving and standardizing DNA isolation techniques (85) and using multiple standardized primer sets have reduced much of the variability in RAPD results between laboratories.

There have been several applications of RAPD analysis to source tracking of food-borne pathogens. For example, RAPD analysis of *Vibrio vulnificus,* a bacterial pathogen associated with seafood consumption, allowed differentiation of isolates from clinical, environmental, and diseased-eel populations as well as different geographic locations (2). In addition, Martinez and colleagues utilized RAPD to type *L. monocytogenes* strains, which also differentiated between clinical, food, and environmental isolates (81).

Repetitive extragenic palindromic (REP)-PCR is similar to RAPD but utilizes primers that target repetitive elements present throughout the bacterial genome. This method utilizes a standard primer set that was initially based on *E. coli* repetitive sequences. These sequences were found to be present in a diverse range of bacterial species and are now utilized to produce fingerprints of many different bacteria (120). REP-PCR has been used to trace potential sources of *Salmonella* contamination during the production of cantaloupe (22).

Multilocus VNTR Analysis

Repetitive DNA elements, such as VNTR, can be utilized to subtype bacterial pathogens, similar to the idea of REP-PCR. VNTR elements are short repetitive DNA sequences that exist in multiple copies. These repeats can be clustered in one area of the genome or dispersed throughout. VNTR elements can serve as targets for genomic events, such as recombination or DNA polymerase slippage, which can result in a change in the number of times the element is repeated. This creates variation between strains at the VNTR sites if it occurs at a relatively high frequency (74). Multilocus VNTR analysis (MLVA) utilizes polymorphisms at multiple VNTR loci around the genome to differentiate strains. The complete genome sequence is required to identify and design PCR primers flanking VNTR sites. PCR products for a particular site can differ in multiples of the number of base pairs within the specific repeat. PCR products can be visualized on an agarose gel or can be adapted to automatic analysis for higher throughput using fluorescent dye-labeled primers and an automated sequencer.

MLVA has been shown to provide comparable or higher discrimination between isolates of *E. coli* O157:H7 than PFGE (65, 87). Additionally, Lindstedt and colleagues (75) have developed an MLVA subtyping scheme for *S. enterica* strains of serotype Typhimurium that utilizes multiplex PCR and an automated capillary electrophoresis system for rapid classification of strains at higher discrimination than PFGE.

Tracing Sources and Transmission of Food-Borne Pathogens

Sources of Microbial Pathogens in Food

There are many similarities between the concepts and methods used to identify sources of microbial contamination of the food chain and those used to track the sources of biological pollution in natural waterways (Fig. 3A). One aim of microbial source tracking in water resource management is to better distinguish between point sources of biological pollution (e.g., sewage treatment plants) and nonpoint contributions (e.g., wildlife). It is also important to elucidate the contribution of various nonpoint or diffuse origins of pollution such as runoff from agricultural sources, including manure from livestock and poultry, and fecal contamination from wildlife (e.g., migrating geese). An objective of molecular subtyping in microbial source tracking is to separate the sources of indicator organisms and potential human pathogens from human sewage and fecal contamination (108). By correctly identifying the sources of microbial contamination, rational decisions can be made to

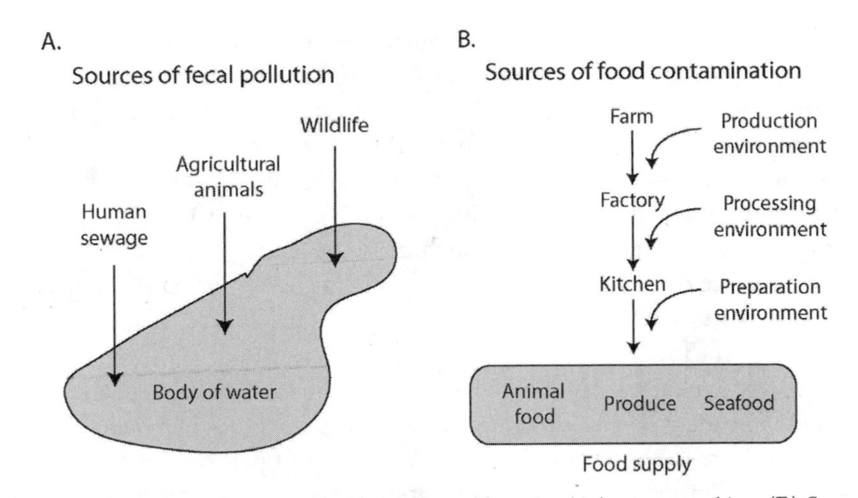

Figure 3 (A) Sources of water pollution addressed by microbial source tracking. (B) Sources of food contamination in the food chain addressed by microbial subtyping.

remediate impairment and to devise long-term strategies to manage water pollution (108).

Typically, monitoring methods used for detecting potential pathogens in natural waters are based upon cultivation and enumeration of fecal indicator bacteria (i.e., fecal coliforms and enterococci). There is a growing number of methods for the molecular characterization of waterborne microorganisms to be used not only for detection, but also for identification of fecal contamination sources. Such molecular subtyping methods are now being applied to help improve our waters by identifying problem sources and determining the effect of implemented remedial solutions. In addition to differentiating groups of indicator organisms, molecular methods are targeting host-specific microorganisms. These methods are useful because they circumvent the need to isolate individual microorganisms and do not require the establishment of reference databases (104).

There are parallel issues in discovering sources of microbial pathogens in the processing of food and in different food products (Fig. 3B). In this case, microbial contamination can enter the food chain at several levels. The first level is the production environment, represented by the farm, orchard, or fishery. The presence of microbial pathogens in livestock at the farm, for example, can be the source of contamination of meat and dairy products as well as fruits and vegetables that often are exposed to fertilizers, manure, or polluted water. Contamination at this preharvest stage is particularly serious for foods that are consumed raw or undercooked, such as green onions or shellfish.

The second level associated with microbial contamination of the food chain is the processing environment represented by the slaughterhouse, cannery, or packing plant (Fig. 3B). Contamination during processing or within a processing plant can come from a variety of sources. In beef processing, for example, intestinal contents from carcasses or fecal material on hides can contaminate meat with human pathogens such as *E. coli* O157:H7 (36). For environmentally adapted bacteria such as *L. monocytogenes*, certain strains colonize surfaces, machinery, and drains and can persist in processing plants for years, where they become virtually a continuous source of food product contamination (56).

The third level at which microbes can enter the food chain is in the preparation environment, either in a kitchen or food preparation service (Fig. 3B). At this level, cross-contamination—the spreading of pathogenic bacteria and viruses between foods, surfaces, and equipment—is a common source of contamination. Infected food handlers are also a contributing source of pathogens during the food preparation stage. Food preparation is an especially important step for transmission of strictly human-adapted pathogens, such as *Shigella* species, serotype Typhi, or hepatitis A. For these host-restricted pathogens, there is no environmental or animal reservoir; pathogens persist only by circulating between infected and uninfected hosts. Infected food handlers, which can be symptomatic or asymptomatic carriers, can contaminate food during preparation particularly if sanitary practices (e.g., hand-washing between tasks) are not followed. A typical scenario of contamination by food preparers was seen in an outbreak of gastrointestinal illness among patrons of a single restaurant in California in January 1998 (117). Eight of the 25 symptomatic patrons and 3 employees had stool specimens that were culture-positive for *Shigella* spp. The link to food preparation was confirmed by PFGE profiling; two isolates from food handlers and three isolates from restaurant patrons had identical PFGE profiles (117).

Environmental sources

Food-borne pathogens present in the natural environment can lead directly to illness, e.g., via swimming in water contaminated with *E. coli* O157:H7 (29), or can indirectly lead to illness by contaminating a food product present in the environment.

Vibrio *spp. in shellfish*

Vibrio parahaemolyticus is one of the leading causes of seafood-associated bacterial gastroenteritis in the United States (84). Prior to 1996, *V. parahaemolyticus* infections were sporadic cases attributed to multiple diverse serotypes. A rapid increase in illness due to serotype O3:K6 was initially observed in Asia in 1996 (26), and this highly virulent clone then spread to other parts of the world. The use of PFGE resolved and differentiated this newly identified serotype from other *V. parahaemolyticus* serotypes (132). Additionally, an outbreak of *V. parahaemolyticus* in Spain was found to be closely related to this recently emerged pandemic clone in Asia and the United States marked by the O3:K6 serotype by PFGE (82).

By contrast, infection with *V. vulnificus* contributes to more severe disease than *V. parahaemolyticus*, as disease is rapid, invasive, and highly lethal, especially in people with underlying chronic illness. *V. vulnificus* infections are typically associated with consumption of raw oysters, and multiple subtypes have been found in a single oyster (18). PFGE profiling of *V. vulnificus* isolates, however, has identified subtypes that are predominantly associated with either environmental or clinical isolates. This suggests that a specific subset of environmental isolates may be responsible for the majority of human disease cases, and, therefore, PFGE profiling may be useful to identify which environmental strains are clinically relevant (113).

Zoonoses and Animal Reservoirs

Pathogenic Escherichia coli *in Cattle*

Beef cattle and dairy cows are the main reservoir for Shiga toxin-producing *E. coli* (STEC) including *E. coli* O157:H7. As a result, there has been much interest in understanding the dynamics of STEC colonization in bovine herds, in elucidating transmission mechanisms between cattle on the same farm and between cattle and the farm environment, and in gaining a better understanding of the bacterial shedding patterns from cattle. Cattle can harbor a diverse assemblage of *E. coli* strains, and individual cows can have completely different subtypes of *E. coli* (61). Because *E. coli* O157:H7 is the most common serotype associated with STEC infections in the United States (84), most studies monitoring STEC populations in cattle have focused specifically on this bacterium. Shere et al. (107) conducted a longitudinal study of *E. coli* O157:H7 on dairy farms in Wisconsin and found that contaminated drinking water was the most likely vehicle for transmission of *E. coli* O157:H7 to cattle. *E. coli* O157:H7 isolates were then classified by PFGE, and identical subtypes were found in the drinking water and the cattle. In addition, when water was free of *E. coli* O157:H7, cattle stopped shedding the bacteria within weeks. As a result of this source tracking study based on molecular subtyping of *E. coli* O157:H7, cattle drinking water has been identified as a logical point for intervention treatments to reduce the spread of this pathogen on farms.

In a study of the diversity and dynamics of *E. coli* O157:H7 in the bovine reservoir, Rice et al. (97) repeatedly sampled 41 cattle herds over a 34-month period to determine the persistence of specific *E. coli* O157:H7 subtypes in herds over time. Strains were isolated from fecal samples and subtyped by PFGE. Subtypes with indistinguishable patterns were shown to persist on four farms for 6 to 24 months. In contrast to those subtypes maintained for long periods, most of the diversity appeared within a particular farm only briefly and most farms had a succession of unique subtypes. Finally, the movement of animals to and from farms did not appear to be related to the diversity of subtypes found on a specific farm.

Microbial subtyping and source tracking also have been used to investigate the transmission of O157 from pre- to postharvest cattle. Fecal and hide prevalence of *E. coli* O157:H7 were significantly correlated with carcass contamination at meat processing plants in the midwestern United States (36). *E. coli* O157:H7 strains isolated from feces, hides, and beef carcasses were then subtyped by PFGE to investigate the potential that the same genotypes were being brought to the processing plant on live cattle and then cross-contaminating carcasses during processing. Because of the relatively low number of *E. coli* O157:H7-positive hides detected, the data from this study were insufficient to provide evidence that either hide or feces were most likely to be the direct source of O157 on carcasses (8).

Commensal *E. coli* also were evaluated to trace transmission from live cattle to postharvest contamination (4). RAPD as well as PCR-based RFLP for the flagellin gene (*fliC*) was utilized to subtype *E. coli* isolated from the same cattle during each step of food processing (i.e., from on the pastures all the way to ground beef) (4). Multiple genetic types of *E. coli* found in the feces of pasture and feedlot cattle were also found on the hides and carcasses at the time of slaughter, indicating that the carcasses were contaminated with *E. coli* of animal origin and that these strains subsequently contaminated the ground beef. Subtype tracing based on RAPD and PFGE patterns in a commercial beef abattoir revealed that the majority of *E. coli* bacteria isolated from hides, carcasses, and conveyors were indistinguishable in their DNA fingerprints from those isolated from ground beef. In addition, few unique subtypes were recovered from ground beef, indicating that most of the contaminants originated from the processing steps prior to grinding (3). This type of transmission would be quite likely for *E. coli* O157:H7 as well.

Salmonella *Transmission via Eggs*

Salmonella enterica serotype Enteritidis is unusual as a food-borne pathogen in that investigations of both outbreak and sporadic cases of infection have identified egg-laying chicken flocks as the primary reservoir and eggs as the major vehicle of transmission (52, 110). Transmission of serotype Enteritidis to humans occurs primarily through contaminated food, most often via uncooked eggs, egg-containing foods, and poultry (Fig. 4). The CDC have reported that 80% of 371 outbreaks of serotype Enteritidis in the United States between 1985 and 1999 were egg associated (91). Serotype Enteritidis-contaminated eggs are estimated to have accounted for approximately 180,000 cases of illness in the United States in 2000 (103). The number of cases of serotype Enteritidis in human gastroenteritis has increased dramatically over the past three decades (119). In the United States, serotype Enteritidis ranks second to serotype Typhimurium as the cause of thousands of cases of salmonellosis each year (1, 84, 91). Several studies also have been conducted to assess the genotypic diversity of serotype Enteritidis from different sources. Liebana and colleagues (72) examined the genotypic diversity from multiple geographic sources using both PS ribotyping and PFGE to resolve 73 subtypes (clones). Interestingly, some of these clones are widespread and occur in several countries and were likely spread as a result of the movement of foodstuffs, animals, or people (72).

As with *E. coli* O157:H7 and *Campylobacter* spp., serotype Enteritidis is of zoonotic origin and can be transmitted to humans via various transmission routes (Fig. 4). Serotype Enteritidis circulates in wild animals and has often been recovered in rodent populations associated with chicken farms (30, 53).

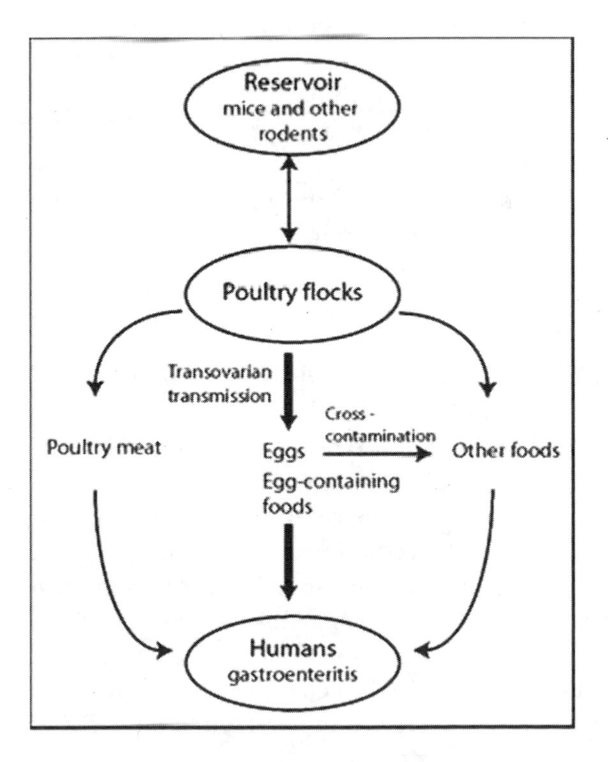

Figure 4 Routes of transmission for serotype Enteritidis (99). (Reprinted from the *Journal of AOAC International*, Volume 89, no. 5, 2006. Copyright 2006 by AOAC International.)

In a study of captured mice from hen houses on farms, for instance, approximately 20 percent of mice had spleens that were culture-positive for serotype Enteritidis (47). It was suggested that the rodent reservoir maintains a diverse population of serotype Enteritidis and only a specialized subpopulation has an ability to infect chickens (46, 99). Molecular subtyping evidence based on ribotyping and PFGE has clearly incriminated multiple wildlife species, including rodents, flies, litter beetles, and foxes, in the maintenance of serotype Enteritidis infection on farms (71).

Serotype Enteritidis contaminates eggs either on the shell surface through contact with fecal material, or by contamination of the internal contents as a result of infection of the reproductive tissue of laying hens, resulting in transovarian transmission (58, 86). The pathogen has been cultured from the albumen (white) of the egg, the vitelline (yolk) membrane, and the yolk (43, 58). The production of eggs with *Salmonella*-infected contents not only serves as the primary vehicle for human disease, but the bacteria also can spread to newborn chicks hatched from contaminated eggs via vertical transmission (58, 114). This route of transmission may contribute to additional outbreaks of infection by serotype Enteritidis that follow consumption of meat from broiler chickens that were initially infected in hatcheries.

Non-egg-containing foods also have been associated with serotype Enteritidis infections. One notable example is an outbreak of salmonellosis in North America in 2001 that was linked to contaminated almonds (25). Subsequent investigation found that a specific phage type (a phenotype based on the sensitivity of strains to lytic bacteriophages) of serotype Enteritidis was detected in raw almonds collected from multiple sources, including environmental swabs from orchards and the associated processing equipment, suggesting that there is diffuse contamination under some environmental conditions (60).

Diversity of Campylobacter *on the Farm*

Kramer and colleagues (66) systematically sampled for *C. jejuni* in the soil and water, wildlife, and livestock in a defined 100-km^2 area of dairy farmland. Using MLST, they resolved 65 sequence types (STs) among 172 isolates. Even on a local scale, the *C. jejuni* population was very diverse genetically: 41 of the 65 (63%) STs were isolated only once. Among the 24 STs that were repeatedly recovered, there was evidence for circulation among sources: 16 of 24 (67%) were recovered from more than one source. These results also support the association between cattle and *C. jejuni* genotypes that are members of a specific clonal complex, that is, a cluster of STs related to ST-61. Furthermore, there was considerable diversity in genotypes from water and environmental samples that were not recovered from livestock or observed in human illness cases. This study highlights the problem of sampling for a genetically diverse population in which most bacterial genotypes are recovered only once. In such cases, sampling must be extensive to determine the frequency and make statistical comparisons between sources. This problem can be rectified to some extent by utilizing methods that permit hierarchical classification that organizes diverse genotypes into related groups, preferably in ways that are biologically meaningful.

Transmission of Human Pathogens to Food Animals via Feed

Because food-producing animals are the primary source of *E. coli* O157:H7 and pathogenic *Salmonella* infections in humans, it follows that bacterial contamination of animal feed contributes to the burden of food-borne illness. Identification of genetically identical isolates from both feed and animals on the same farm supports the hypothesis that animal feed represents an important vehicle for introducing *Salmonella* and *E. coli* O157:H7 into herds. Support for this hypothesis comes from a study by Davis and colleagues (31), who cultured cattle feed and cattle fecal specimens from multiple farms for *E. coli* O157:H7 and serotype Typhimurium and subtyped isolates using PFGE. No clear identity between feed and fecal *E. coli* O157:H7 was observed, but PFGE revealed a close genetic relationship between feed and fecal isolates on one farm. Similarly, serotype Typhimurium strains were isolated from several feed and fecal samples and found to be genetically identical by PFGE.

Contamination during Food Processing

In order to further prevent food-borne illnesses in humans, it is also important to determine where contaminating agents are entering the food supply. Bacteria can enter food processing facilities via raw products, and contamination of finished products can result from the presence and persistence of these bacteria in the processing environment of the plant. Microbial subtyping can add valuable information for discriminating these sources, in contrast to simply isolating and enumerating microorganisms at specific stages of processing (126). It is critical to determine which of these scenarios is the case for any particular setting, as the methods developed to mediate or control the spread of pathogens would be different in each case.

Of particular concern in food processing plants is *L. monocytogenes*, as it has been shown to thrive in processing environments, can form biofilms to become resistant to sanitizers, and is psychrotrophic. *L. monocytogenes* is particularly problematic for ready-to-eat (RTE) foods, as the consumer typically eats these foods without further heating or preparation. Because of the severity of disease caused by *L. monocytogenes* infection, the current U.S. regulations have a zero-tolerance policy for this organism in RTE food products.

This stringent standard has resulted in many recalls and economic losses (131); therefore, understanding contamination patterns through the entire processing procedure is extremely important to prevent contamination of finished products.

Smoked Fish Processing

The patterns of *L. monocytogenes* contamination and spread in seafood processing plants, particularly those that produce smoked or frozen RTE foods, have been studied intensively. For example, a combination of PFGE and RAPD were used to identify sources of contamination in a shrimp processing facility (33). Subtypes of *L. monocytogenes* were found to be specific to particular areas of the plant; some isolates were exclusive to water, whereas others were exclusive to the processing area. This study also found a large number of subtypes on the raw product, but only two of these same subtypes were present on the finished frozen product, suggesting that specific subtypes may survive processing better than others. In another study of two smoked fish processing facilities, ribotyping was used to subtype *L. monocytogenes* isolates (56). A single ribotype was found to account for 32.1% of the environmental isolates from a particular plant, and it persisted for over 2 years. The subtypes present in raw fish, however, did not seem to contribute to the persistent, plant-specific *L. monocytogenes* microflora.

Longitudinal studies have been conducted in smoked fish processing plants to identify persistent subtypes of *L. monocytogenes* in the plant environment and to assess how these subtypes may spread throughout the plant. For example, Thimothe and colleagues (115) sampled four smoked fish processing plants monthly for 1 year and isolated *L. monocytogenes* from raw product, food contact surfaces, non-food-contact surfaces (floors, drains, etc.), and finished product. The isolates were characterized by ribotyping. Plant-specific ribotypes were identified from plant environmental samples, while *L. monocytogenes* on raw materials did not represent the source of finished product contamination. Persistent subtypes identified in the environment were also isolated from the finished product or on food contact or employee contact surfaces. Thus, in these processing plants, the food processing environment appears to be the most common source of *L. monocytogenes* contamination of RTE foods.

In a follow-up to this study, intervention methods were designed for each plant and the same areas were sampled monthly for a year to examine the effect of the intervention program (68). Ribotyping was then used to identify key equipment and methods that were contributing to the spread of *L. monocytogenes* through a specific plant (conveyors, food handlers, etc.). While the prevalence of some specific persistent ribotypes decreased after the intervention in some plants, some persistent types were still present following the intervention.

In a similar longitudinal study, Autio et al. (5) sampled a smoked trout processing plant for *L. monocytogenes*. Two production lots of fish were sampled at every step during processing, while food contact surfaces and employees were also sampled. The isolates were characterized using PFGE. *L. monocytogenes* from the plant environment, particularly from the recirculating brine water, appeared to be the primary source of contamination in the finished product. This result was used to devise an eradication program for the plant, and the plant and product were then sampled for a 5-month period. During this time, no samples were positive for *L. monocytogenes*. Another study that utilized PFGE to characterize *L. monocytogenes* was conducted in Iceland at a smoked salmon processing plant. The study found that subtypes in the raw product were an important source of final product contamination in some smokehouses (48). Markkula et al. (79) also found identical PFGE profiles for *L. monocytogenes* in both the raw and finished smoked fish.

In the studies highlighted, both ribotyping and PFGE provided similar evidence that *L. monocytogenes* contamination of the finished product resulted from contamination by plant-specific environmental isolates. Because of these data, appropriate intervention strategies could be devised to decrease the probability of food contamination. Furthermore, these findings highlight the importance of first tracking an organism through the plant environment before developing a plant-specific intervention strategy.

Processing of Latin-Style Cheeses

Another food product implicated in outbreaks of *L. monocytogenes* is Latin-style fresh cheese. In studies of three cheese-processing facilities, multiple samples obtained over a 6-month period included finished product, food contact surfaces, and environmental samples. *L. monocytogenes* isolates were subtyped using a combination of ribotyping and PCR-RFLP for two virulence genes (64). Persistent contamination of a particular ribotype was found in one plant; it was isolated from finished product as well as from the processing environment and food contact surfaces. No ribotypes persisted for more than two visits at the second plant, suggesting that persistent *L. monocytogenes* contamination was successfully prevented in this plant.

Tracking *Salmonella enterica* during Meat Processing

Salmonella enterica is an enteric pathogen that commonly colonizes the intestinal tract of mammals and can contaminate meat and poultry during processing. A sampling study undertaken at local swine slaughterhouses in Belgium identified incoming pigs as a likely source for cross-contamination during processing, as an average of 28% of incoming pigs were positive for *Salmonella* (16). In this study, a significant association was found between the number of animals that carried *Salmonella* and the number of contaminated carcasses at the end of the slaughter line. In order to trace cross-contamination throughout the slaughterhouse, live pigs and carcasses at various stages of processing were sampled for *Salmonella* and isolates were subtyped by PFGE (15). In one slaughterhouse, samples from live pigs matched those from the carcasses, and in another plant, aerosols generated during carcass washing matched samples from the carcasses.

In France, Giovannacci and colleagues (44) found further evidence for cross-contamination of pig carcasses and the resulting pork cuts during slaughterhouse processing by utilizing PFGE to type *Salmonella* isolated from pigs, carcasses, meat cuts, and the slaughterhouse environment. PFGE was used to characterize matched fecal and carcass samples in a swine processing plant to determine the source of carcass contamination (130). Swine were found to carry multiple genotypes of *Salmonella*, and the feces or carcasses from other animals were found to contaminate approximately half of the carcasses. Taken together, these studies have provided ample evidence for the major sources of *Salmonella* contamination in pork products: the GI tracts of the entering pigs and poor control of the plant environment facilitate cross-contamination between carcasses. These findings have identified specific areas to target for implementation and prevention measures aimed at reducing and eliminating the occurrence and spread of *Salmonella* in the processing environment.

Surveillance, Outbreak Investigation, and Molecular Databases

Microbial Subtyping Databases and Networks Directed at Food Safety

PulseNet. PulseNet began in 1995 as a response to a large outbreak of *E. coli* O157:H7 disease in the western United States and initially included the state public health laboratories from four states (Fig. 5). These laboratories were the beginning of a national molecular subtyping network for food-borne bacterial disease surveillance (111). Currently, all states participate in PulseNet, with the objectives to allow for real-time communication among state and local health departments and international partners and to facilitate early identification of common-source outbreaks. State laboratories are responsible for subtyping food-borne pathogens isolated within their states using standardized PFGE typing and pattern analysis technology. Each PulseNet lab has all the equipment required to generate the PFGE patterns and the capability to normalize the patterns, compare them with others, and maintain a local database of PFGE patterns for each pathogen of interest. These laboratories can query the national PFGE data-

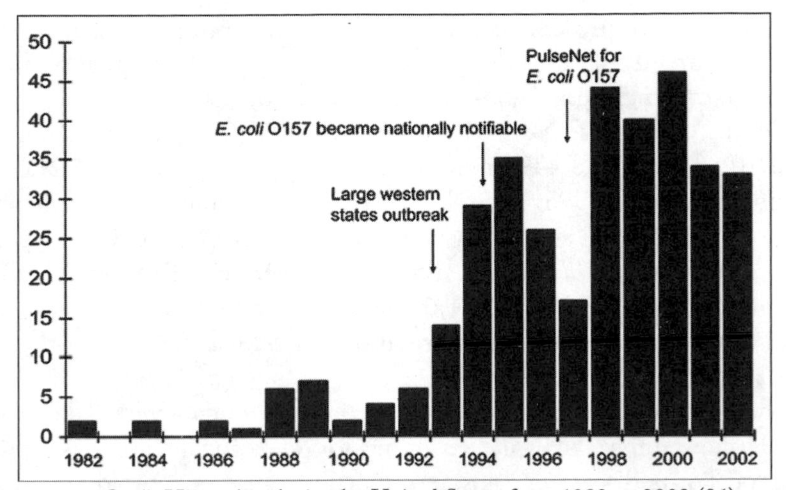

Figure 5 *E. coli* O157:H7 outbreaks in the United States from 1982 to 2002 (96). *y* axis is number of outbreaks.

base to find matches to the PFGE types that were generated. This type of query initiates a recent-match message if two or more laboratories have submitted identical or closely related patterns within a specified timeframe. This alert provides an early warning to the local laboratories about possible multistate food-borne disease outbreaks.

PulseNet has facilitated the identification of numerous multistate food-borne disease outbreaks, many of which would have gone otherwise unnoticed at first because of the temporal differences between cases. *Listeria* outbreaks are difficult to trace because of the long incubation time and the limited susceptible populations (e.g., pregnant women and the elderly). Centralized food production also increases the difficulty associated with identifying outbreaks, as cases may be scattered over a wide geographic area. The routine subtyping of *L. monocytogenes*, for example, was critical to the identification of a diffuse multistate outbreak linked to delicatessen turkey meat (89), which had previously been shown to be responsible for an outbreak of febrile gastroenteritis (42).

FoodNet. FoodNet was initiated in 1995 as a population-based, active disease surveillance network with the purpose of estimating the burden of food-borne illness in the United States. This system was designed to investigate the sources of infection in outbreaks and sporadic cases and to build up the public health infrastructure for better handling of emerging food-borne disease issues. In addition, another function of FoodNet is to attribute the proportions of food-borne diseases to specific food items. This is done mainly by conducting case-control studies of laboratory-confirmed illness within the surveillance area (51). There are currently 10 sites that participate in FoodNet, including California, Colorado, Connecticut, Georgia, Maryland, Minnesota, New Mexico, New York, Oregon, and Tennessee. These 10 sites include 44.1 million people, which represents 15.2% of the U.S. population. In 1996, FoodNet began active, population-based surveillance for laboratory-diagnosed cases of *Campylobacter*, STEC O157, *Listeria*, *Salmonella*, *Shigella*, *Vibrio*, and *Yersinia*. In 1997, FoodNet added surveillance of *Cryptosporidium*, *Cyclospora*, and hemolytic uremic syndrome cases, and in 2000 began collecting information on STEC serotypes other than O157:H7 (non-O157 serotypes).

Although a comparison of the 1996 through 1998 data with data collected for 2004 showed that the incidence for most food-borne diseases has decreased (23), food-borne disease remains an important cause of morbidity and mortality in many individuals. Therefore, continued surveillance of food-borne illness is critical, as the prevalence of any given pathogen can change significantly over time.

There are, however, several limitations to the health burden of food-borne illness based on the FoodNet surveillance, as noted by the CDC (23). The network relies on 10 sites whose population varies in constitution and disease incidence. As a result, the catchment areas for some pathogens may be larger in some sites than others. This system also relies exclusively on laboratory diagnosis, and many, if not most, food-borne illnesses are not laboratory diagnosed. In addition, the source of acute gastroenteritis for a reported illness might include other non-food-borne routes of transmission, such as contaminated water or person-to-person contact, and finally, protocols for enteric pathogen isolation and identification can vary across sites.

Pathogen Tracker. Pathogen Tracker (www.pathogentracker.net) is an interactive internet-accessible database designed to compare bacterial strains with regard to epidemiologic, phenotypic, and genotypic strain characteristics. The database interface allows subscribers to compare isolates of interest with an online database comprising isolates from multiple sources and geographic and temporal distributions. This database currently allows access to genetic, phenotypic, and source information from a collection of food-borne and zoonotic pathogens and food spoilage organisms. Major organisms represented in this database include *L. monocytogenes*, *V. parahaemolyticus*, *Pseudomonas* spp., and *Streptococcus agalactiae*. The database contains DNA subtyping information (e.g., ribotype images), DNA sequence information, and phenotypic information (e.g., Biolog data, serotypes) for bacterial isolates. Martin Wiedmann, Kathryn Boor, and collaborators at Cornell

University developed the strain collection and database. The USDA, Dairy Management Inc., and the American Meat Institute Foundation have provided funding for the strain collection and database development. The website for Pathogen Tracker 2.0 is hosted by the Cornell Theory Center, which provides computational resources and support for the project.

PubMLST. PubMLST (www.pubmlst.org) is a publicly accessible database specifically designed for accessing and analyzing MLST data. The database contains genotypic profiles and information on isolates from multiple bacterial species, several of which are important etiological agents of food-borne diseases. MLST data on *Campylobacter* species comprise several thousand strains of bacteria, many of them originally isolated from clinical cases and animal reservoirs. The *C. jejuni* MLST website (http://pubmlst.org/campylobacter/) was developed by Keith Jolley and Man-Suen Chan, at the University of Oxford, with support from the Wellcome Trust (63).

STEC Center. The STEC Center website (www.shigatox.net) is designed to integrate research on the Shiga toxin-producing *E. coli* strains, including *E. coli* O157:H7, by providing a standard reference collection of well-characterized strains and central online-accessible databases. Genetic and phenotypic information is available to the scientific community through several online databases, comprised primarily of data on pathogenic *Escherichia* and *Shigella* strains associated with diarrheal disease. Molecular subtyping is based on allele profiles determined by multilocus enzyme electrophoresis and genotyping using MLST and extended MLST, as well as virulence gene profiling. EcMLST is the online database for MLST and DNA polymorphisms detected by RFLP (95). The STEC Center is based at the National Food Safety and Toxicology Center at Michigan State University. The project is supported by the National Institute of Allergy and Infectious Diseases through the Food and Waterborne Diseases Integrated Research Network, Microbiology Research Unit.

MLVA Web service. The MLVA bacterial genotyping site (http://bacterial-genotyping.igmors.u-psud.fr/) is a gateway to publicly accessible multilocus VNTR analysis databases and software. The aim of the site is to facilitate pathogenic bacterial strain genotyping, essentially for epidemiological purposes. Molecular subtyping is based on polymorphic tandem repeat variation at multiple loci in the genome. The databases include genotypic information on several bacterial pathogens, including food-borne *S. enterica* strains. The website allows for the comparison of locally produced data with genotypes present in the database. The primary MLVA site is hosted at the Institut de Génétique et Microbiologie, Université d'Orsay Paris XI, was developed by Patrick Bouchon, and is maintained by Ibtissem GRISSA.

Norovirus molecular epidemiology database. The norovirus database contains more than 1,000 RNA polymerase and capsid gene sequences of norovirus strains and associated epidemiological data. The database can be searched by gene sequence or on the basis of specific epidemiological criteria (age, country, and outbreak setting). This database is curated by the staff of the Enteric Virus Unit and the Bioinformatics Group of the Health Protection Agency, United Kingdom (http://www.hpa.org.uk/srmd/bioinformatics/norwalk/norovirus.htm).

Outbreak Investigations

One of the principal applications of molecular subtyping methods, such as PFGE, is to investigate outbreaks or clusters of disease cases. This application is often triggered by a sudden increase in reported cases of a syndrome or disease. The subtyping results will enable epidemiologists or laboratory personnel to identify cases linked to a given outbreak and distinguish those from cases that are not. Coupled with epidemiological information, suspected food vehicles and potential sources are often further investigated. If the same subtype of bacterium is isolated from an implicated food item and from disease cases, then this provides a direct link to the source of infection. In this case, the molecular subtyping results provide a genotypic signature of the source. By contrast, patients with isolates representing distinct subtypes are assumed to have acquired the infection from a different source and are often classified as sporadic cases.

There are several underlying assumptions to be considered when using molecular subtyping to investigate outbreaks for the purpose of tracing and identifying the source of infection. One critical assumption is that a sole source of infection carries a single subtype. If multiple types occur within a particular food vehicle, then this can obscure the effective tracing of the source. A second assumption is that the genetic profiles remain stable during the course of an outbreak. It is possible, however, that microevolution can occur during the course of a large outbreak as the pathogen spreads temporally and geographically. The loss or acquisition of mobile elements can result in new mismatched genotypes that would be excluded from the outbreak cases.

Despite these limitations, molecular subtyping methods have been critical for tracing food-borne outbreaks to their sources. Most outbreaks in recent years have first been detected through the PulseNet system and then followed up by epidemiological investigations at the local level. Sometimes the epidemiological evidence leads to a particular food, which can then be tested for the presence of the pathogen. In many cases, however, the pathogen cannot be identified in food or the implicated food item is not available. Therefore, epidemiologists often rely solely on the epidemiologic and molecular subtyping data associated with the clinical cases to make conclusions regarding an association with a given outbreak. Molecular subtyping, for example, is critical to link outbreaks across the country that may be associated with centralized food production. Examples of linking food sources to *E. coli* O157:H7 outbreaks using PFGE include linking two temporally distinct outbreaks to a common alfalfa seed distributor (17, 38) and linking a multistate outbreak to mesclun lettuce (55). In another study, epidemiological investigations implicated the ground beef in tacos prepared at a fast-food chain as a source of *E. coli* O157:H7 from patients in Arizona, California, and Nevada (62), but definitive links to the beef could not be made, as none of the suspect product was left. Finally, one of the largest waterborne outbreaks of *E. coli* O157:H7 occurred at a county fair in New York, where patient isolates had similar PFGE types to isolates obtained from the well water at the fair (14). This outbreak was also unusual in that some patients were coinfected with *C. jejuni*, which also was traced to isolates from the well water by PFGE typing.

Other challenges also can hamper outbreak investigation efforts. For instance, outbreaks of infection by *L. monocytogenes* present an additional complication, as incubation time before the onset of disease can be as long as 70 days. There are many challenges associated with such a long incubation period (e.g., meat is rarely still available in these situations, patients cannot recall food-eating history, etc.). Therefore, given such a rare disease incidence as with *Listeria*, PulseNet subtyping activities are very important to assess whether other cases with similar profiles have been identified elsewhere. Molecular subtyping is essential for distinguishing outbreak cases that appear over a long period of time from the sporadic cases that make up the background rate of disease. A multistate outbreak of infections with *L. monocytogenes* that lasted 8 months, for example, was first identified via PFGE and then linked to contaminated delicatessen meat via epidemiological investigation and isolation of *L. monocytogenes* with the identical PFGE type from the implicated meat (89). Ribotyping and PFGE were used in combination to identify an outbreak of *L. monocytogenes* in Australia and to trace back to contaminated chicken; the outbreak ended after a recall of the product was issued (59). PFGE and an epidemiological investigation identified raw-milk Mexican-style cheese in an outbreak of *L. monocytogenes* that was traced back to the raw-milk supplier (76).

Food Attribution

E. coli *O157:H7 Outbreaks in the United States*

From 1982 to 2002, a total of 350 outbreaks of *E. coli* O157:H7 infection were reported in the United States (96). Of these outbreaks, 183 (52%) were transmitted via food. Rangel and colleagues were then able to attribute the known food-borne outbreaks to specific food vehicles. Data from outbreaks over the past 20 years show an increase in outbreaks attributed to fresh produce; however, it is unclear whether this increase is due to enhanced surveillance activities. It is possible, for example, that older questionnaire data may not have included produce item questions. Nevertheless, these data suggest that safety measures should be addressed for these commodities.

Attributing outbreaks to specific food vehicles is useful to allow regulators and industry to implement and evaluate control measures to prevent further outbreaks. For example, outbreak investigations that implicated fast-food hamburgers led to major improvements in fast-food meat safety in the United States. Despite this, ground beef continues to be frequently implicated in outbreaks of *E. coli* O157:H7. The conclusion from the 2003 Food Attribution Data Workshop necessitates the association of food-borne illnesses to specific food vehicles in order to make informed decisions about food safety interventions (9).

Salmonellosis in Denmark

Accurate food attribution requires intensive monitoring of all relevant food animals, foods, and human cases, as well as efficient methods for subtyping potential pathogens. Denmark established the Danish Zoonosis Center in 1994 as an epidemiological surveillance and research unit. The Zoonosis Center collects data from all national surveillance and control programs on zoonoses and conducts an ongoing assessment of sources of human food-borne disease.

Hald and colleagues (50) developed a mathematical model for quantifying the contribution of each of the major animal-food sources to human salmonellosis based on the data from the integrated Danish *Salmonella* surveillance in 1999. The model was set up to calculate the number of domestic and sporadic cases caused by different *Salmonella* serotypes and phage types as a function of the prevalence of these *Salmonella* types in the animal-food sources and the amount of the food source consumed (Fig. 6). The principle behind the model is to compare the number of human cases caused by different *Salmonella* types with the prevalence of the types isolated from different food sources weighted by the amount of the food source consumed. This approach assumes that bacterial-dependent factors, such as differences

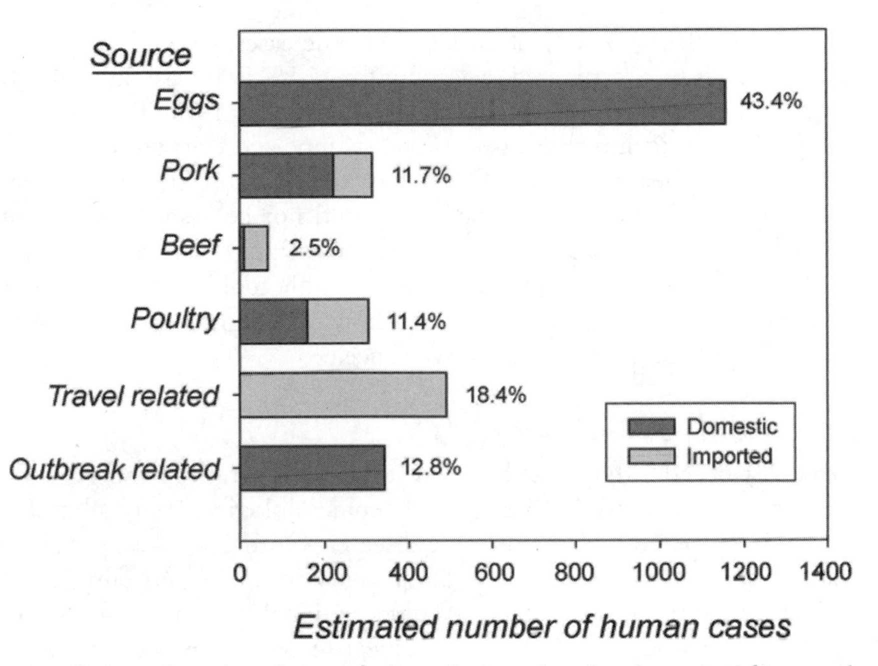

Figure 6 Estimated number of cases of salmonellosis attributed to domestic and imported food sources based on a Bayesian model (50).

between serotypes in survival rates in foods, and food source-dependent factors, such as differences between food types in bacterial load, are negligible.

The bacterial surveillance data were modeled using nine animal-food sources, including domestic and imported cuts of pork and beef, as well as poultry and eggs. For the model, the prevalences of identified *Salmonella* types (serotypes and phage types) in pork, beef, and imported meat were calculated from the percentages of positive samples. The prevalence for the poultry data was estimated from the percentage of positive flocks, and it was assumed that the number of contaminated table eggs produced by infected flocks was proportional to the flock size. The amount of food consumed for different meat types and shell eggs was estimated from national statistics of food availability in the Danish market. The joint posterior distribution was estimated by fitting the model to the reported number of domestic and sporadic cases per *Salmonella* type in a Bayesian framework using a Markov chain Monte Carlo simulation. The numbers of domestic and sporadic cases were obtained by subtracting the estimated numbers of travel- and outbreak-associated cases from the total number of reported cases.

The most important food sources were found to be table eggs and domestically produced pork, comprising 47.1% (95% credibility interval [CI], 43.3

to 50.8%) and 9% (95% CI, 7.8 to 10.4%) of the cases, respectively (Fig. 6). Imported foods were estimated to account for 11.8% (95% CI, 5.0 to 19.0%) of the cases. Imported eggs were not considered a contributing source of serotype Enteritidis infections because most imported eggs are heat treated and used in further processing. Other food sources considered had only a minor impact, whereas 25% of the cases could not be associated with any source. This approach of quantifying the contributions of the various sources to human salmonellosis has proved to be a valuable tool in risk management in Denmark and provides an example of how to integrate quantitative risk assessment with zoonotic disease surveillance.

PERSPECTIVE

In the past decade, there has been outstanding progress in the growth of molecular methods and databases for the epidemiological study of etiological agents associated with food-borne diseases. Outbreak investigation and epidemiological tracing have been the primary incentives encouraging the development of the PulseNet and FoodNet national systems for food-borne disease surveillance. In the future, we expect to see the application of new molecular genotyping systems with rapid throughput, such as those based on single-nucleotide polymorphisms (13), to generate high-quality genotypic data for microbial source tracking. These high-throughput systems can be expanded beyond outbreak investigations to include isolates from sporadic illness as well as from food and animal sources. In this way, the molecular subtyping of microorganisms will open the door for the precise quantification of the various sources of hazards and risks in food safety.

ACKNOWLEDGMENTS

We thank Shannon Manning for critical reading of an earlier version of this chapter and Robert Tauxe for sharing his ideas on the principal sources of contamination in the food chain.

We acknowledge support from the USDA, the NIH, the National Center for Food Protection and Defense, and the Michigan Agricultural Experiment Station. The STEC Center is part of the NIAID Food and Waterborne Diseases Intergrated Research Network and is supported under NIH Research Contract number N01-AI-30058.

REFERENCES

1. **Angulo, F. J., and D. L. Swerdlow.** 1998. *Salmonella enteritidis* infections in the United States. *J. Am. Vet. Med. Assoc.* **213:**1729–1731.
2. **Arias, C. R., M. J. Pujalte, E. Garay, and R. Aznar.** 1998. Genetic relatedness among environmental, clinical, and diseased-eel *Vibrio vulnificus* isolates from different geographic regions by ribotyping and randomly amplified polymorphic DNA PCR. *Appl. Environ. Microbiol.* **64:**3403–3410.

3. **Aslam, M., G. G. Greer, F. M. Nattress, C. O. Gill, and L. M. McMullen.** 2004. Genotypic analysis of *Escherichia coli* recovered from product and equipment at a beef-packing plant. *J. Appl. Microbiol.* **97**:78–86.
4. **Aslam, M., F. Nattress, G. Greer, C. Yost, C. Gill, and L. McMullen.** 2003. Origin of contamination and genetic diversity of *Escherichia coli* in beef cattle. *Appl. Environ. Microbiol.* **69**:2794–2799.
5. **Autio, T., S. Hielm, M. Miettinen, A. M. Sjoberg, K. Aarnisalo, J. Bjorkroth, T. Mattila-Sandholm, and H. Korkeala.** 1999. Sources of *Listeria monocytogenes* contamination in a cold-smoked rainbow trout processing plant detected by pulsed-field gel electrophoresis typing. *Appl. Environ. Microbiol.* **65**:150–155.
6. **Ayling, R. D., M. J. Woodward, S. Evans, and D. G. Newell.** 1996. Restriction fragment length polymorphism of polymerase chain reaction products applied to the differentiation of poultry campylobacters for epidemiological investigations. *Res. Vet. Sci.* **60**:168–172.
7. **Baloga, A. O., and S. K. Harlander.** 1991. Comparison of methods for discrimination between strains of *Listeria monocytogenes* from epidemiological surveys. *Appl. Environ. Microbiol.* **57**:2324–2331.
8. **Barkocy-Gallagher, G. A., T. M. Arthur, G. R. Siragusa, J. E. Keen, R. O. Elder, W. W. Laegreid, and M. Koohmaraie.** 2001. Genotypic analyses of *Escherichia coli* O157:H7 and O157 nonmotile isolates recovered from beef cattle and carcasses at processing plants in the Midwestern states of the United States. *Appl. Environ. Microbiol.* **67**:3810–3818.
9. **Batz, M. B., M. P. Doyle, G. Morris, Jr., J. Painter, R. Singh, R. V. Tauxe, M. R. Taylor, and D. M. Wong.** 2005. Attributing illness to food. *Emerg. Infect. Dis.* **11**:993–999.
10. **Baumler, A. J., R. M. Tsolis, T. A. Ficht, and L. G. Adams.** 1998. Evolution of host adaptation in *Salmonella enterica*. *Infect. Immun.* **66**:4579–4587.
11. **Beltran, P., J. M. Musser, R. Helmuth, J. J. Farmer III, W. M. Frerichs, I. K. Wachsmuth, K. Ferris, A. C. McWhorter, J. G. Wells, A. Cravioto, and R. K. Selander.** 1988. Toward a population genetic analysis of *Salmonella*: genetic diversity and relationships among strains of serotypes *S. choleraesuis, S. derby, S. dublin, S. enteritidis, S. heidelberg, S. infantis, S. newport*, and *S. typhimurium*. *Proc. Natl. Acad. Sci. USA* **85**:7753–7757.
12. **Bender, J. B., C. W. Hedberg, J. M. Besser, D. J. Boxrud, K. L. MacDonald, and M. T. Osterholm.** 1997. Surveillance by molecular subtype for *Escherichia coli* O157:H7 infections in Minnesota by molecular subtyping. *N. Engl. J. Med.* **337**:388–394.
13. **Best, E. L., A. J. Fox, J. A. Frost, and F. J. Bolton.** 2005. Real-time single-nucleotide polymorphism profiling using Taqman technology for rapid recognition of *Campylobacter jejuni* clonal complexes. *J. Med. Microbiol.* **54**:919–925.
14. **Bopp, D. J., B. D. Sauders, A. L. Waring, J. Ackelsberg, N. Dumas, E. Braun-Howland, D. Dziewulski, B. J. Wallace, M. Kelly, T. Halse, K. A. Musser, P. F. Smith, D. L. Morse, and R. J. Limberger.** 2003. Detection, isolation, and molecular subtyping of *Escherichia coli* O157:H7 and *Campylobacter jejuni* associated with a large waterborne outbreak. *J. Clin. Microbiol.* **41**:174–180.
15. **Botteldoorn, N., L. Herman, N. Rijpens, and M. Heyndrickx.** 2004. Phenotypic and molecular typing of *Salmonella* strains reveals different contamination sources in two commercial pig slaughterhouses. *Appl. Environ. Microbiol.* **70**:5305–5314.
16. **Botteldoorn, N., M. Heyndrickx, N. Rijpens, K. Grijspeerdt, and L. Herman.** 2003. Salmonella on pig carcasses: positive pigs and cross contamination in the slaughterhouse. *J. Appl. Microbiol.* **95**:891–903.

17. **Breuer, T., D. H. Benkel, R. L. Shapiro, W. N. Hall, M. M. Winnett, M. J. Linn, J. Neimann, T. J. Barrett, S. Dietrich, F. P. Downes, D. M. Toney, J. L. Pearson, H. Rolka, L. Slutsker, and P. M. Griffin.** 2001. A multistate outbreak of *Escherichia coli* O157:H7 infections linked to alfalfa sprouts grown from contaminated seeds. *Emerg. Infect. Dis.* **7**:977–982.

18. **Buchrieser, C., V. V. Gangar, R. L. Murphree, M. L. Tamplin, and C. W. Kaspar.** 1995. Multiple *Vibrio vulnificus* strains in oysters as demonstrated by clamped homogeneous electric field gel electrophoresis. *Appl. Environ. Microbiol.* **61**:1163–1168.

19. **Cai, S., D. Y. Kabuki, A. Y. Kuaye, T. G. Cargioli, M. S. Chung, R. Nielsen, and M. Wiedmann.** 2002. Rational design of DNA sequence-based strategies for subtyping *Listeria monocytogenes*. *J. Clin. Microbiol.* **40**:3319–3325.

20. **Call, D. R., M. K. Borucki, and T. E. Besser.** 2003. Mixed-genome microarrays reveal multiple serotype and lineage-specific differences among strains of *Listeria monocytogenes*. *J. Clin. Microbiol.* **41**:632–639.

21. **Call, D. R., M. K. Borucki, and F. J. Loge.** 2003. Detection of bacterial pathogens in environmental samples using DNA microarrays. *J. Microbiol. Methods* **53**:235–243.

22. **Castillo, A., I. Mercado, L. M. Lucia, Y. Martinez-Ruiz, J. Ponce de Leon, E. A. Murano, and G. R. Acuff.** 2004. Salmonella contamination during production of cantaloupe: a binational study. *J. Food Prot.* **67**:713–720.

23. **Centers for Disease Control and Prevention.** 2005. Preliminary FoodNet data on the incidence of infection with pathogens transmitted commonly through food—10 sites, United States, 2004. *Morb. Mortal. Wkly. Rep.* **54**:352–356.

24. **Centers for Disease Control and Prevention.** 2005. Summary of notifiable diseases—United States, 2003. *Morb. Mortal. Wkly. Rep.* **52**:1–85.

25. **Chan, E. S., J. Aramini, B. Ciebin, D. Middleton, R. Ahmed, M. Howes, I. Brophy, I. Mentis, F. Jamieson, F. Rodgers, M. Nazarowec-White, S. C. Pichette, J. Farrar, M. Gutierrez, W. J. Weis, L. Lior, A. Ellis, and S. Isaacs.** 2002. Natural or raw almonds and an outbreak of a rare phage type of *Salmonella enteritidis* infection. *Can. Commun. Dis. Rep.* **28**:97–99.

26. **Chiou, C. S., S. Y. Hsu, S. I. Chiu, T. K. Wang, and C. S. Chao.** 2000. *Vibrio parahaemolyticus* serovar O3:K6 as cause of unusually high incidence of food-borne disease outbreaks in Taiwan from 1996 to 1999. *J. Clin. Microbiol.* **38**:4621–4625.

27. **Chizhikov, V., A. Rasooly, K. Chumakov, and D. D. Levy.** 2001. Microarray analysis of microbial virulence factors. *Appl. Environ. Microbiol.* **67**:3258–3263.

28. **Clark, C. G., T. M. A. C. Kruk, L. Bryden, Y. Hirvi, R. Ahmed, and F. G. Rodgers.** 2003. Subtyping of *Salmonella enterica* serotype Enteritidis strains by manual and automated *Pst*I-*Sph*I ribotyping. *J. Clin. Microbiol.* **41**:27–33.

29. **Craun, G. F., R. L. Calderon, and M. F. Craun.** 2005. Outbreaks associated with recreational water in the United States. *Int. J. Environ. Health Res.* **15**:243–262.

30. **Davies, R. H., and C. Wray.** 1995. Mice as carriers of *Salmonella enteritidis* on persistently infected poultry units. *Vet. Rec.* **137**:337–341.

31. **Davis, M. A., D. D. Hancock, D. H. Rice, D. R. Call, R. DiGiacomo, M. Samadpour, and T. E. Besser.** 2003. Feedstuffs as a vehicle of cattle exposure to *Escherichia coli* O157:H7 and *Salmonella enterica*. *Vet. Microbiol.* **95**:199–210.

32. **Desai, M., E. J. Threlfall, and J. Stanley.** 2001. Fluorescent amplified-fragment length polymorphism subtyping of the *Salmonella enterica* serovar Enteritidis phage type 4 clone complex. *J. Clin. Microbiol.* **39:**201–206.

33. **Destro, M. T., M. F. Leitao, and J. M. Farber.** 1996. Use of molecular typing methods to trace the dissemination of *Listeria monocytogenes* in a shrimp processing plant. *Appl. Environ. Microbiol.* **62:**705–711.

34. **Dingle, K. E., F. M. Colles, D. R. Wareing, R. Ure, A. J. Fox, F. E. Bolton, H. J. Bootsma, R. J. Willems, R. Urwin, and M. C. Maiden.** 2001. Multilocus sequence typing system for *Campylobacter jejuni*. *J. Clin. Microbiol.* **39:**14–23.

35. **Duim, B., T. M. Wassenaar, A. Rigter, and J. Wagenaar.** 1999. High-resolution genotyping of *Campylobacter* strains isolated from poultry and humans with amplified fragment length polymorphism fingerprinting. *Appl. Environ. Microbiol.* **65:**2369–2375.

36. **Elder, R. O., J. E. Keen, G. R. Siragusa, G. A. Barkocy-Gallagher, M. Koohmaraie, and W. W. Laegreid.** 2000. Correlation of enterohemorrhagic *Escherichia coli* O157 prevalence in feces, hides, and carcasses of beef cattle during processing. *Proc. Natl. Acad. Sci. USA* **97:**2999–3003.

37. **Feil, E. J., B. C. Li, D. M. Aanensen, W. P. Hanage, and B. G. Spratt.** 2004. eBURST: inferring patterns of evolutionary descent among clusters of related bacterial genotypes from multilocus sequence typing data. *J. Bacteriol.* **186:**1518–1530.

38. **Ferguson, D. D., J. Scheftel, A. Cronquist, K. Smith, A. Woo-Ming, E. Anderson, J. Knutsen, A. K. De, and K. Gershman.** 2005. Temporally distinct *Escherichia coli* O157 outbreaks associated with alfalfa sprouts linked to a common seed source—Colorado and Minnesota, 2003. *Epidemiol. Infect.* **133:**439–447.

39. **Finlay, B. B.** 2001. Microbiology. Cracking Listeria's password. *Science* **292:**1665–1667.

40. **Fitzgerald, C., L. O. Helsel, M. A. Nicholson, S. J. Olsen, D. L. Swerdlow, R. Flahart, J. Sexton, and P. I. Fields.** 2001. Evaluation of methods for subtyping *Campylobacter jejuni* during an outbreak involving a food handler. *J. Clin. Microbiol.* **39:**2386–2390.

41. **Foley, S. L., S. Simjee, J. Meng, D. G. White, P. F. McDermott, and S. Zhao.** 2004. Evaluation of molecular typing methods for *Escherichia coli* O157:H7 isolates from cattle, food, and humans. *J. Food Prot.* **67:**651–657.

42. **Frye, D. M., R. Zweig, J. Sturgeon, M. Tormey, M. LeCavalier, I. Lee, L. Lawani, and L. Mascola.** 2002. An outbreak of febrile gastroenteritis associated with delicatessen meat contaminated with *Listeria monocytogenes*. *Clin. Infect. Dis.* **35:**943–949.

43. **Gast, R. K., and P. S. Holt.** 2001. Multiplication in egg yolk and survival in egg albumen of *Salmonella enterica* serotype Enteritidis strains of phage types 4, 8, 13a, and 14b. *J. Food Prot.* **64:**865–868.

44. **Giovannacci, I., S. Queguiner, C. Ragimbeau, G. Salvat, J. L. Vendeuvre, V. Carlier, and G. Ermel.** 2001. Tracing of *Salmonella* spp. in two pork slaughter and cutting plants using serotyping and macrorestriction genotyping. *J. Appl. Microbiol.* **90:**131–147.

45. **Graves, L. M., B. Swaminathan, M. W. Reeves, S. B. Hunter, R. E. Weaver, B. D. Plikaytis, and A. Schuchat.** 1994. Comparison of ribotyping and multilocus enzyme electrophoresis for subtyping of *Listeria monocytogenes* isolates. *J. Clin. Microbiol.* **32:**2936–2943.

46. **Guard-Bouldin, J., R. K. Gast, T. J. Humphrey, D. J. Henzler, C. Morales, and K. Coles.** 2004. Subpopulation characteristics of egg-contaminating *Salmonella enterica* serovar

Enteritidis as defined by the lipopolysaccharide O chain. *Appl. Environ. Microbiol.* **70**:2756–2763.

47. **Guard-Petter, J., D. J. Henzler, M. M. Rahman, and R. W. Carlson.** 1997. On-farm monitoring of mouse-invasive *Salmonella enterica* serovar Enteritidis and a model for its association with the production of contaminated eggs. *Appl. Environ. Microbiol.* **63**:1588–1593.
48. **Gudmundsdottir, S., B. Gudbjornsdottir, H. L. Lauzon, H. Einarsson, K. G. Kristinsson, and M. Kristjansson.** 2005. Tracing *Listeria monocytogenes* isolates from cold-smoked salmon and its processing environment in Iceland using pulsed-field gel electrophoresis. *Int. J. Food Microbiol.* **101**:41–51.
49. **Hahm, B. K., Y. Maldonado, E. Schreiber, A. K. Bhunia, and C. H. Nakatsu.** 2003. Subtyping of foodborne and environmental isolates of *Escherichia coli* by multiplex-PCR, rep-PCR, PFGE, ribotyping and AFLP. *J. Microbiol. Methods* **53**:387–399.
50. **Hald, T., D. Vose, H. C. Wegener, and T. Koupeev.** 2004. A Bayesian approach to quantify the contribution of animal-food sources to human salmonellosis. *Risk Anal.* **24**:255–269.
51. **Hardnett, F. P., R. M. Hoekstra, M. Kennedy, L. Charles, and F. J. Angulo.** 2004. Epidemiologic issues in study design and data analysis related to FoodNet activities. *Clin. Infect. Dis.* **38**:S121–S126.
52. **Hedberg, C. W., M. J. David, K. E. White, K. L. MacDonald, and M. T. Osterholm.** 1993. Role of egg consumption in sporadic *Salmonella enteritidis* and *Salmonella typhimurium* infections in Minnesota. *J. Infect. Dis.* **167**:107–111.
53. **Henzler, D. J., and H. M. Opitz.** 1992. The role of mice in the epizootiology of *Salmonella enteritidis* infection on chicken layer farms. *Avian Dis.* **36**:625–631.
54. **Herikstad, H., S. Yang, T. J. Van Gilder, D. Vugia, J. Hadler, P. Blake, V. Deneen, B. Shiferaw, and F. J. Angulo.** 2002. A population-based estimate of the burden of diarrhoeal illness in the United States: FoodNet, 1996–7. *Epidemiol. Infect.* **129**:9–17.
55. **Hilborn, E. D., J. H. Mermin, P. A. Mshar, J. L. Hadler, A. Voetsch, C. Wojtkunski, M. Swartz, R. Mshar, M. A. Lambert-Fair, J. A. Farrar, M. K. Glynn, and L. Slutsker.** 1999. A multistate outbreak of *Escherichia coli* O157:H7 infections associated with consumption of mesclun lettuce. *Arch. Intern. Med.* **159**:1758–1764.
56. **Hoffman, A. D., K. L. Gall, D. M. Norton, and M. Wiedmann.** 2003. *Listeria monocytogenes* contamination patterns for the smoked fish processing environment and for raw fish. *J. Food Prot.* **66**:52–60.
57. **Hu, H., R. Lan, and P. R. Reeves.** 2002. Fluorescent amplified fragment length polymorphism analysis of *Salmonella enterica* serovar Typhimurium reveals phage-type-specific markers and potential for microarray typing. *J. Clin. Microbiol.* **40**:3406–3415.
58. **Humphrey, T. J.** 1994. Contamination of egg shell and contents with Salmonella enteritidis: a review. *Int. J. Food Microbiol.* **21**:31–40.
59. **Inglis, T. J., A. Clair, J. Sampson, L. O'Reilly, S. Vandenberg, K. Leighton, and A. Watson.** 2003. Real-time application of automated ribotyping and DNA macrorestriction analysis in the setting of a listeriosis outbreak. *Epidemiol. Infect.* **131**:637–645.
60. **Isaacs, S., J. Aramini, B. Ciebin, J. A. Farrar, R. Ahmed, D. Middleton, A. U. Chandran, L. J. Harris, M. Howes, E. Chan, A. S. Pichette, K. Campbell, A. Gupta, L. Y. Lior, M. Pearce, C. Clark, F. Rodgers, F. Jamieson, I. Brophy, and A. Ellis.** 2005. An international outbreak of salmonellosis associated with raw almonds contaminated with a rare phage type of *Salmonella enteritidis*. *J. Food Prot.* **68**:191–198.

61. **Jarvis, G. N., M. G. Kizoulis, F. Diez-Gonzalez, and J. B. Russell.** 2000. The genetic diversity of predominant *Escherichia coli* strains isolated from cattle fed various amounts of hay and grain. *FEMS Microbiol. Ecol.* **32**:225–233.
62. **Jay, M. T., V. Garrett, J. C. Mohle-Boetani, M. Barros, J. A. Farrar, R. Rios, S. Abbott, R. Sowadsky, K. Komatsu, R. Mandrell, J. Sobel, and S. B. Werner.** 2004. A multistate outbreak of *Escherichia coli* O157:H7 infection linked to consumption of beef tacos at a fast-food restaurant chain. *Clin. Infect. Dis.* **39**:1–7.
63. **Jolley, K. A., M. S. Chan, and M. C. Maiden.** 2004. mlstdbNet—distributed multi-locus sequence typing (MLST) databases. *BMC Bioinformatics* **5**:86.
64. **Kabuki, D. Y., A. Y. Kuaye, M. Wiedmann, and K. J. Boor.** 2004. Molecular subtyping and tracking of *Listeria monocytogenes* in Latin-style fresh-cheese processing plants. *J. Dairy Sci.* **87**:2803–2812.
65. **Keys, C., S. Kemper, and P. Keim.** 2005. Highly diverse variable number tandem repeat loci in the E. coli O157:H7 and O55:H7 genomes for high-resolution molecular typing. *J. Appl. Microbiol.* **98**:928–940.
66. **Kramer, J. M., J. A. Frost, F. J. Bolton, and D. R. Wareing.** 2000. *Campylobacter* contamination of raw meat and poultry at retail sale: identification of multiple types and comparison with isolates from human infection. *J. Food Prot.* **63**:1654–1659.
67. **Kudva, I. T., P. S. Evans, N. T. Perna, T. J. Barrett, F. M. Ausubel, F. R. Blattner, and S. B. Calderwood.** 2002. Strains of *Escherichia coli* O157:H7 differ primarily by insertions or deletions, not single-nucleotide polymorphisms. *J. Bacteriol.* **184**:1873–1879.
68. **Lappi, V. R., J. Thimothe, K. K. Nightingale, K. Gall, V. N. Scott, and M. Wiedmann.** 2004. Longitudinal studies on *Listeria* in smoked fish plants: impact of intervention strategies on contamination patterns. *J. Food Prot.* **67**:2500–2514.
69. **Lecuit, M., S. Vandormael-Pournin, J. Lefort, M. Huerre, P. Gounon, C. Dupuy, C. Babinet, and P. Cossart.** 2001. A transgenic model for listeriosis: role of internalin in crossing the intestinal barrier. *Science* **292**:1722–1725.
70. **Lee, T. M., L. L. Chang, C. Y. Chang, J. C. Wang, T. M. Pan, T. K. Wang, and S. F. Chang.** 2000. Molecular analysis of *Shigella sonnei* isolated from three well-documented outbreaks in school children. *J. Med. Microbiol.* **49**:355–360.
71. **Liebana, E., L. Garcia-Migura, C. Clouting, F. A. Clifton-Hadley, M. Breslin, and R. H. Davies.** 2003. Molecular fingerprinting evidence of the contribution of wildlife vectors in the maintenance of Salmonella Enteritidis infection in layer farms. *J. Appl. Microbiol.* **94**:1024–1029.
72. **Liebana, E., L. Garcia-Migura, J. Guard-Petter, S. W. McDowell, S. Rankin, H. M. Opitz, F. A. Clifton-Hadley, and R. H. Davies.** 2002. Salmonella enterica serotype Enteritidis phage types 4, 7, 6, 8, 13a, 29 and 34: a comparative analysis of genomic fingerprints from geographically distant isolates. *J. Appl. Microbiol.* **92**:196–209.
73. **Lindsay, J. A.** 1997. Chronic sequelae of foodborne disease. *Emerg. Infect. Dis.* **3**:443–352.
74. **Lindstedt, B. A.** 2005. Multiple-locus variable number tandem repeats analysis for genetic fingerprinting of pathogenic bacteria. *Electrophoresis* **26**:2567–2582.
75. **Lindstedt, B. A., T. Vardund, L. Aas, and G. Kapperud.** 2004. Multiple-locus variable-number tandem-repeats analysis of Salmonella enterica subsp. enterica serovar Typhimurium using PCR multiplexing and multicolor capillary electrophoresis. *J. Microbiol. Methods* **59**:163–172.

76. **MacDonald, P. D., R. E. Whitwam, J. D. Boggs, J. N. MacCormack, K. L. Anderson, J. W. Reardon, J. R. Saah, L. M. Graves, S. B. Hunter, and J. Sobel.** 2005. Outbreak of listeriosis among Mexican immigrants as a result of consumption of illicitly produced Mexican-style cheese. *Clin. Infect. Dis.* **40:**677–682.
77. **Madden, R. H., L. Moran, and P. Scates.** 1998. Frequency of occurrence of *Campylobacter* spp. in red meats and poultry in Northern Ireland and their subsequent subtyping using polymerase chain reaction-restriction fragment length polymorphism and the random amplified polymorphic DNA method. *J. Appl. Microbiol.* **84:**703–708.
78. **Maiden, M. C., J. A. Bygraves, E. Feil, G. Morelli, J. E. Russell, R. Urwin, Q. Zhang, J. Zhou, K. Zurth, D. A. Caugant, I. M. Feavers, M. Achtman, and B. G. Spratt.** 1998. Multilocus sequence typing: a portable approach to the identification of clones within populations of pathogenic microorganisms. *Proc. Natl. Acad. Sci. USA* **95:**3140–3145.
79. **Markkula, A., T. Autio, J. Lunden, and H. Korkeala.** 2005. Raw and processed fish show identical *Listeria monocytogenes* genotypes with pulsed-field gel electrophoresis. *J. Food Prot.* **68:**1228–1231.
80. **Martin, I. E., S. D. Tyler, K. D. Tyler, R. Khakhria, and W. M. Johnson.** 1996. Evaluation of ribotyping as epidemiologic tool for typing *Escherichia coli* serogroup O157 isolates. *J. Clin. Microbiol.* **34:**720–723.
81. **Martinez, I., L. M. Rorvik, V. Brox, J. Lassen, M. Seppola, L. Gram, and B. Fonnesbech-Vogel.** 2003. Genetic variability among isolates of *Listeria monocytogenes* from food products, clinical samples and processing environments, estimated by RAPD typing. *Int. J. Food Microbiol.* **84:**285–297.
82. **Martinez-Urtaza, J., A. Lozano-Leon, A. DePaola, M. Ishibashi, K. Shimada, M. Nishibuchi, and E. Liebana.** 2004. Characterization of pathogenic *Vibrio parahaemolyticus* isolates from clinical sources in Spain and comparison with Asian and North American pandemic isolates. *J. Clin. Microbiol.* **42:**4672–4678.
83. **McLauchlin, J., G. Ripabelli, M. M. Brett, and E. J. Threlfall.** 2000. Amplified fragment length polymorphism (AFLP) analysis of Clostridium perfringens for epidemiological typing. *Int. J. Food Microbiol.* **56:**21–28.
84. **Mead, P. S., L. Slutsker, V. Dietz, L. F. McCaig, J. S. Bresee, C. Shapiro, P. M. Griffin, and R. V. Tauxe.** 1999. Food-related illness and death in the United States. *Emerg. Infect. Dis.* **5:**607–625.
85. **Micheli, M. R., R. Bova, E. Pascale, and E. D'Ambrosio.** 1994. Reproducible DNA fingerprinting with the random amplified polymorphic DNA (RAPD) method. *Nucleic Acids Res.* **22:**1921–1922.
86. **Nakamura, M., N. Nagamine, M. Norimatsu, S. Suzuki, K. Ohishi, M. Kijima, Y. Tamura, and S. Sato.** 1993. The ability of Salmonella enteritidis isolated from chicks imported from England to cause transovarian infection. *J. Vet. Med. Sci.* **55:**135–136.
87. **Noller, A. C., M. C. McEllistrem, A. G. Pacheco, D. J. Boxrud, and L. H. Harrison.** 2003. Multilocus variable-number tandem repeat analysis distinguishes outbreak and sporadic *Escherichia coli* O157:H7 isolates. *J. Clin. Microbiol.* **41:**5389–5397.
88. **Noller, A. C., M. C. McEllistrem, O. C. Stine, J. G. Morris, Jr., D. J. Boxrud, B. Dixon, and L. H. Harrison.** 2003. Multilocus sequence typing reveals a lack of diversity among *Escherichia coli* O157:H7 isolates that are distinct by pulsed-field gel electrophoresis. *J. Clin. Microbiol.* **41:**675–679.
89. **Olsen, S. J., M. Patrick, S. B. Hunter, V. Reddy, L. Kornstein, W. R. MacKenzie, K. Lane, S. Bidol, G. A. Stoltman, D. M. Frye, I. Lee, S. Hurd, T. F. Jones, T. N.**

LaPorte, W. Dewitt, L. Graves, M. Wiedmann, D. J. Schoonmaker-Bopp, A. J. Huang, C. Vincent, A. Bugenhagen, J. Corby, E. R. Carloni, M. E. Holcomb, R. F. Woron, S. M. Zansky, G. Dowdle, F. Smith, S. Ahrabi-Fard, A. R. Ong, N. Tucker, N. A. Hynes, and P. Mead. 2005. Multistate outbreak of *Listeria monocytogenes* infection linked to delicatessen turkey meat. *Clin. Infect. Dis.* **40**:962–967.

90. **Owen, R. J., and S. Leeton.** 1999. Restriction fragment length polymorphism analysis of the flaA gene of *Campylobacter jejuni* for subtyping human, animal and poultry isolates. *FEMS Microbiol. Lett.* **176**:345–350.
91. **Patrick, M. E., P. M. Adcock, T. M. Gomez, S. F. Altekruse, B. H. Holland, R. V. Tauxe, and D. L. Swerdlow.** 2004. *Salmonella enteritidis* infections, United States, 1985–1999. *Emerg. Infect. Dis.* **10**:1–7.
92. **Petersen, L., and D. G. Newell.** 2001. The ability of Fla-typing schemes to discriminate between strains of *Campylobacter jejuni*. *J. Appl. Microbiol.* **91**:217–224.
93. **Piffaretti, J. C., H. Kressebuch, M. Aeschbacher, J. Bille, E. Bannerman, J. M. Musser, R. K. Selander, and J. Rocourt.** 1989. Genetic characterization of clones of the bacterium *Listeria monocytogenes* causing epidemic disease. *Proc. Natl. Acad. Sci. USA* **86**:3818–3822.
94. **Pupo, G. M., R. Lan, and P. R. Reeves.** 2000. Multiple independent origins of Shigella clones of *Escherichia coli* and convergent evolution of many of their characteristics. *Proc. Natl. Acad. Sci. USA* **97**:10567–10572.
95. **Qi, W., D. W. Lacher, A. C. Bumbaugh, K. E. Hyma, L. M. Ouellette, T. M. Large, C. L. Tarr, and T. S. Whittam.** 2004. *Ec*MLST: an online database for multilocus sequence typing of pathogenic *Escherichia coli*. *Proceedings of the 2004 IEEE Computational Systems Bioinformatics Conference.*
96. **Rangel, J. M., P. H. Sparling, C. Crowe, P. M. Griffin, and D. L. Swerdlow.** 2005. Epidemiology of *Escherichia coli* O157:H7 outbreaks, United States, 1982–2002. *Emerg. Infect. Dis.* **11**:603–609.
97. **Rice, D. H., K. M. McMenamin, L. C. Pritchett, D. D. Hancock, and T. E. Besser.** 1999. Genetic subtyping of *Escherichia coli* O157 isolates from 41 Pacific Northwest USA cattle farms. *Epidemiol. Infect.* **122**:479–484.
98. **Ross, I. L., and M. W. Heuzenroeder.** 2005. Use of AFLP and PFGE to discriminate between *Salmonella enterica* serovar Typhimurium DT126 isolates from separate food-related outbreaks in Australia. *Epidemiol. Infect.* **133**:635–644.
99. **Saeed, A. M., S. T. Walk, M. M. Arshad, and T. S. Whittam.** 2006. Clonal structure and variation in virulence of *Salmonella* Enteritidis isolated from mice, chickens, and humans. *J. Assoc. Off. Anal. Chem.* **89**:504–511.
100. **Sails, A. D., B. Swaminathan, and P. I. Fields.** 2003. Utility of multilocus sequence typing as an epidemiological tool for investigation of outbreaks of gastroenteritis caused by *Campylobacter jejuni*. *J. Clin. Microbiol.* **41**:4733–4739.
101. **Salama, N., K. Guillemin, T. K. McDaniel, G. Sherlock, L. Tompkins, and S. Falkow.** 2000. A whole-genome microarray reveals genetic diversity among Helicobacter pylori strains. *Proc. Natl. Acad. Sci. USA* **97**:14668–14673.
102. **Sauders, B. D., K. Mangione, C. Vincent, J. Schermerhorn, C. M. Farchione, N. B. Dumas, D. Bopp, L. Kornstein, E. D. Fortes, K. Windham, and M. Wiedmann.** 2004. Distribution of *Listeria monocytogenes* molecular subtypes among human and food isolates from New York State shows persistence of human disease-associated *Listeria monocytogenes* strains in retail environments. *J. Food Prot.* **67**:1417–1428.

103. **Schroeder, C. M., A. L. Naugle, W. D. Schlosser, A. T. Hogue, F. J. Angulo, J. S. Rose, E. D. Ebel, W. T. Disney, K. G. Holt, and D. P. Goldman.** 2005. Estimate of illnesses from *Salmonella* enteritidis in eggs, United States, 2000. *Emerg. Infect. Dis.* **11**:113–115.

104. **Scott, T. M., T. M. Jenkins, J. Lukasik, and J. B. Rose.** 2005. Potential use of a host associated molecular marker in *Enterococcus faecium* as an index of human fecal pollution. *Environ. Sci. Technol.* **39**:283–287.

105. **Selander, R. K., P. Beltran, N. H. Smith, R. Helmuth, F. A. Rubin, D. J. Kopecko, K. Ferris, B. D. Tall, A. Cravioto, and J. M. Musser.** 1990. Evolutionary genetic relationships of clones of *Salmonella* serovars that cause human typhoid and other enteric fevers. *Infect. Immun.* **58**:2262–2275.

106. **Selander, R. K., J. M. Musser, D. A. Caugant, M. N. Gilmour, and T. S. Whittam.** 1987. Population genetics of pathogenic bacteria. *Microb. Pathog.* **3**:1–7.

107. **Shere, J. A., K. J. Bartlett, and C. W. Kaspar.** 1998. Longitudinal study of *Escherichia coli* O157:H7 dissemination on four dairy farms in Wisconsin. *Appl. Environ. Microbiol.* **64**: 1390–1399.

108. **Simpson, J. M., J. W. Santo Domingo, and D. J. Reasoner.** 2002. Microbial source tracking: state of the science. *Environ. Sci. Technol.* **36**:5279–5288.

109. **Spratt, B. G., and M. C. Maiden.** 1999. Bacterial population genetics, evolution and epidemiology. *Philos. Trans. R. Soc. Lond. B Biol. Sci.* **354**:701–710.

110. **St. Louis, M. E., D. L. Morse, M. E. Potter, T. M. DeMelfi, J. J. Guzewich, R. V. Tauxe, and P. A. Blake.** 1988. The emergence of grade A eggs as a major source of Salmonella enteritidis infections. New implications for the control of salmonellosis. *JAMA* **259**: 2103–2107.

111. **Swaminathan, B., T. J. Barrett, S. B. Hunter, and R. V. Tauxe.** 2001. PulseNet: the molecular subtyping network for foodborne bacterial disease surveillance, United States. *Emerg. Infect. Dis.* **7**:382–389.

112. **Swaminathan, B., S. B. Hunter, P. M. Desmarchelier, P. Gerner-Smidt, L. M. Graves, S. Harlander, R. Hubner, C. Jacquet, B. Pedersen, K. Reineccius, A. Ridley, N. A. Saunders, and J. A. Webster.** 1996. WHO-sponsored international collaborative study to evaluate methods for subtyping *Listeria monocytogenes*: restriction fragment length polymorphism (RFLP) analysis using ribotyping and Southern hybridization with two probes derived from *L. monocytogenes* chromosome. *Int. J. Food Microbiol.* **32**:263–278.

113. **Tamplin, M. L., J. K. Jackson, C. Buchrieser, R. L. Murphree, K. M. Portier, V. Gangar, L. G. Miller, and C. W. Kaspar.** 1996. Pulsed-field gel electrophoresis and ribotype profiles of clinical and environmental *Vibrio vulnificus* isolates. *Appl. Environ. Microbiol.* **62**:3572–3580.

114. **Thiagarajan, D., A. M. Saeed, and E. K. Asem.** 1994. Mechanism of transovarian transmission of *Salmonella enteritidis* in laying hens. *Poult. Sci.* **73**:89–98.

115. **Thimothe, J., K. K. Nightingale, K. Gall, V. N. Scott, and M. Wiedmann.** 2004. Tracking of *Listeria monocytogenes* in smoked fish processing plants. *J. Food Prot.* **67**:328–341.

116. **Todd, E. C.** 1989. Costs of acute bacterial foodborne disease in Canada and the United States. *Int. J. Food Microbiol.* **9**:313–326.

117. **Trevejo, R. T., S. L. Abbott, M. I. Wolfe, J. Meshulam, D. Yong, and G. R. Flores.** 1999. An untypeable *Shigella flexneri* strain associated with an outbreak in California. *J. Clin. Microbiol.* **37**:2352–2353.

118. **Urisman, A., K. F. Fischer, C. Y. Chiu, A. L. Kistler, S. Beck, D. Wang, and J. L. DeRisi.** 2005. E-Predict: a computational strategy for species identification based on observed DNA microarray hybridization patterns. *Genome Biol.* **6:**R78.

119. **Velge, P., A. Cloeckaert, and P. Barrow.** 2005. Emergence of Salmonella epidemics: the problems related to *Salmonella enterica* serotype Enteritidis and multiple antibiotic resistance in other major serotypes. *Vet. Res.* **36:**267–288.

120. **Versalovic, J., T. Koeuth, and J. R. Lupski.** 1991. Distribution of repetitive DNA sequences in eubacteria and application to fingerprinting of bacterial genomes. *Nucleic Acids Res.* **19:**6823–6831.

121. **Welsh, J., and M. McClelland.** 1990. Fingerprinting genomes using PCR with arbitrary primers. *Nucleic Acids Res.* **18:**7213–7218.

122. **Whittam, T. S.** 1996. Genetic variation and evolutionary processes in natural populations of *Escherichia coli*, p. 2708–2720. *In* F. C. Neidhardt, R. Curtiss III, J. L. Ingraham, E. C. C. Lin, K. B. Low, B. Magasanik, W. S. Reznikoff, M. Riley, M. Schaechter, and H. E. Umbarger (ed.), *Escherichia coli and Salmonella: Cellular and Molecular Biology*, 2nd ed. American Society for Microbiology, Washington, D.C.

123. **Whittam, T. S., I. K. Wachsmuth, and R. A. Wilson.** 1988. Genetic evidence of clonal descent of *Escherichia coli* O157:H7 associated with hemorrhagic colitis and hemolytic uremic syndrome. *J. Infect. Dis.* **157:**1124–1133.

124. **Whittam, T. S., and R. A. Wilson.** 1988. Genetic relationships among pathogenic *Escherichia coli* of serogroup O157. *Infect. Immun.* **56:**2467–2473.

125. **Wick, L. M., W. Qi, D. W. Lacher, and T. S. Whittam.** 2005. Evolution of genomic content in the stepwise emergence of *Escherichia coli* O157:H7. *J. Bacteriol.* **187:**1783–1791.

126. **Wiedmann, M.** 2002. Subtyping of bacterial foodborne pathogens. *Nutr. Rev.* **60:**201–208.

127. **Wiedmann, M., J. L. Bruce, C. Keating, A. E. Johnson, P. L. McDonough, and C. A. Batt.** 1997. Ribotypes and virulence gene polymorphisms suggest three distinct *Listeria monocytogenes* lineages with differences in pathogenic potential. *Infect. Immun.* **65:**2707–2716.

128. **Williams, J. G., A. R. Kubelik, K. J. Livak, J. A. Rafalski, and S. V. Tingey.** 1990. DNA polymorphisms amplified by arbitrary primers are useful as genetic markers. *Nucleic Acids Res.* **18:**6531–6535.

129. **Wilson, W. J., C. L. Strout, T. Z. DeSantis, J. L. Stilwell, A. V. Carrano, and G. L. Andersen.** 2002. Sequence-specific identification of 18 pathogenic microorganisms using microarray technology. *Mol. Cell. Probes* **16:**119–127.

130. **Wonderling, L., R. Pearce, F. M. Wallace, J. E. Call, I. Feder, M. Tamplin, and J. B. Luchansky.** 2003. Use of pulsed-field gel electrophoresis to characterize the heterogeneity and clonality of *Salmonella* isolates obtained from the carcasses and feces of swine at slaughter. *Appl. Environ. Microbiol.* **69:**4177–4182.

131. **Wong, S., D. Street, S. I. Delgado, and K. C. Klontz.** 2000. Recalls of foods and cosmetics due to microbial contamination reported to the U.S. Food and Drug Administration. *J. Food Prot.* **63:**1113–1116.

132. **Yeung, P. S., M. C. Hayes, A. DePaola, C. A. Kaysner, L. Kornstein, and K. J. Boor.** 2002. Comparative phenotypic, molecular, and virulence characterization of *Vibrio parahaemolyticus* O3:K6 isolates. *Appl. Environ. Microbiol.* **68:**2901–2909.

133. **Yuki, N., K. Susuki, M. Koga, Y. Nishimoto, M. Odaka, K. Hirata, K. Taguchi, T. Miyatake, K. Furukawa, T. Kobata, and M. Yamada.** 2004. Carbohydrate mimicry between human ganglioside GM1 and *Campylobacter jejuni* lipooligosaccharide causes Guillain-Barre syndrome. *Proc. Natl. Acad. Sci. USA* **101:**11404–11409.

134. **Zhang, W., B. M. Jayarao, and S. J. Knabel.** 2004. Multi-virulence-locus sequence typing of *Listeria monocytogenes*. *Appl. Environ. Microbiol.* **70:**913–920.

Microbial Source Tracking
Edited by Jorge W. Santo Domingo and Michael J. Sadowsky

Shellfish and Microbial Source Tracking

5

John Scott Meschke and David Boyle

Molluscan shellfish (oysters, clams, mussels, and scallops) are a valuable commodity both as a food source and as the sustaining product of an important coastal industry. Consequently, fecal contamination of shellfish growing waters can create both public health and economic concerns. Indicator microorganisms are commonly used to determine the sanitary quality of food and water. The prevalence of these organisms serves as an indication of the incidence of fecal contamination and the possible presence of pathogenic organisms. However, the detection of these organisms alone provides no indication of the source of the contamination and little utility towards the management of fecal contamination sources. Microbial source tracking (MST) refers to a broad class of microbiological techniques used to determine the source(s) of fecal contamination. These methods, although still under development, have already been widely applied to the management of fecally contaminated water bodies in the establishment of bacterial total maximum daily loads. These techniques may also provide a potentially important management tool for the shellfish industry and the regulation of the sanitary quality of shellfish. This chapter will review the impact of fecal contamination on the shellfish industry, summarize the legal framework for regulation of the sanitary quality of shellfish, and examine the use of MST techniques as applied to shellfish growing waters.

John Scott Meschke, School of Public Health and Community Medicine, University of Washington, Seattle, WA 98105-6099. David Boyle, Washington State Dept. of Health, 1610 N.E. 150th Street, Public Health Laboratories, Shoreline, WA 98155-7224.

Shellfish as a Commodity

The U.S. Shellfish Industry

The commercial shellfishery (oysters, clams, mussels, and scallops) of the United States is a valuable resource. Commercial shellfish landings accounted for nearly 50% of the value of the overall commercial fishery, although they are less than 20% of the overall volume of commercial fishery landings from 1984 to 1993 (124). The 1995 U.S. commercial harvest of oysters, clams, and mussels had a dockside value of $200 million (5).

In the 1995 Federal Shellfish Register, 87% of the marine and estuarine waters in the contiguous U.S. coastal areas (over 33,000 square miles) were classified for shellfish harvest under the National Shellfish Sanitation Program (http://issc.org/documents/issc%20analysis%20of%20classified%20waters%201985-2003.pdf) (5). These areas were each classified as approved, conditionally approved, restricted, conditionally restricted, or prohibited. The remaining 13% of marine and estuarine waters are inactive; that is, they are part of the state shellfish programs but are neither monitored for water quality nor do they have a sanitary survey maintained, and they are thus effectively prohibited from commercial harvest. Of the over 4,000 growing areas classified for shellfish harvesting, approximately 69% were approved for harvest, 8% conditionally approved, 10% restricted, <1% conditionally restricted, and 13% prohibited from harvest. The relative acreage of classified shellfish-growing waters in the 21 contiguous coastal states in 1995 is summarized in Table 1. A 2003 survey of state shellfish managers (http://issc.org/documents/issc%20analysis%20of%20classified%20waters%201985-2003.pdf) found slight improvements in the classified growing waters, with approximately 70% of growing waters classified as approved, 15% as conditionally approved, 5% as restricted, less than 1% as conditionally restricted, and only 10% as prohibited.

The shellfish harvest from classified state shellfish waters fall into three distinct categories: commercial harvest of wild shellfish stock, shellfish aquaculture, and recreational harvest. Aquaculture, particularly of oysters, is a growing sector in the overall shellfish harvest. For example, the National Marine Fisheries Service reported that oyster aquaculture increased from 43% in 1985 to 72% of the overall oyster harvest in 1995 (123). This is important as states like Washington, which have a comparatively small classified acreage for harvest but significant aquaculture, may play a significant role in the overall shellfish harvest economy.

The shellfish growing waters of the contiguous United States can be divided into five regions: North Atlantic (ME, NH, MA, CT, and RI), Mid-Atlantic (NY, NJ, DE, MD, and VA), South Atlantic (NC, SC, GA, and part of FL), Gulf of Mexico (AL, MS, LA, TX, and part of FL), and West Coast (CA, OR, and WA).

Table 1 Classified U.S. shellfish-growing waters for 1995[a]

State	Classified acres (10^3)	% Harvest limited
Maine	1,852	16
New Hampshire	61	10
Massachusetts	1,548	27
Rhode Island	284	14
Connecticut	369	66
New York	1,134	14
New Jersey	737	29
Delaware	326	23
Maryland	1,440	9
Virginia	1,650	6
North Carolina	2,803	15
South Carolina	783	10
Georgia	187	63
Florida	1,445	79
Alabama	292	100
Mississippi	431	36
Louisiana	3,962	46
Texas	1,620	49
California	24	96
Oregon	91	89
Washington	308	36

[a]Reproduced from NOAA State of the Coast Report, 1998, http://state_of_coast.noaa.gov/bulletins/html/sgw_04/sfw.html.

North Atlantic Coast. The important shellfish-growing waters of the North Atlantic Coast region include Penobscot Bay and parts of Long Island Sound and Delaware Bay (5).

Mid-Atlantic Coast. The most important estuary for shellfish harvest in the United States is the Chesapeake Bay, with a classified acreage of >2.5 million acres in Maryland and Virginia (5).

South Atlantic Coast. The second most important estuary for shellfish harvest is the Albemarle/Pamlico Sound region of North Carolina, with nearly 2 million classified acres (5).

Gulf Coast. The Gulf of Mexico region has the largest fraction of classified shellfish-growing waters (i.e., more than a third), with the majority being located in the intertidal areas of Louisiana (5). The Breton/Chandeleur and Mississippi Sounds of Louisiana are important estuarine harvesting areas, with approximately 1.1 million acres each classified for harvest. Other important gulf estuarine harvest areas include Atchafalaya/Vermillion (500,000 classified acres) and the Terrebone/Timbalier (410,000 classified acres) Bay regions of Louisiana and the Lower Laguna Madre region of Texas (399,000 classified acres) (5). Unfortunately, the growing waters in this region are also among the most harvest-limited (i.e., conditionally open or closed to harvest) (5). In 1995, 100% of Alabama, 36% of Mississippi, 46% of Louisiana, 49% of Texas, and 79% of Florida growing waters were harvest-limited. In total over half of the growing waters in the Gulf region are harvest-limited.

West Coast. The West Coast region has the smallest fraction of shellfish-growing waters in the United States. Additionally, the West Coast region has the highest percentage of waters prohibited to harvesting (~1/3 as compared to 13% of Gulf waters and 7% of Mid-Atlantic waters) (5). However, Washington State is the leading producer of commercially farmed bivalve shellfish in the United States. More than 86 million pounds of shellfish worth $76 million are harvested commercially each year in the state of Washington, and the recreational harvest is estimated at 700,000 pounds of clams and 900,000 pounds of oysters per year (http://www.psat.wa.gov/Publications/manplan00/18_shellfish.pdf). Washington is also a leading producer of geoduck clams. Puget Sound is one of the richest shellfish-growing areas in the West Coast Region. Other important shellfish-growing areas in the West Coast region include Grays Harbor in Washington, Tillamook Bay and Coos Bay in Oregon, and Morro Bay in California.

Types of Shellfish

The types of bivalve molluscs harvested in the United States include clams, mussels, oysters, and scallops. Oysters, mussels, and most clams are typically harvested in near-shore coastal or estuarine regions. Most scallops and a significant fraction of clams are harvested further offshore (more than 3 miles). Oysters account for half of the annual harvest (weight and value), followed by clams (40% weight, 46% value), mussels (9% weight, 2% value) and scallops (less than 1% weight and value) (5).

Oysters

The major types of oysters commercially harvested in the United States are the Eastern oyster (*Crassostrea virginica*), harvested primarily from the Gulf of Mexico and Atlantic regions, and the Pacific oyster (*Crassostrea gigas*), harvested primarily on the west coast. However, the Pacific oyster is not native to the west coast of North America. Originally, the Olympia oyster, *Ostrea lurida*, was the native species. However, numbers dwindled at the

Table 2 Examples of state contribution to oyster harvest in 1995[a]

State	Poundage of oysters harvested
Louisiana	13.8
Washington	8
Texas	5.4
Connecticut	5

[a]Adapted from reference 7 in millions of pounds.

turn of the century and the American oyster, *Crassostrea virginica*, was introduced from the east coast. Unfortunately, this species did not survive well, and by the 1920s large quantities of Pacific oyster seed were being imported from Japan. *Crassostrea gigas* has adapted well to various environments and is now the most widespread and ubiquitous oyster species in the world (22). The relative fractions of the oyster harvest in the United States are summarized in Table 2. Although Louisiana produces more poundage of oysters, Washington produced nearly eight times the number of oysters per acre as Louisiana through intensive aquaculture (5).

Clams

Clams are a broad grouping of bivalve molluscs, consisting of several taxonomic genera. Most clams are oval shaped with two symmetrical shells. An exception is the razor clam (*Siliqua patula*), which has an elongated shell that suggests an old-fashioned straight razor. The majority of clams harvested annually in the United States come from the Atlantic coast, with the North Atlantic and Mid-Atlantic regions contributing 80% of the harvested clams (Table 3). The most common types of clams commercially harvested on the Atlantic coast include the surf clams (*Spisula solidissima*), quahog clams (*Mercenaria mercenaria* and *Arctica islandica*), and softshell clams (*Mya arenaria*). On the west coast, Manila clams (*Ruditapes philippinarum*) and the geoduck clams (*Panopea abrupta* or *Panope generosa*) are the predominant commercially harvested types.

Table 3 Examples of state contributions to clam harvest in 1995[a]

State	% of clams harvested
New Jersey	35
Massachusetts	11
Washington	10
New York	8

[a]Adapted from reference 7.

Mussels

The commercially important species of mussels in the United States are from the family *Mytillidae*, the saltwater mussels. The main commercial species is the common or blue mussel (*Mytilus edulis*). Mussels are bivalve molluscs which are sedentary as adults and secrete byssal threads to secure them to their substrate. Mussels may be found naturally in the intertidal zone and are commercially cultivated. The mussel harvest in the United States comes primarily from classified waters of Maine.

Scallops

Scallops are bivalve molluscs of the family *Pectinidae*. They are active swimmers and, as a result, have a central adductor muscle that is larger and more developed than that of other bivalves. Scallops have a fan-shaped shell with fluted edges. The New England sea scallop (*Placepten magellinacus*) is the most important commercial scallop species in the United States.

Shellfish as Filter Feeders

Shellfish do not need clean water to grow, but they must be grown in clean water to be safely consumed. Shellfish are filter feeders that efficiently remove particulate matter from their growing waters as a food source. As a result, they may incidentally bioaccumulate pathogenic microbes to concentrations greater than present in the water column (12, 57). Levels of these pathogens in the shellfish can be 100 times or more higher than those found in the water column (http://www.cfsan.fda.gov/~ear/nss2-toc.html). This creates a potential health risk for consumers ingesting raw or undercooked shellfish (68).

Shellfish may filter 6 to 60 liters of seawater per day, depending on size, species, and other factors (http://www.psat.wa.gov/Programs/shellfish/fact_sheets/ecology_web1.pdf). Both nutrient level and temperature are critical elements in the filtration rate of bivalves, which is thought to be under physiological control. Uptake and depuration rates may vary for different types of shellfish. Further, shellfish may demonstrate selective accumulation of particular pathogens (15, 139, 140). Uptake may also vary seasonally, with shellfish preferentially accumulating microorganisms during periods of low water temperature (between 11.5 and 21.5°C), which results in higher incidence of human disease during these periods (16, 68). This tendency to preferentially accumulate pathogens, combined with improved survival of pathogens in cooler water conditions, may explain much of the seasonality associated with shellfish-borne infectious disease (68). Other studies have shown differences in the elimination rate for different pathogen types from shellfish tissues, with viruses tending to be more gradually removed than fecal coliforms and other bacteria (36–38, 96, 120, 151). Two commercial practices used to allow contaminated shellfish to purge themselves of microbial contaminants are relaying and depuration. Relaying involves the relocation of shellfish from a fecally impacted site to an approved harvest site for several months. In contrast, depuration typically involves relocating contaminated shellfish for approximately 48 hours to a land-based facility supplied with UV-disinfected water.

Shellfish-Borne Illness

Food-borne diseases from consumption of shellfish are a significant public health concern, especially since rates of shellfish consumption continue to increase worldwide (134). Over 2,000 cases of shellfish-associated illnesses were reported from 1991 to 1998 in the United States (156). As with most other food- or waterborne illnesses, this number is believed to represent only the tip of the iceberg. The vast majority of cases are thought to go unreported, as the most common associated illness is a relatively mild gastroenteritis (113). The total number of food-borne illnesses is estimated to be 76 million cases of illness per year in the United States (113). Shellfish are thought to be responsible for roughly 6% of total food-borne illness, or 4.5 million cases annually (113, 173).

The practice of consuming raw or partially cooked shellfish increases the risk of contracting shellfish-related illnesses (32, 53, 90). While cooking may significantly reduce the risks of consuming shellfish, cooking of shellfish meats does not ensure that virus contamination is eliminated (89, 112). Tests on poliovirus have shown that up to 7% of the initial inoculum to oyster tissues may remain viable following steaming for 30 minutes with peak temperature of the surrounding tissue reaching 94°C (34). Other studies have shown that noroviruses may be more heat resistant than poliovirus, and, given their low infectious dose, ingesting even a small amount following cooking could cause disease (35). Interestingly, Kirkland et al. found that the amount of shellfish consumed was the main predictor of illness, not whether it was raw or steamed (89). This study also makes it clear that the heat required to render an oyster "done" in culinary terms may not be enough to inactivate all of the viruses present within the tissue.

Outbreaks of illness from shellfish are typically attributed to either viruses or bacteria. However, illnesses associated with marine algal toxins have also been reported (41). Of the reported cases of shellfish-borne illness in the United States, viruses (primarily hepatitis A virus and noroviruses) were conclusively identified in more than 2,200 cases, and an additional 8,000 cases were likely caused by viruses, based on symptomology, onset, and duration of illness. The reason for the large number of inconclusively identified cases is that the methods for adequate virus detection in shellfish are relatively new and emerging. (143). By comparison, confirmed bacterial cases of shellfish illness total less than 1,000, with the exception of *Salmonella* (which has not been a significant problem in North America since 1950) (143). Broadly, the microbial contaminants associated with shellfish-borne disease may be classified as either (i) being naturally occurring (autochthonous) in shellfish-growing waters or (ii) related to fecal contamination of the growing waters (allochthonous). The major contaminants in each group are described below.

Naturally Occurring Microbial Contaminants

Of the pathogenic microbes associated with shellfish-borne illness, only *Vibrio* spp. and *Aeromonas* spp. are indigenous to the marine environment. *Vibrio* spp. are halophilic, gram-negative, motile rod-shaped bacteria, many of which are slightly curved. *Vibrio* spp. are commonly associated with the water column, phytoplankton, sediments, and certain shellfish species. Due to the estuarine ecology of these bacteria, their occurrence does not correlate well with indicator microorganisms often associated with fecal pollution and, therefore, cannot be controlled by water quality control measures such as wastewater treatment (29). At least 10 different *Vibrio* spp. have been implicated in shellfish-borne illness (45) (Table 4). Of the *Vibrio* species, *V. cholerae,* the etiologic agent of cholera, has historically been of greatest concern to public health. *Vibrio cholerae* was also the first organism to be linked with an outbreak associated with eating contaminated shellfish (134).

Cholera continues to present a serious problem to public health in developing countries. The prevalence of the disease in developing countries is primarily linked to a lack of effective water supply and sanitation systems (167). The disease is sporadically reported in the developed world, primarily due to vacationers returning from holiday or to the eating of shellfish imported from affected areas, although *V. cholerae* is endemic to the Gulf of Mexico coast (59, 132). It is possible for *V. cholerae* to be introduced to pristine waters, as it has been reported to be transmitted in the bilge water of marine vessels (111) and in cyanobacteria and algae (42). Once ingested, the bacterium colonizes the intestinal epithelium and releases a toxin, the cholera toxin, which produces a profuse, watery diarrhea (classically known as "rice water stool"), with symptoms occurring within 24 h of infection and lasting for 5 to 60 h (118). Loss of appetite and occasional vomiting accompany the diarrhea. The severity of the disease is strain-dependent, with 75% of infected patients being asymptomatic or experiencing only mild symptoms (55). In extreme forms of cholera (cholera gravis) diarrheal emissions can be over 1 liter/h and may rapidly lead to death by circulatory collapse, primarily as a result of dehydration, if left untreated. *V. cholerae* can be segregated into three

Table 4 Pathogenic microorganisms found to accumulate in bivalve molluscs

Organism type(s)	Source
Viruses	
Calicivirus (norovirus, sapovirus)	Fecal
Hepatitis A	Fecal
Hepatitis E	Fecal
Astrovirus	Fecal
Enterovirus	Fecal
Rotavirus	Fecal
Coronavirus	Fecal
Toroviruses	Fecal
Picobirnaviruses	Fecal
Pestiviruses	Fecal
Bacteria	
Vibrio spp.	Naturally occurring
V. cholerae, V. parahaemolyticus, V. vulnificus (Biogroup 1), *V. mimicus, V. fluvialis, V. furnissii, V. hollisae, V. damsela, V. metschnikovii,* and *V. cincinnatiensis*	
Aeromonas spp.	Naturally occurring
A. hydrophila and *A. caviae*	
Clostridium difficile and *C. botulinum*	Naturally occurring
Enterobacteriaceae	Fecal
Salmonella spp., *Escherichia coli, Shigella* spp. (*S. flexneri, S. sonnei,* and *S. dysenteriae*), *Plesiomonas shigelloides,* and *Yersinia enterocolitica*	
Listeria monocytogenes	Fecal
Campylobacter spp.	Fecal
C. jejuni, C. coli, C. lari, and *C. fetus*	
Parasites	
Giardia lamblia (G. duodenalis, G. intestinalis)	Fecal
Cryptosporidium parvum	Fecal

major subgroups: *V. cholerae* O1, *V. cholerae* O139, and *V. cholerae* non-O1. The *Vibrio cholerae* O1 subgroup has been responsible for seven previous pandemics. *V. cholerae* O139 was first identified in Madras, India in 1992 and does not react with the O1 antisera. It has been speculated that the latter subgroup may be responsible for the eighth pandemic (44, 71). The final subgroup is *V. cholerae* non-O1, members of which are biochemically indistinct from both types O1 and O139 but do not possess the gene for cholera toxin production. Instead, *V. cholerae* non-O1 produces a different toxin which causes a severe choleralike disease (72).

Vibrio parahaemolyticus is the most common *Vibrio* sp. identified with shellfish-borne disease in the United States. It was first isolated from food poisoning outbreaks in Japan in the early 1950s and is also the most common cause of outbreaks there (104). Infection by *V. parahaemolyticus* is primarily associated with the consumption of raw or undercooked oysters; the incubation takes from 6 to 96 hours (104, 164). The symptoms include vomiting, nausea, diarrhea, abdominal cramps, headaches, fever, and chills. The diarrhea is watery and very occasionally bloody. The symptoms are usually mild and self-limiting. This disease is typically not fatal, although complications may arise in compromised patients. Typically *V. parahaemolyticus* needs elevated water temperatures to grow, and therefore, outbreaks or individual infections occur primarily in the summer months or when water temperatures are ≥15°C (31, 71, 83). Interestingly, most environmental strains of *V. parahaemolyticus* are avirulent but all isolates from infected patients carry at least two markers associated with virulence: the thermostable direct hemolysin and the thermostable direct hemolysin-related hemolysin (4). Why there are discrepancies between the environmental and clinical isolates is not understood.

The third most common *Vibrio* species associated with shellfish-borne illness is *V. vulnificus*. There are three biotypes, of which only biotype 1 is of concern as an agent of gastroenteritis. The number of reported outbreaks of disease from *V. vulnificus* is limited, but the etiology of this disease is very significant as it is often acquired by patients who have chronic liver disease and results in a mortality rate of approximately 50%. One route for infection is via infected seafood, in which the infection manifests itself as primary septicemia. Long-term care is often associated with the disease as the result of multiple organ damage caused by acute toxic shock (118). Other *Vibrio* species which have been implicated with shellfish-borne disease include *V. mimicus*, *V. fluvialis*, *V. furnissii*, *V. hollisae*, *V. damsela*, *V. metschnikovii*, and *V. cincinnatiensis*.

The opportunistic pathogens of the genus *Aeromonas*, once classified as *Vibrio* spp., are now in the family *Aeromonadaceae* (82). *Aeromonas hydrophila* has been associated with outbreaks of shellfish-related illnesses (1, 10, 114). Like *Vibrio* spp., they are a common component of the microflora in estuarine and marine environments. *A. hydrophila* and *A. caviae* are strains commonly associated with diarrheal diseases, typically in the young, elderly, or immunocompromised. Toxicity is due to several factors including hemolysins, exotoxins, and lipopolysaccharide endotoxins (114, 179). A type III secretion system has been recently identified and characterized from an *A. hydrophila* diarrheal isolate (155).

Clostridium species are frequently associated with fecal contamination, but these spore-forming anaerobes can be easily isolated from many environmental sources without the presence of an obvious fecal contamination due to their considerably environmentally resistant spore form. *Clostridium perfringens* and *C. botulinum* are often isolated from marine shellfish and sediments (15, 174). Although *C. perfringens* is an enteric pathogen, there are no reported outbreaks of related infection associated with shellfish consumption. The toxin from *C. botulinum* is a highly toxic neurotoxin, and outbreaks of poisoning from seafood have been previously reported (141, 173). The presence of *C. botulinum* or the toxin has also been reported in blue-crab products in the United States (85). The symptoms of botulism poisoning include nausea and vomiting followed by a number of neurological signs and symptoms related to flaccid paralysis: visual impairment (blurred or double vision), loss of normal mouth and throat functions, weakness or total paralysis, and death by respiratory failure. Treatment of affected patients requires the use of antitoxins (40).

Though not pathogenic, a variety of marine algae can cause toxic blooms, during which a variety of toxins can be accumulated in shellfish tissues and can result in poisoning (http://www.cfsan.fda.gov/~mow/intro.html). Paralytic shellfish poisoning is caused by a variety of derivatives of saxitoxins produced by toxic dinoflagellates. Amnesiac shellfish poisoning is caused by domoic acid, a neurotoxin produced by certain strains of a marine diatom, *Pseudo-nitzschia* (78). Diarrhetic shellfish poisoning is presumably caused by a group of high-molecular-weight polyethers (including okadaic acid, the dinophysis toxins, the pectenotoxins, and yessotoxin) chiefly produced by *Dinophysis* spp. (http://www.cfsan.fda.gov/~mow/intro.html). Despite severity of illness, shellfish-associated toxins have accounted for only a few hundred reported cases of illness in the United States (26).

Fecal Pollution-Related Microbial Contaminants

Fecal pollution from nonhuman (pets, domestic livestock, and wildlife) and human sources is frequently a major factor associated with urbanization that contributes to the degradation of water quality (175). Viral illnesses are the most commonly encountered cause of shellfish-related illness (91). There are over 100 different viruses present in human fecal material, and multiple virus types have been reported in the same shellfish specimen or batch of shellfish (88, 163). Virus particles introduced to shellfish-growing waters by fecal contamination are more persistent than most vegetative bacterial species and may persist in seawater for extended periods of time (121). During that time they may be removed from the water column by filter-feeding bivalves. Viral contamination of shellfish presents unique problems, since they adhere directly to tissues within the shellfish and have been shown to persist for up to several months. In addition, viral particles can also adhere to particles and lie dormant in sediments which can be later absorbed into shellfish after disturbances of the seabed. There is no correlation between viral load and bacterial concentration in bivalves (58, 79, 120, 146). Of the known enteric viruses, noroviruses and hepatitis A virus pose the greatest risk to public health (92, 134).

Noroviruses are spherical, nonenveloped particles 27 nm to 32 nm in diameter, with a single-stranded, positive-sense RNA genome surrounded by a protein capsid. The classical symptomology of noroviral illness involves nausea, vomiting, abdominal cramps, diarrhea, low-grade fever, and headaches (35, 131). Gastroenteritis symptoms typically develop in 24 to 48 hours (median 33 to 36 hours) and have a common duration of 1 to 3 days. Viral shedding (up to 10^7 particles/gram of feces) typically subsides with symptoms. However, some individuals may continue to shed noroviruses for more than 3 weeks after the onset of illness (145). Immunity to the virus is limited due to constant mutation of the viral antigens. The infectious dose for noroviruses is not known but is thought to be as low as 10 virus particles. Host susceptibility to noroviruses may vary between individuals and is related to their Lewis blood group antigens (70, 74, 75, 103, 108).

Noroviruses are estimated to be the leading cause of food-borne illness attributable to a known agent (30, 43). In the United States, noroviruses cause an estimated 23 million episodes of illness, 50,000 hospitalizations, and 3,000 deaths annually (113). Noroviruses are also thought to be the most common source of all gastroenteritis associated with shellfish. The disease is generally nonfatal, with only elderly or debilitated patients being at risk of mortality (131). Many norovirus outbreaks associated with shellfish are reported annually worldwide (3, 21, 54, 107, 119, 154, 156, 164, 171, 173). For example, of the 11 outbreaks of seafood-borne illness that occurred between 2001 and 2004 in the state of Washington, 5 were determined to be caused by consumption of oysters containing noroviruses. A sixth norovirus outbreak was traced to a "mixed seafood" salad. A seventh outbreak was attributed to consumption of oysters but no agent was specified. However, based on the description of illness, a norovirus was the most likely responsible pathogenic agent (S. Dreitzler, personal communication). A recent survey of viral contamination of oysters from 11 countries over 3 years found that 10% were positive for norovirus RNA (21).

Hepatitis A virus (HAV), of the family *Picornaviridae,* genus *Hepatovirus*, is the primary etiologic agent of infectious hepatitis. HAV is a nonenveloped, icosahedral virus (27 to 32 nm in diameter) with a single-stranded, positive-sense genome. Unlike noroviruses, which have only one capsid protein, HAV has four. This difference makes HAV even more persistent in the environment than noroviruses. Infectious dose is low and estimated to be between 10 and 100 viral particles. HAV spreads primarily via the fecal-oral route, and poor sanitation in developing nations means that nearly all of the population are immune due to exposure as young children. The disease is usually asymptomatic in infants and children <6 years old (anicteric hepatitis), so HAV presents a more dangerous problem to developed countries, where exposure occurs at an older age when the disease is more likely to be symptomatic. In shellfish-borne HAV infection, the disease is not generally severe. In symptomatic patients, the incubation period of the disease is 15 to 50 days (mean, 30 days). The early disease symptoms prior to affecting liver function (the icteric phase) include anorexia, nausea, vomiting, fatigue, fever, and abdominal cramps (14). Jaundice generally occurs 5 to 7 days following the onset of gastrointestinal symptoms; however, in 15% of reported jaundice cases, it was not preceded by gastrointestinal symptoms. The icteric phase is recognizable by golden-brown urine due to the increased excretion of conjugated bilirubin. The illness is generally self-limiting, lasting up to several months and infrequently causing fulminant disease (14, 18). In the United States, the estimated case fatality rate among reported cases of hepatitis A is <1.3%, with a fatality rate of 1.8% in persons older than 50 years (110). HAV is endemic throughout much of the world and is often associated with shellfish consumption (14, 23, 25, 27, 51, 53, 93, 95, 98, 101, 180). HAV was responsible for the largest recorded outbreak of disease linked to shellfish, with an outbreak in China in 1988 linked to the consumption of clams, where over 292,301 people were affected (69).

Noroviruses and HAV are the most common enteric viruses implicated with gastroenteritis from shellfish. However, other pathogenic viruses have been isolated from shellfish tissues and may be associated with illness following shellfish consumption. The enteroviruses, unlike what their name suggests, are not common enteric pathogens. Enteroviruses are often excreted in feces, but clinically relevant enteroviruses are more likely to manifest as respiratory or meninge-type diseases and have not been associated with the consumption of shellfish (134). Enteroviruses (polioviruses, coxsackieviruses, and echoviruses) are members of the family *Picornaviridae*. Polioviruses, classically the most notorious of the enteroviruses, are the causative agents of paralytic poliomyelitis. Polioviruses are no longer significant in terms of global health impact, due to widespread vaccination programs in most nations. However, screening of shellfish in France has shown that up to 45% of samples tested positive for enteroviruses which may represent a significant risk or may be at least an indicator of fecal pollution (99). Shellfish screened for enterovirus have frequently tested positive, even in relatively pristine waters (49).

Hepatitis E virus (HEV) is not related to HAV or the other hepatitis agents (137). This positive-strand RNA virus was initially classed as a calicivirus based on virion size and shape and the presence of three open reading frames in its genome (138). However, the organization of these open reading frames has led it to be reclassified into its own group, the "hepatitis E-like viruses" (135). HEV-associated disease is rare in developed countries but still persists in much of Africa, India, Asia, and South America (2, 76, 100, 106). Though the disease is frequently asymptomatic, symptoms caused by HEV are similar to those caused by HAV. Also like HAV, HEV rarely causes chronic liver damage but is distinctly different, resulting in fatalities of over 20% in pregnant women due to fulminant liver failure caused by HEV infection. Despite transmission by the fecal-oral route, shellfish-borne transmission of HEV has not been documented.

The rotaviruses and astroviruses are clinically significant viruses, as they are responsible for diarrheal disease in infants and young children. While these viruses have been detected in shellfish, transmission via this food type has not been established as a route of infection. Enteric adenoviruses are DNA viruses of the family *Adenoviridae.* Adenovirus types 40 and 41 have been associated with acute gastroenteritis (177). They are commonly found in shellfish harvested from water contaminated with human sewage, but epidemiological links to illness from shellfish consumption are lacking (50, 84). Other viral agents such as the enteric coronaviruses, sapoviruses, toroviruses, pestiviruses, and picobirnaviruses have been identified as causative agents of gastroenteritis (115, 177), and as they are expelled with feces they may also be associated with shellfish-borne disease, but there are no reported outbreaks of infections from these viruses linked with shellfish consumption.

In addition to the many viral types that may be found in shellfish, numerous enteric bacteria have also been isolated. Several members of the *Enterobacteriaceae* are present in many fresh and marine water environments due to fecal contamination from animal and human reservoirs and may survive for several months. Some studies suggest that *Salmonella* spp. may be ubiquitous in tropical aquaculture environments (29). Several species of *Enterobacteriaceae*, *Salmonella* spp., *Shigella*, *Escherichia coli*, and *Yersinia enterocolitica*, have been directly identified as enteric pathogens (44). Of these, *Salmonella* spp., *Shigella* spp., and *Plesiomonas* spp. have each been associated with food-borne disease outbreaks linked with shellfish (67, 109, 136, 147, 160, 166). Diseases associated with these bacteria are rarely fatal and typically are self-limiting diarrheal illnesses; however, *Salmonella enterica* serovar Typhi and certain strains of *Shigella* have been associated with more severe illness. *E. coli* has classically been used as an indicator of fecal pollution, but several strains of *E. coli* are clinically relevant (for example, verocytoxin *E. coli*, which may cause illnesses with potentially severe complications—hemolytic uremic syndrome causing serious kidney damage in children and thrombocytopenic purpura in the elderly). *Yersinia enterocolitica* is generally associated with mild to moderate diarrheal disease. Although a potential risk exists from consumption of shellfish harvested from fecally contaminated growing waters, no published cases of verocytoxin *E. coli* or *Y. enterocolitica* infection have been directly linked with the consumption of contaminated shellfish (142).

Several species of the genus *Campylobacter* have been identified in shellfish in the United Kingdom and The Netherlands (40, 178). In one study, 44% of all shellfish samples tested contained *Campylobacter* species. *Campylobacter jejuni* and *C. coli* are the two most common gastrointestinal pathogens within this genus, but *C. lari* has also been isolated from infected patients and identified in shellfish (40). Typically, patients can be asymptomatic to severely ill, with the disease lasting up to 1 week. Symptoms can include fever, abdominal cramps, and diarrhea, and the disease is normally self-limiting, with 10% of patients relapsing (119). Several outbreaks of gastroenteritis by *Listeria* spp. have been linked to eating contaminated shellfish (13, 142, 172). *Listeria monocytogenes* is commonly identified in shellfish (33, 117, 142), although in comparison to some other pathogens there have been relatively few cases directly associated with shellfish consumption (150). Listeriosis can be a very serious disease, most commonly presenting as either septicemia or a central nervous system attack, and may be fatal (149). Pregnant women, newborn babies, immunocompromised patients, and the elderly are particularly at risk. *Cryptosporidium parvum* and *Giardia* spp. are important enteric protozoan parasites. *Cryptosporidium parvum* results in diarrheal disease that is self-limiting and typically persists for about 2 weeks in immunocompetent hosts. Protozoan cysts and oocysts are environmentally stable, and individuals may become infected by consuming as few as 10 to 100 cysts or oocysts. *Cryptosporidium* oocysts have been demonstrated to survive and remain infectious in marine water (46, 122). Immunocompromised patients are susceptible to chronic *Cryptosporidium* infection since there is no effective prophylactic control of this disease. Further, *Cryptosporidium* strains have been isolated from oysters and mussels from Chesapeake Bay (46, 47, 65). *Cryptosporidium* was identified in nearly 20% of oysters examined in one study, with the greatest occurrence rates corresponding to times of greatest weekly and monthly rainfall (48). Infection by *Giardia* spp. is the most common form of gastroenteric parasitosis worldwide. *Giardia* is a flagellar parasite that infects both humans and animals (94). *Giardia* has been isolated from clams taken from Chesapeake Bay (66). Despite isolation of *C. parvum* oocysts and *Giardia* cysts from shellfish in Europe and the United States, there have been to date no reported outbreaks of protozoan diseases directly associated with consumption of molluscan shellfish (60–62, 66).

Regulation of Shellfish Quality

Use of Fecal Indicators

With the exception of certain international regulatory programs for monitoring of *Vibrio* spp., *Salmonella* spp., and some algal toxins in shellfish tissues, pathogens are typically not directly monitored in the regulation of the microbial quality of shellfish (41). Rather, indicator microorganisms (e.g., total coliforms, fecal coliforms, and *E. coli*) are relied upon to indicate the presence of fecal contamination in either the growing water or shellfish tissues. The indicators relied upon and the site of monitoring (water or tissues) vary by location. One notable exception to the strict reliance on indicator organisms is the recent move by some countries (e.g., Hong Kong) to monitor for noroviruses or HAV in imported shellfish lots.

The term "indicator organism" can lead to confusion. Three classes of indicator organisms (general or process indicator organisms, fecal indicator organisms, and index/model organisms) are now recognized, which helps to eliminate any ambiguity suggested by this term (7). Of these classifications, fecal indicator organisms and index/model organisms are important in the regulation of shellfish quality. Fecal indicators are those organisms that indicate the presence of fecal contamination but not necessarily the presence or absence of fecal pathogens. This class of indicator includes bacterial indicators of sanitary quality (e.g., total coliforms, fecal coliforms, and *E. coli*), which have historically been used in shellfish monitoring. The index/model organisms are surrogates that are indicative of pathogen presence and behavior. Examples of an index/model organism might be *E. coli* as an indicator of *Salmonella* spp. or coliphage as an indicator of human enteric viruses. A benefit of index/model organisms is that they may be able to be used to track a source of fecal contamination.

The total coliform bacteria are one of the most commonly utilized groups of fecal indicator organisms. The group is empirically defined to consist of all aerobic and facultative anaerobic, gram-negative, non-spore-forming rod-shaped bacteria that ferment lactose with gas and acid formation within 48 h at 35°C. This fecal indicator is widely used in food and water industries. Due to shortcomings in the feces specificity of total coliforms, the fecal or thermotolerant coliforms have recently been used as an alternate indicator group. Fecal coliforms are defined as those bacteria in the total coliform group that ferment lactose with production of gas within 24 h at an elevated temperature of 44.5°C. *Escherichia coli* is a common type of fecal coliform bacteria that lives in the lower intestines of warm-blooded animals (including birds and mammals). Each day, humans as well as other warm-blooded animals will pass a significant number of *E. coli* cells per gram of feces.

Escherichia coli has historically been considered of fecal origin, although environmental reservoirs have recently been identified (19, 77, 144, 159).

The enterococcus species, *E. faecium* and *E. faecalis*, inhabit the intestines of both humans and animals. *Enterococcus faecalis* colonizes the large intestine and approximately 10^7 organisms are shed per gram of feces, while *E. faecium* colonizes the same site but normally in lesser numbers. Enterococci have been chosen as an indicator of fecal contamination for recreational waters, in part because they persist for longer in the marine environment than *E. coli* (52). Another study has demonstrated that the enterococcus group was a more robust indicator despite weather conditions or location of sampling (127).

The use of a bacterial standard (e.g., fecal coliform) for determination of harvest area restrictions is questionable because of the differences in survival and resistance to environmental stresses (e.g., disinfection, sunlight, salt tolerance, etc.) between bacteria and human enteric viruses. This is particularly true in growing waters receiving chlorinated municipal wastes, as vegetative bacteria are typically more susceptible to chlorination than viruses (9, 64, 105). Studies have documented that fecal indicator bacteria, the organisms currently used to measure sanitary quality of shellfish-growing area water, may be inadequate as indicators of enteric viruses and *Vibrio* spp. in shellfish and water (36, 38, 68, 97, 176).

F+ and somatic coliphages have been suggested as good indicators for viral contamination in oysters due to their similar pattern of persistence in the environment (17). Specifically, because these are bacterial viruses that have similar size, shape, and nucleic acid composition to human enteric viruses, they have become common surrogates for enteric viruses. Bacteriophages have been shown to be preferentially accumulated by shellfish during winter and spring months, the period during which human viral contamination of growing waters is thought to be highest (16, 68, 99). Therefore, coliphages may be a good indicator of viral contamination.

U.S. Standards

The major regulatory program affecting the microbial quality of shellfish in the United States is the National Shellfish Sanitation Program (NSSP). The NSSP is a cooperative program composed of state officials, the shellfish industry, and federal agencies. The overall purpose of the NSSP is to establish uniform guidelines and facilitate the production and processing of shellfish under sanitary conditions. The NSSP arose from a series of public health initiatives formulated in the mid-1920s following a widespread outbreak of typhoid fever (1,500 cases and 150 deaths) (143). To accomplish its goal, the

NSSP describes a Model Ordinance for state or local regulatory authorities to follow in the regulation of shellfish.

Although the U.S. Food and Drug Administration coordinates and administers the NSSP, one or more regulatory agencies in each participating state manage the sanitation programs for domestic and imported shellfish (http://www.cfsan.fda.gov/~ear/nss2-toc.html). Under the NSSP, shellfish-growing waters are classified as approved, conditionally approved, conditionally restricted, restricted, or prohibited. The state regulating authority must conduct sanitary surveys of shellfish-growing areas prior to harvest of shellfish stock for human consumption or for classification of a growing area to a classification other than "prohibited." Sanitary surveys are written evaluations of all environmental factors, including actual and potential pollution sources, which have a bearing on the water quality in a shellfish-growing area. At a minimum, a sanitary survey includes the following: a shoreline survey; a survey of the bacteriological quality of the water; an evaluation of the effect of any meteorological, hydrodynamic, and geographic characteristics on the growing area; an analysis of the data from the shoreline survey and the bacteriological, hydrodynamic, meteorological, and geographic evaluations; and a determination of the appropriate growing-area classification.

The NSSP allows for a growing area to be classified using either a total or fecal coliform standard. Sample collection strategies authorized by the NSSP for the application of the total or fecal coliform standard include (i) adverse pollution condition (for areas impacted by known or potential point sources) and (ii) systematic random sampling (for areas affected by nonpoint source contamination). Approved growing areas must have a median or geometric mean fecal coliform concentration of not more than 14 per 100 ml and not more than 10% of samples exceeding a concentration of 43 per 100 ml for a five-tube decimal dilution test. Shellfish may be directly marketed from approved waters following harvest. Conditionally approved waters must meet approved standards using the same bacterial standards under predictable conditions. Shellfish may be harvested and directly marketed from conditionally approved waters as long as the water-quality standards are met, and the harvest areas are closed under other conditions. Restricted waters must not have a median or geometric mean fecal coliform concentration of greater than 88 per 100 ml, with not more than 10% of samples exceeding a most-probable number of 260 cells per 100 ml in a five-tube decimal dilution test. Shellfish may be harvested from restricted areas but must undergo a suitable postharvest purification procedure, like depuration or relaying, prior to direct marketing. Conditionally restricted waters are waters that sometimes meet "restricted" criteria (same bacterial standard). Shellfish may be harvested from conditionally restricted waters as long as the water quality meets the "restricted" criteria and the shellfish are suitably purified prior to marketing. Prohibited waters are waters from which shellfish may not be harvested under any conditions. Harvest restrictions (conditional, restricted, or prohibited classifications) are determined primarily on the concentration of fecal coliform bacteria associated with human sewage and animal wastes. However, blooms of toxic algae and industrial contaminants can also trigger harvest area restrictions.

International Shellfish Standards

The E.U. shellfish standards differ from the U.S. standards (Table 5) in that they are based on 100 g of shellfish tissue compared to 100 ml of growing water for those in the United States (6, 96). In the E.U., shellfish are classified into categories A, B, and C (6). For category A shellfish, all samples must have less than 300 fecal coliforms or 230 *E. coli* bacteria per 100 g of tissue. Category A shellfish may be directly used for human consumption. For category B shellfish, a 90th-percentile compliance of $<$6,000 fecal coliforms or $<$4,600 *E. coli* bacteria per 100 g of tissue must be met. Category C shellfish must have less than 60,000 fecal coliforms (46,000 *E. coli* bacteria). Categories B and C both require postharvest treatment prior to sale for human use. Category B shellfish must be relayed or depurated prior to market, and category C shellfish require protected relaying for $>$2 months or heat treatment prior to market (6, 96). In Australia and New Zealand, guidelines for shellfish waters stipulate a median fecal coliform concentration of less than 14 CFU/100 ml in order to be approved for harvest. Similarly, shellfish tissues with greater than 5 CFU/gram of tissue are not to be harvested (57).

Fecal Contamination Sources

While the quality of shellfish is dependent upon many factors, clean growing water is undeniably one of the most important. There are generally two source types of fecal contamination that can adversely impact shellfish-growing waters: point and nonpoint sources. An example of the two pollution-source types simultaneously affecting shellfish water quality is that of Puget Sound and its tributaries, where impacts range from poorly maintained septic tanks, poorly managed runoff from farm and wild animals, storm water runoff, and at least 109 municipal treatment plants. A clear example of a point source of fecal contamination impacting shellfish-growing waters is the outfall of a wastewater treatment plant. This type of point source can be a particular concern in shellfish-growing waters, since most of the indicator bacteria used for monitoring of the microbial quality of the water are removed or inactivated during wastewater treatment, while enteric viruses may not be adequately removed or inactivated (168). Studies have demonstrated that municipal

Table 5 Overview of shellfish harvest area classification criteria used in the European Union and the U.S.[a]

Criterion	European Union	United States
Number of harvest area categories and designations	3: "A," "B," "C"	4: "Approved," "Conditionally Approved," "Restricted," "Prohibited"
Growing area classification method	Indicator organism density in shellfish meats	Sanitary survey and density of indicator organisms in growing area water samples
Indicator organism(s)	Fecal coliforms or *E. coli*	Total coliforms or fecal coliforms
Growing area categories, applicable indicator organism densities, and intended use of shellfish	Category "A": all shellfish meat samples must contain ≤300 Fc/100 g; shellfish are for direct human consumption.	"Approved" ("Conditionally Approved"): geometric mean or median Fc densities in water samples must be ≤14/100 ml and no greater than 10% of these water samples can have an Fc density of >43/100 ml during adverse pollution conditions; shellfish are for direct human consumption.
	Category "B": 90% of shellfish meat samples must contain ≤6,000 Fc/100 g; shellfish require depuration or relay prior to consumption.	"Restricted": geometric mean of Fc densities in water samples must be ≤88/100 ml and no greater than 10% of these water samples can have Fc densities of >260/100 ml; shellfish require depuration or relay prior to consumption.
	Category "C": all shellfish samples must contain ≤60,000 Fc/100 g; shellfish require prolonged relay (>2 mos).	"Prohibited": water samples exceed "Restricted" growing area standards; no shellfish harvesting allowed for human consumption.
Sampling frequency per yr	Varies by member states	"Approved" and "Conditionally Approved": during adverse conditions, 5 times per location or by systematic random sampling

[a]Fc, fecal coliforms.

sewage has contaminated shellfish with viral pathogens (8, 157). Sources of viral outbreaks involving oysters have included harvesting within a prohibited zone around an oil rig with a malfunctioning sewage facility (8). However, from a shellfish management standpoint, point sources of fecal contamination are easier to deal with than nonpoint sources, as growing waters in the immediate vicinity of a point source require classification as prohibited.

Fecal contamination can be from human, domestic animal, or wildlife origin. Human nonpoint sources potentially affecting shellfish-growing waters include the following: on-site waste disposal (septic) systems, leaking municipal wastewater conveyances, land-applied biosolids, urban storm water runoff, and passenger or fishing vessels. Animal nonpoint sources of fecal contamination include runoff from livestock operations, land application of agricultural wastes, runoff containing domestic animal (e.g., canine), terrestrial, and marine wildlife wastes, and avian waste sources. Studies have examined rainfall as a contributing factor to shellfish and growing-water contamination (58, 63, 116). Outbreaks of noroviruses have also implicated overboard dumping of fecal wastes and vomitus by oyster harvesters in Louisiana (8, 20, 39, 90).

Shellfish often grow best where the waters are sheltered and the nutrient levels are high, typically estuarine environments. Areas of coastline with this type of geography/topography are also often heavily populated with humans, and therefore, human sewage is released to varying degrees into the environment. The rapid "suburbanization" of rural watersheds and shoreline has led to increased use of on-site septic systems for sewage treatment and disposal. Failing on-site waste treatment systems are one of the leading contributors to fecal contamination of shellfish-growing waters. It should be no surprise that since the early 1980s, nonpoint source fecal pollution has become a key factor in closure of shellfish beds and failed on-site septic systems have been implicated as a major source of pollution (http://www.psat.wa.gov/Publications/StateSound2004/PSATSOS2004.pdf). The problem is exemplified in the Puget Sound (PS) region of Washington, which has experienced a population growth of over 40 percent since 1980, with nearly two-thirds of the state's 6 million people residing around the shores of the sound (http://www.psat.wa.gov/Programs/shellfish/fact_sheets/resource_web1.pdf). Much of the fastest population growth affecting PS has occurred in rural, shellfish-rich areas. Unfortunately, one of the major ecological impacts of population growth around PS is human wastewater discharge. The near-shore ecosystems of PS comprise one of the primary growing areas of shellfish in the west. Since 1980, approximately one-quarter of the area classified for commercial shellfish harvesting has been downgraded and taken out of production, primarily because of fecal pollution from humans and animals (http://www.psat.wa.gov/Publications/manplan00/18_shellfish.pdf). In the state of Washington, failing septic systems were a contributing factor in 8 out of 10 shellfish-growing area restrictions in 1995.

APPLICATION OF MST METHODS TO SHELLFISH MANAGEMENT

The sanitary quality of water is typically established based on the levels of fecal indicator bacterial species therein. Although these organisms often indicate the presence of fecal contamination, their presence alone does not provide much insight as to the source of the contamination or ways to mitigate it. Often, the likely source of fecal contamination is obvious (e.g., a cow standing in a stream, the location of wastewater outfall, etc.). However, in other instances, the source of fecal pollution can be difficult to determine. Consequently, there is a keen interest in techniques that will allow determination of the source of fecal contamination in addition to indication of the presence of fecal contamination. While no current MST method is accepted as a regulatory tool, these methods can play a valuable role in identification and management of fecal contamination sources. Knowing whether a pollution source is human or animal may indicate to environmental managers entirely different strategies for risk management.

Traditionally, human fecal waste has been considered more likely to contain pathogens of importance to humans (human-specific or human-adapted), and thus was considered to pose a greater hazard than animal wastes (http://www.state.nj.us/dep/dsr/research/mar-rps.pdf). As a result, early studies suggested a need to identify the fecal source in order to adequately assess and minimize the risk of illness (129). However, several recent human outbreaks have been reportedly due to domestic animal waste contamination (28). Accordingly, the Environmental Protection Agency (EPA) has recently reversed their position regarding waterways impacted by animal sources of fecal contamination, including wildlife. The EPA's previous policy, as stated in its 1994 Water Quality Standards Handbook, allowed states and authorized tribes to justify a decision not to apply the bacteriological criteria to particular recreational waters when high concentrations of bacteria were found to be of animal origin (169). The EPA's current position is that in these situations, it is inappropriate to conclude that these sources present no risk to human health from waterborne pathogens, based on the ability of warm-blooded animals to harbor and shed human pathogens. Consequently, states and authorized tribes should not use broad exemptions from the bacteriological criteria for waters designated for primary contact recreation based on the presumption that high levels of bacteria resulting from nonhuman fecal contamination present no risk to human health. Identification of fecal source is still a potentially valuable tool for management of fecal contamination.

Several methods of MST have been described for identification of fecal pollution sources in contaminated water (158, 170). These methods may be divided broadly into two distinct classes, library-dependent and libraryless methods. The techniques may be further classified as based on phenotypic or genotypic characterization of microbial (typically bacterial) isolates. Unfortunately, few studies on the application of microbial source tracking methods to shellfish management have as yet been published. These MST methods have been sponsored by state and local governments, shellfish growers, and Native American tribes to identify fecal pollution sources in waters used for both commercial and recreational shellfish harvesting. Thorough reviews of the available MST methods and their application to total maximum daily loads and water quality management are available elsewhere and are beyond the scope of this chapter (158, 170).

Overview of Methods Currently Applied

Among the concerns for MST methods applied to shellfish management are a need for a library, the level of pollution-source discrimination, potential "age" biasing of fecal indicators isolated from environmental waters, the differential stability of the phenotypes and genotypes of the fecal indicators, the characteristics of the isolated population, and the differential survival rates of indicator isolates from different sources.

DNA Fingerprinting of E. coli

Perhaps the earliest application of a source tracking technique to management of shellfish-growing waters involved the use of DNA fingerprinting (http://lakes.chebucto.org/H-2/bst.html). Fecal contamination was documented in a tidal creek used for culture of clams. Fearing the downgrading of the growing area, a clam grower contacted George M. Simmons (at the Virginia Polytechnic Institute and State University in Blacksburg, Virginia), who used a library-based DNA fingerprinting technique to compare isolates of *E. coli* from local fecal sources (human and animal) to isolates obtained from the growing waters. The initial hypothesis was that the most likely source of the fecal contamination was nonpoint-source pollution from on-site waste disposal systems. However, the analysis of 88 unknown isolates compared to a library of over 200 DNA fingerprints found that only 8 of the isolated strains matched a human source and the majority of identified isolates were consistent with wildlife sources (e.g., deer and raccoon). Reduction of wildlife on properties adjacent to the shellfish waters resulted in a 2-log reduction in the magnitude of fecal coliform concentrations and prevention of the downgrading of the growing waters.

Ribotyping

The currently most-applied MST method for determination of fecal contamination sources in shellfish-growing waters is the molecular technique of ribotyping. This technique is a particular type of DNA fingerprinting that targets exclusively the 16S rRNA genes. As with other fingerprinting methods, this method requires an extensive library. The level of source discrimination and the confidence of the method's results are directly related to the size of the library. Ribotyping methods have been used in a number of studies of fecal contamination of shellfish-growing waters (150). Ribotyping has been an effective epidemiologic tool applied to a variety of bacterial types that may be found in shellfish, including *Vibrio* spp., *E. coli*, and *Salmonella* spp. (128, 130, 133, 161, 165). Discriminate analysis of ribotype profiles for *E. coli* isolates from sewage and animal sources in Apalachicola National Estuarine Research Reserve has demonstrated the ability to differentiate human and nonhuman sources of fecal contamination (130). A study examining the geographical variation in ribotype profiles found that a single restriction enzyme was unable to discriminate between animal species, but it was able to differentiate *E. coli* isolates as human or nonhuman over a large geographic region (153).

In one case, ribotyping was used in a study of the fecal contamination of Grays Harbor, a bay on the central coast of Washington (148). Growing waters in the outer harbor were conditionally approved with a closure-triggering event of fecal coliform increase in the effluent of an industrial wastewater plant used to treat pulp mill waste. The plant currently receives no fecal material but was initially seeded with cattle manure approximately 40 years ago. The ribotyping study found that none of the isolates from the presumably affected shellfish beds matched strains isolated from the industrial wastewater plant effluent. Rather, isolates from the shellfish beds were found to match human, wildlife, and livestock strains. The conclusion of the study was that an increase in the fecal coliform levels of the wastewater plant effluent as a triggering event was faulty and should be reconsidered.

Ribotyping methods were also applied in two studies in Hampton/Seabrook Harbor in New Hampshire (81). The harbor contains the state's largest beds of softshell clams (*Mya arenaria*) and was conditionally approved for recreational harvest in the absence of significant rainfall. Initial expectations were that birds were the most significant source of fecal contamination in the harbor. However, in the first study, comparison of 390 isolates from harbor water samples found humans to be the most frequent source of the contamination. In the second study of two storm water pipes leading to the harbor, avian sources were identified as the most common source of contamination. However, human isolates were found in both pipes and in one pipe at a frequency similar to those of the birds.

An additional example of the use of ribotyping for source identification occurred at the Isle of Palms, a barrier island community in South Carolina. In this instance, the waterways surrounding the island were frequently contaminated with high levels of fecal coliforms. As a result of the bacterial contamination, the waters have been closed for shellfish on numerous occasions (152). Possible sources of the bacterial contamination include septic systems, wild animals, and domesticated animals. To determine the source of the contamination, ribotyping of *E. coli* isolates was performed. In this study, the average rate of correct classification (ARCC), as determined by running each isolate from the library as an unknown against the entire database, was 92%. The findings of this study suggest that temporal and spatial diversity may not be significant factors in broad source classification of *E. coli* (152). In addition, it was determined that 88% of isolates examined were classified as being derived from animal sources, while only 11% were derived from human sources. Deer and dog were presumed to contribute significantly to the contamination.

The Three Bays area of Barnstable, Massachusetts had high seasonal levels of fecal coliform bacteria initially assumed be caused by animal contamination (http://www.3bays.org/Reportspdf/2001-DNAReport.pdf). However, the pattern of occurrence in 1999 suggested a potential for human contamination. A DNA fingerprinting (ribotyping) study was performed to determine the source of the fecal contamination. Samples from two coves confirmed a high percentage of contribution from human sources.

The Webhannet watershed in Wells, Maine, is affected by chronic bacterial contamination from unidentified sources, resulting in classification of the shellfish harvest areas as restricted (http://www.seagrant.umaine.edu/MST/Webh_Report.pdf). Ribotyping was performed on isolates from the most contaminated regions of the watershed. Wildlife sources as a group contributed the most significant fraction of the bacterial contamination, but human contamination was the most significant individual source of contamination. Domestic pets and livestock sources were also identified. These findings were then used to arrive at recommended management practices to address the contamination levels. However, in this study, roughly one-third of the isolates were unclassifiable due to deficiencies in library size. This highlights a significant shortcoming of library-based methods.

Routine bacterial water-quality monitoring indicated that the shellfish-growing waters of Barkley and Clayoquot Sounds were negatively impacted by seasonal rainfall. Closure of the areas to shellfish harvesting during long periods of the year would be devastating to aquaculture and the commercial shellfisheries. In an attempt to mitigate potential harvest restrictions, a bacterial source-tracking study was performed to identify the sources affecting the shellfish waters. Ribotyping of 212 *E. coli* isolates implicated eight source groups in the contamination. The study had a high classification rate of 86.3%, with just 27 isolates being unclassifiable. The leading sources of contamination were attributed to unknown avian sources, marine mammals, and gulls. Human waste was determined to not be a source of contamination in any of the samplings.

The Nisqually Reach, Nisqually River, and McAllister Creek do not meet the Washington water quality standards for fecal coliform bacteria and are listed as §303(d) impaired water bodies. In 2000, the Washington Department of Health downgraded 74 acres of commercial shellfish beds in Nisqually Reach from conditionally approved to restricted. This downgrade was one of several recent downgrades of the classification for shellfish-growing water in the area. A ribotyping study was performed on nearly 1,000 isolates collected from sampling locations classified as agricultural, rural tributary, marine water, residential, storm water, and wildlife refuge to determine the relative contribution of fecal source types to the bacterial load of the water. Over 230 clonal types of *E. coli* were found in 946 isolates. These clonal types were then compared to a library developed for the region. Bird sources were found to make up 43% of the isolates but only 25% of the clonal types. Almost 30% of clonal types were a previously unknown type, but they represented only 11% of the total isolates. Overall, the study had a high rate of matched isolates (i.e., approximately 89%) compared to the expected 60 to 80 percent. Bovine sources were the second most common at agricultural land use sites, while human sources were the second most common at residential sites. Canine and rodent sources were also significant at some sampling locations. For the isolates collected from the shellfish beds (marine water), avian and rodent sources were the most frequently isolated, while human sewage and bovine source types were isolated in five out of nine sampling events (http://www.co.thurston.wa.us/health/ehrp/pdf/Nisqually/Nisqually%20DNA%20Edited%20Final%20document.pdf).

Morro Bay Estuary in California has been monitored for fecal coliforms for the protection of shellfish harvesting since 1953. Recent elevated levels of fecal coliform bacteria required downgrading of a portion of the growing areas from conditionally approved to prohibited, while other portions of the growing area were downgraded with rainfall restrictions. Potential sources of fecal contamination included agricultural runoff, domestic animal waste, leaking/failing septic systems, terrestrial and marine wildlife, leaking lift stations, and faulty wastewater treatment plant (WWTP) operations. To identify the potential sources of contamination, ribotyping was performed on more than 1,600 *E. coli* isolates from three bay sites, both creek mouths, two seeps, bay sediment, and oyster tissue (http://www.swrcb.ca.gov/rwqcb3/TMDL/documents/MBDNAFinalReport.pdf). The ribotyping results were compared to a library containing over 75,000 ribotypes isolated from various known fecal sources. In addition, the library was augmented with isolates from known fecal sources including wild mammals, humans, domestic mammals, and total birds collected from the location of the study. The study had a classification rate of 74.4% for isolates, with 424 of the isolates not matching ribotypes in the library. Overall, the largest fraction of isolates came from bird (22%), human (17%), bovine (14%), and canine sources (9%). Birds were the largest source of *E. coli* for all sites, including oysters, other than one of the creek mouths (bovine) and one of the seeps (human). The vast majority of unknown isolates were unique to the oyster samples, and the majority of ribotypes attributed to birds were also unique to the oyster samples (http://www.swrcb.ca.gov/rwqcb3/TMDL/documents/MBDNAFinalReport.pdf). Upon further investigation it was determined that *E. coli* attributed to birds entered the bay by a direct route, while bovine, canine, and human sources entered through less direct pathways, e.g., seeps, creek mouths, and rainwater runoff.

One of the most detailed studies utilizing MST techniques to evaluate fecal pollution sources in shellfish waters was performed in Puget Sound (http://www.co.thurston.wa.us/shellfish/pdf/Final%20DNA%20report%2002.pdf). Henderson Inlet is in the south of Puget Sound, northeast of Olympia, WA. The shoreline of Henderson Inlet is not densely housed (14 houses per mile of shoreline), but the watershed of the inlet contains high-density commercial and residential development. The population of the watershed has increased, and since 1984, the water quality at the south end of the outlet has been in a state of decline for shellfish waters. Since 1984, 180 acres of shellfish-harvestable water were first downgraded from approved to conditionally approved, and then 120 acres graded conditionally approved were downgraded to prohibited. It was determined that nonpoint source pollution associated with significant rainfall events was responsible for the downgrades in harvest area quality. The conditional approval for the area required closure of the area for 5 days following rainfall events of greater than 0.75 inches in a 24-hour period. In a 1998 review of water quality data, violations of the bacterial standard were observed with rainfall events of less than 0.75 inches. The conditions for approval were then adjusted to 0.5 inches of rainfall in 24 hours. Despite efforts to control nonpoint source pollution, water quality continued to decline, and the acreage downgraded to conditionally approved and to prohibited increased to 360 and 128 acres respectively. There are three commercial growers and a substantial recreational harvest in Henderson Inlet. Sources of nonpoint source pollution include failing near-shore on-site waste disposal systems, high-density animal farming, and storm water runoff. Ribotyping of *E. coli* isolates from the impacted areas was performed to determine the source of the contamination. Twenty-seven sources were identified from 943 isolates, including human and a variety of domestic and wild animals. The percentage of matches identified for the isolates to the DNA library was 86%. Source identification of isolates from two creeks draining to Henderson Inlet was consistent with the type of land use observed in each of the creek's watersheds, both in terms of the number of sampling events in which a particular source was identified and in terms of the number of clonal types identifies for each source type. Marine water was found to reflect the complexity of the overall watershed, with the majority of isolates being typed as avian, human, rodent, or canine. A survey of isolates from shellfish tissues found human sources of contamination, although the number was too limited to allow meaningful conclusions. A concurrent pathogen survey using PCR methods of detection identified 4 of 45 oyster tissue samples containing human pathogens. Three contained *Vibrio* spp., while the fourth contained *Campylobacter jejuni*. Overall, the results of the source tracking study and the bacterial pathogen scan supported the State Department of Health's decision to downgrade the shellfish beds.

Host-Specific Marker PCR

Host marker PCR methods have been successfully applied to water samples collected from Tillamook Bay, Oreg. (11). Fecal contamination resulted in closures of Tillamook Bay shellfish beds to harvest for over 100 days per year. A PCR method examining host-specific 16S rRNA gene markers from *Bacteroides* and *Prevotella* species was able to detect a host-specific marker in 17 of 22 samples. Detection of ruminant markers was most often associated with isolates from rural areas, and detection of the human marker was most often associated with isolates from an urban setting or those impacted by a sewage outfall. Ten samples were positive for human-specific markers, while one or both of two ruminant-specific markers were detected in 14 samples. The sources of contamination to the Bay were found to include farm animal waste, sewage effluent, and on-site waste systems. Seven samples contained both human and ruminant markers.

Animal enteroviruses (e.g., bovine enteroviruses in cattle and deer feces) have also been used as host-specific indicators for animal fecal pollution (80, 102). Fecal sources are determined based on the detection of host-specific viral pathogens (e.g., enteroviruses and adenoviruses for humans, canine parvovirus for dogs, and bovine enterovirus for cattle). Quantitative reverse transcription-PCR methods for viral pathogen detection have been used as a means to trace fecal contamination sources in North Carolina shellfish-harvesting and beach waters (125, 126).

Coliphage Typing

Coliphages are viruses of coliform bacteria. F+ RNA coliphage is morphologically similar to human enteric viruses. These viruses are more similar in resistance to treatment processes and persistence in the environment to human enteric viruses (e.g., noroviruses) than to bacterial indicators. As a result, various types of coliphages have been proposed as indicators of viral contamination. Additionally, serotypic and genotypic analyses of F+ RNA coliphages have been proposed as MST methods. These methods do not require the use of libraries, although the level of source discrimination is typically limited to human versus animal and is incapable of discriminating between animal sources. F+ RNA coliphages are prevalent in sewage and other fecal wastes of humans and animals. There are four known antigenically distinct types of F+ RNA coliphage, with groups II and III predominating in humans and types I and IV predominating in animals. Thus, it may be possible to distinguish between human and animal fecal sources using F+ RNA coliphage genotyping or serotyping. One study using a coliphage-serotyping method has been applied to F+ coliphages isolated from shellfish tissues harvested from North Carolina waters (24). This study concluded that serotype II F+ RNA coliphages were reliable human-specific indicators for human enteric viruses. In another study which examined 200 bacteriophage isolates from a variety of locations, >99% of phages could be serotyped into one of four groups, and >96% could be classified by genotyping (73). Of 32 oyster samples collected from Calico Creek, NC, all but one (unknown genotype) were genotyped as either type II (31) or III (1), indicating a likely human source of contamination.

In a study in France, a coliphage-genotyping method was used to determine the source of fecal contamination in an estuary used for shellfish harvest (M. Gourmelon et al., unpublished results). This study used a previously described genotyping method, standardized it against fecal samples, and then applied it to estuarine waters in comparison to a host-specific-marker PCR method (targeting the 16S rRNA genes of *Bacteroides* spp.). This study detected human fecal pollution in most samples for waters under urban influence, but characterization of water under agricultural influence was more difficult, since only one-third of samples were positive for animal contamination by PCR for a cattle-specific marker or by the presence of a few coliphages of types I and IV.

In another recent study, genotyping F+ RNA phages was used to source-type contamination in Greek shellfish waters (A. Vantarakis et al., unpublished results). Group IV phages were the most frequently isolated, indicating possible animal sources of contamination. Group II and III phages, indicating human contamination, were present in low levels. The study concluded that the F+ RNA phage methods showed great promise for tracking of fecal pollution sources. Genotyping of F+ RNA phages was recently coupled with the detection of adenovirus to address shellfish quality problems by microbial source tracking (R. Rangdale et al., unpublished results). The study determined that these viral methods could be used in the analysis of shellfish tissue and that genotyping methods of F+ RNA phages were sufficiently developed whereas PCR detection of adenovirus was limited by uncertainties in excretion rates, ambiguous specificity, and molecular methodology.

Antibiotic Resistance and Multiple Antibiotic Resistance (MAR) Analyses
Antibiotic resistance/MAR analysis is a phenotypic technique examining antibiotic resistance profiles for isolated fecal indicators. MAR analysis has been applied to *E. coli* isolated from the Manasquan River Estuary in New Jersey (http://www.state.nj.us/dep/dsr/research/mar-rps.pdf). The Manasquan River Estuary is an important shellfish resource. However, due to high levels of coliform bacteria, a significant portion of the area has been closed to harvesting since 1996, and another portion is classified as "special restricted." *Escherichia coli* isolates were grown in the presence of 12 different antibiotics, and their antibiotic resistance profiles were generated and compared to a library of over 4,000 profiles for *E. coli* strains isolated from known fecal sources. This method was hindered by overlap between the profiles for different categories (e.g., human, nonavian wild animal, avian wild animal, etc.) and by a high rate of unclassifiable profiles at some sites. Most sites tested in this study found fecal contamination inputs from multiple classes of organisms (humans, pets, farm animals, nonavian wild animals, and avian wild animals). The ARCC ranged from 65 to 94% depending on the group classification.

MAR analysis and regression analysis have been used to evaluate fecal contamination in Murrel's Inlet in South Carolina (87). Comparison of frequency and patterns of antibiotic resistance to 10 antibiotics in 321 *E. coli* isolates from 23 surface-water sites and 9 sewage system lift stations suggested that the majority of sources to the estuary were nonhuman, including isolates from areas with high-density septic systems. One exception was contamination at a site in close proximity to a sewage lift station and several large pleasure boats (86). An MAR study of 765 *E. coli* isolates from point and nonpoint sources in Apalachicola National Estuarine Research Reserve, a significant shellfish harvest area, found that human source-type profiles clustered with point source profiles, while animal source profiles clustered among the nonpoint sources (129).

In another MAR study on the Broad Creek and Okatee River watersheds in South Carolina, *E. coli* isolates were evaluated for resistance to a panel of 10 antibiotics (175). A general trend of more resistance associated with urbanized areas than with less-developed areas was observed. Water isolates were compared to isolates from WWTPs impacting the waters. Cluster analysis showed that the antibiotic resistance patterns of most surface-water isolates demonstrating resistance matched composite human sources (WWTPs).

Port Stephens is one of the largest oyster-producing estuaries in New South Wales, Australia, accounting for over $2 million annually (56, 57). It is also just south of another estuary implicated in a large HAV outbreak from contaminated oysters. Bacterial analysis indicates localized fecal pollution, particularly following significant precipitation. Possible sources of contamination in the watershed include urban runoff, animal wastes, and a large number of failing septic systems. As a result of the HAV outbreak, a study was commissioned by the shellfish industry to investigate land use practices that may have negatively affected the productivity and health of shellfish-growing waters. Antibiotic resistance patterns of fecal streptococci against four antibiotics (at four concentrations) were used to investigate the source of contamination in the estuary. Profiles of 199 isolates from unknown sources were compared to profiles of 166 isolates from known sources. The overall ARCC was 71%. No single source was identified as the most important regarding fecal contamination of affected oyster leases. Based on the analysis, cattle, human, and chicken sources were identified as probable sources of contamination.

FUTURE DIRECTIONS AND RESEARCH NEEDS

Shellfish-growing waters must be essentially free of significant levels of bacterial, viral, and protozoan human pathogens. As filter feeders, bivalve shellfish concentrate microbial contaminants from their growing waters and thus reflect the quality of the water. Current methods for monitoring the quality of growing waters rely on bacterial indicators of fecal contamination. These organisms are general indicators of sanitation, not of the occurrence of pathogens, and knowledge of their presence alone may not allow adequate management of the shellfish-growing waters. Identification of fecal sources is an important tool for management of shellfish-growing waters and mitigation of fecal contamination; e.g., knowing whether a pollution source is human or animal may indicate to environmental managers entirely different strategies for risk management. Several methods have been applied to shellfish-growing waters for the identification of fecal contamination sources. Library-based methods (e.g., ribotyping) have been most frequently utilized and offer the best source discrimination. However, they are cumbersome and relatively expensive considering the number of isolates needed to develop robust libraries. Library-independent methods (e.g., phage genotyping or host-specific virus detection) offer a more efficient and cost-effective alternative but as yet do not individually offer adequate discriminatory power between animal sources. No one MST method has been approved for regulation of shellfish-growing water quality. Additional development is needed to improve the discrimination of libraryless methods for fecal source identification. Until individual MST methods are developed to the point of being accepted as both regulatory and management tools, one way to overcome the limitations of any one method is to perform multiple methods concurrently. Although the performance of multiple methods may lead to increased costs, linking the results of multiple methods may allow sources of contamination to be identified by a preponderance of the evidence.

ACKNOWLEDGMENT

We thank Capt. William Burkhardt III for providing Table 5 and assisting with manuscript review.

REFERENCES

1. **Abeyta, C., C. A. Kaysner, M. M. Wekell, J. J. Sullivan, and G. N. Stelma.** 1986. Recovery of *Aeromonas hydrophila* from oysters implicated in an outbreak of foodborne illness. *J. Food Prot.* **49:**643–646.
2. **Aggarwal, R., K. A. McCaustland, J. B. Dilawari, S. D. Sinha, and B. H. Robertson.** 1999. Genetic variability of hepatitis E virus within and between three epidemics in India. *Virus Res.* **59:**35–48.
3. **Akihara, S., T. G. Phan, T. A. Nguyen, G. Hansman, S. Okitsu, and H. Ushijima.** 2005. Existence of multiple outbreaks of viral gastroenteritis among infants in a day care center in Japan. *Arch. Virol.* **150:**2061–2075.
4. **Alam, M. J., S. Miyoshi, and S. Shinoda.** 2003. Studies on pathogenic *Vibrio parahaemolyticus* during a warm weather season in the Seto Inland Sea, Japan. *Environ. Microbiol.* **5:**706–710.
5. **Alexander, C. E.** 1998. Classified shellfish growing waters. *In NOAA's State of the Coast Report*. NOAA, Silver Spring, Md.
6. **Anonymous.** 1998. *The Food Safety (Fishery Products and Live Shellfish) (Hygiene) Regulations 1998.* Statutory Instrument 1998 no. 994. [Online.] http://www.opsi.gov.uk/si/si1998/19980994.htm.
7. **Ashboldt, N. J., W. O. Grabow, and M. Snozzi.** 2001. Indicators of microbial water quality. *In* L. Fewtrell and J. Bartram (ed.), *Water Quality: Guidelines, Standards and Health.* IWA Publishing, London, United Kingdom.
8. **Berg, D. E., M. A. Kohn, T. A. Farley, and L. M. McFarland.** 2000. Multi-state outbreaks of acute gastroenteritis traced to fecal-contaminated oysters harvested in Louisiana. *J. Infect. Dis.* **181:**S381–S386.

9. **Berg, G., D. R. Dahling, G. A. Brown, and D. Berman.** 1978. Validity of fecal coliforms, total coliforms, and fecal streptococci as indicators of viruses in chlorinated primary sewage effluents. *Appl. Environ. Microbiol.* **36:**880–884.
10. **Bernardeschi, P., I. Bonechi, and G. Cavallini.** 1988. *Aeromonas hydrophila* infection after cockles ingestion. *Haematologica* 73:548–549.
11. **Bernhard, A. E., T. Goyard, M. T. Simonich, and K. G. Field.** 2003. Application of a rapid method for identifying fecal pollution sources in a multi-use estuary. *Water Res.* **37:**909–913.
12. **Bouchriti, N., and S. M. Goyal.** 1993. Methods for the concentration and detection of human enteric viruses in shellfish: a review. *New Microbiol.* **16:**105–113.
13. **Brett, M. S., P. Short, and J. McLauchlin.** 1998. A small outbreak of listeriosis associated with smoked mussels. *Int. J. Food Microbiol.* **43:**223–229.
14. **Brown, E. A., and J. T. Stapleton.** 2003. Hepatitis A Virus, p. 1452–1463. *In* P. R. Murray, E. J. Baron, M. A. Pfaller, F. C. Tenover, and R. H. Yolken (ed.), *Manual of Clinical Microbiology*, 8th ed. ASM Press, Washington, D.C.
15. **Burkhardt, W., III, and K. R. Calci.** 2000. Selective accumulation may account for shellfish-associated viral illness. *Appl. Environ. Microbiol.* **66:**1375–1378.
16. **Burkhardt, W., III, W. D. Watkins, and S. R. Rippey.** 1992. Seasonal effects on accumulation of microbial indicator organisms by *Mercenaria mercenaria*. *Appl. Environ. Microbiol.* **58:**826–831.
17. **Burkhardt, W., III, W. D. Watkins, and S. R. Rippey.** 1992. Survival and replication of male-specific bacteriophages in molluscan shellfish. *Appl. Environ. Microbiol.* **58:**1371–1373.
18. **Butt, A. A., K. E. Aldridge, and C. V. Sanders.** 2004. Infections related to the ingestion of seafood. Part I: viral and bacterial infections. *Lancet Infect. Dis.* **4:**201–212.
19. **Byappanahalli, M., and R. Fujioka.** 2004. Indigenous soil bacteria and low moisture may limit but allow faecal bacteria to multiply and become a minor population in tropical soils. *Water Sci. Technol.* **50:**27–32.
20. **Centers for Disease Control and Prevention.** 1997. Viral gastroenteritis associated with eating oysters—Louisiana, December 1996–January 1997. *Morb. Mortal. Wkly. Rep.* **46:** 1109–1112.
21. **Cheng, P. K., D. K. Wong, T. W. Chung, and W. W. Lim.** 2005. Norovirus contamination found in oysters worldwide. *J. Med. Virol.* **76:**593–597.
22. **Chew, K. K.** 1990. Global bivalve shellfish introductions: implications for sustaining a fishery or strong potential for economic gain? *World Aquacult.* **21:**9–22.
23. **Chironna, M., C. Germinario, D. De Medici, A. Fiore, S. Di Pasquale, M. Quarto, and S. Barbuti.** 2002. Detection of hepatitis A virus in mussels from different sources marketed in Puglia region (South Italy). *Int. J. Food Microbiol.* **75:**11–18.
24. **Chung, H., L.-A. Jaykus, G. Lovelace, and M. D. Sobsey.** 1998. Bacteriophages and bacteria as indicators of enteric viruses in oysters and their harvest waters. *Water Sci. Technol.* **38:**37–44.
25. **Coelho, C., A. P. Heinert, C. M. Simoes, and C. R. Barardi.** 2003. Hepatitis A virus detection in oysters (*Crassostrea gigas*) in Santa Catarina State, Brazil, by reverse transcription-polymerase chain reaction. *J. Food Prot.* **66:**507–511.

26. **Committee on Evaluation of the Safety of Fishery Products, Food and Nutrition Board, Institute of Medicine.** 1991. *Seafood Safety.* National Academic Press, Washington, D.C.
27. **Conaty, S., P. Bird, G. Bell, E. Kraa, G. Grohmann, and J. M. McAnulty.** 2000. Hepatitis A in New South Wales, Australia from consumption of oysters: the first reported outbreak. *Epidemiol. Infect.* **124:**121–130.
28. **Craun, G. F., R. L. Calderon, and M. F. Craun.** 2004. Waterborne outbreaks caused by zoonotic pathogens in the U.S.A. *In* J. A. Cotruvo, A. Dufour, G. Rees, J. Bartram, R. Carr, D. O. Cliver, G. F. Craun, R. Fayer, and V. P. J. Gannon (ed.), *Waterborne Zoonoses.* IWA Publishing, London, United Kingdom.
29. **Dalsgaard, A.** 1998. The occurrence of human pathogenic *Vibrio* spp. and *Salmonella* in aquaculture. *Int. J. Food Sci. Technol.* **33:**127–138.
30. **Deneen, V. C., J. M. Hunt, C. R. Paule, R. I. James, R. G. Johnson, M. J. Raymond, and C. W. Hedberg.** 2000. The impact of foodborne calicivirus disease: the Minnesota experience. *J. Infect. Dis.* **181**(Suppl. 2):S281–S283.
31. **DePaola, A., J. L. Nordstrom, J. C. Bowers, J. G. Wells, and D. W. Cook.** 2003. Seasonal abundance of total and pathogenic *Vibrio parahaemolyticus* in Alabama oysters. *Appl. Environ. Microbiol.* **69:**1521–1526.
32. **Desenclos, J. C., K. C. Klontz, M. H. Wilder, O. V. Nainan, H. S. Margolis, and R. A. Gunn.** 1991. A multistate outbreak of hepatitis A caused by the consumption of raw oysters. *Am. J. Public Health* **81:**1268–1272.
33. **Dhanashree, B., S. K. Otta, and I. Karunasagar.** 2003. Typing of *Listeria monocytogenes* isolates by random amplification of polymorphic DNA. *Indian J. Med. Res.* **117:**19–24.
34. **DiGirolamo, R., J. Liston, and J. R. Matches.** 1970. Survival of virus in chilled, frozen, and processed oysters. *Appl. Microbiol.* **20:**58–63.
35. **Dolin, R., N. R. Blacklow, H. DuPont, R. F. Buscho, R. G. Wyatt, J. A. Kasel, R. Hornick, and R. M. Chanock.** 1972. Biological properties of Norwalk agent of acute infectious nonbacterial gastroenteritis. *Proc. Soc. Exp. Biol. Med.* **140:**578–583.
36. **Dore, W. J., K. Henshilwood, and D. N. Lees.** 2000. Evaluation of F-specific RNA bacteriophage as a candidate human enteric virus indicator for bivalve molluscan shellfish. *Appl. Environ. Microbiol.* **66:**1280–1285.
37. **Dore, W. J., and D. N. Lees.** 1995. Behavior of *Escherichia coli* and male-specific bacteriophage in environmentally contaminated bivalve molluscs before and after depuration. *Appl. Environ. Microbiol.* **61:**2830–2834.
38. **Dore, W. J., M. Mackie, and D. N. Lees.** 2003. Levels of male-specific RNA bacteriophage and *Escherichia coli* in molluscan bivalve shellfish from commercial harvesting areas. *Lett. Appl. Microbiol.* **36:**92–96.
39. **Dowell, S. F., C. Groves, K. B. Kirkland, H. G. Cicirello, T. Ando, Q. Jin, J. R. Gentsch, S. S. Monroe, C. D. Humphrey, C. Slemp, D. M. Dwyer, R. A. Meriwether, and R. I. Glass.** 1995. A multistate outbreak of oyster-associated gastroenteritis: implications for interstate tracing of contaminated shellfish. *J. Infect. Dis.* **171:**1497–1503.
40. **Eastaugh, J., and S. Shepherd.** 1989. Infectious and toxic syndromes from fish and shellfish consumption. A review. *Arch. Intern. Med.* **149:**1735–1740.
41. **Endtz, H. P., J. S. Vliegenthart, P. Vandamme, H. W. Weverink, N. P. van den Braak, H. A. Verbrugh, and A. van Belkum.** 1997. Genotypic diversity of *Campylobacter lari* isolated from mussels and oysters in The Netherlands. *Int. J. Food Microbiol.* **34:**79–88.

42. **Epstein, P. R.** 1993. Algal blooms in the spread and persistence of cholera. *Biosystems* **31**:209–221.

43. **Evans, H. S., P. Madden, C. Douglas, G. K. Adak, S. J. O'Brien, T. Djuretic, P. G. Wall, and R. Stanwell-Smith.** 1998. General outbreaks of infectious intestinal disease in England and Wales: 1995 and 1996. *Commun. Dis. Public Health* **1**:165–171.

44. **Farmer, J. J., III.** 2003. *Enterobacteriaceae*: introduction and identification, p. 636–653. *In* P. Murray, E. J. Baron, J. H. Jorgensen, M. A. Pfaller, and R. H. Yolken (ed.), *Manual of Clinical Microbiology*, 8th ed. ASM Press, Washington, D.C.

45. **Farmer, J. J., III, M. Janda, and K. Kirkhead.** 2003. *Vibrio*, p. 706–718. *In* P. Murray, E. J. Baron, J. H. Jorgensen, M. A. Pfaller, and R. H. Yolken (ed.), *Manual of Clinical Microbiology*, 8th ed. ASM Press, Washington, D.C.

46. **Fayer, R., T. K. Graczyk, E. J. Lewis, J. M. Trout, and C. A. Farley.** 1998. Survival of infectious *Cryptosporidium parvum* oocysts in seawater and Eastern oysters (*Crassostrea virginica*) in the Chesapeake Bay. *Appl. Environ. Microbiol.* **64**:1070–1074.

47. **Fayer, R., E. J. Lewis, J. M. Trout, T. K. Graczyk, M. C. Jenkins, J. Higgins, L. Xiao, and A. A. Lal.** 1999. *Cryptosporidium parvum* in oysters from commercial harvesting sites in the Chesapeake Bay. *Emerg. Infect. Dis.* **5**:706–710.

48. **Fayer, R., J. M. Trout, E. J. Lewis, L. Xiao, A. Lal, M. C. Jenkins, and T. K. Graczyk.** 2002. Temporal variability of *Cryptosporidium* in the Chesapeake Bay. *Parasitol. Res.* **88**:998–1003.

49. **Formiga-Cruz, M., A. K. Allard, A. C. Conden-Hansson, K. Henshilwood, B. E. Hernroth, J. Jofre, D. N. Lees, F. Lucena, M. Papapetropoulou, R. E. Rangdale, A. Tsibouxi, A. Vantarakis, and R. Girones.** 2003. Evaluation of potential indicators of viral contamination in shellfish and their applicability to diverse geographical areas. *Appl. Environ. Microbiol.* **69**:1556–1563.

50. **Formiga-Cruz, M., A. Hundesa, P. Clemente-Casares, N. Albinana-Gimenez, A. Allard, and R. Girones.** 2005. Nested multiplex PCR assay for detection of human enteric viruses in shellfish and sewage. *J. Virol. Methods* **125**:111–118.

51. **Formiga-Cruz, M., G. Tofino-Quesada, S. Bofill-Mas, D. N. Lees, K. Henshilwood, A. K. Allard, A. C. Conden-Hansson, B. E. Hernroth, A. Vantarakis, A. Tsibouxi, M. Papapetropoulou, M. D. Furones, and R. Girones.** 2002. Distribution of human virus contamination in shellfish from different growing areas in Greece, Spain, Sweden, and the United Kingdom. *Appl. Environ. Microbiol.* **68**:5990–5998.

52. **Fujioka, R. S., and B. S. Yoneyama.** 2002. Sunlight inactivation of human enteric viruses and fecal bacteria. *Water Sci. Technol.* **46**:291–295.

53. **Furuta, T., M. Akiyama, Y. Kato, and O. Nishio.** 2003. A food poisoning outbreak caused by purple Washington clam contaminated with norovirus (Norwalk-like virus) and hepatitis A virus. *Kansenshogaku Zasshi* **77**:89–94. (In Japanese.)

54. **Gallimore, C. I., J. S. Cheesbrough, K. Lamden, C. Bingham, and J. J. Gray.** 2005. Multiple norovirus genotypes characterized from an oyster-associated outbreak of gastroenteritis. *Int. J. Food Microbiol.* **103**:323–330.

55. **Gangerosa, E. J., and W. H. Mosely.** 1974. Epidemiology and surveillance of cholera, p. 381–403. *In* D. Barua and W. Burrows (ed.), *Cholera*. W. B. Saunders, Philadelphia, Pa.

56. **Geary, P. M.** 2003. *On-site Treatment System Failure and Shellfish Contamination in Port Stephens, N.S.W.* University of New Castle, New Castle, Australia.

57. **Geary, P. M., and C. M. Davies.** 2003. Bacterial source tracking and shellfish contamination in a coastal catchment. *Water Sci. Technol.* **47:**95–100.

58. **Gerba, C. P., S. M. Goyal, R. L. LaBelle, I. Cech, and G. F. Bodgan.** 1979. Failure of indicator bacteria to reflect the occurrence of enteroviruses in marine waters. *Am. J. Pub. Health* **69:**1116–1119.

59. **Gergatz, S. J., and L. M. McFarland.** 1989. Cholera on the Louisiana Gulf Coast: historical notes and case report. *J. Louisiana State Med. Soc.* **141:**29–34.

60. **Gomez-Bautista, M., L. M. Ortega-Mora, E. Tabares, V. Lopez-Rodas, and E. Costas.** 2000. Detection of infectious *Cryptosporidium parvum* oocysts in mussels (*Mytilus galloprovincialis*) and cockles (*Cerastoderma edule*). *Appl. Environ. Microbiol.* **66:**1866–1870.

61. **Gomez-Couso, H., F. Freire-Santos, C. F. Amar, K. A. Grant, K. Williamson, M. E. Ares-Mazas, and J. McLauchlin.** 2004. Detection of *Cryptosporidium* and *Giardia* in molluscan shellfish by multiplexed nested-PCR. *Int. J. Food Microbiol.* **91:**279–288.

62. **Gomez-Couso, H., F. Mendez-Hermida, J. A. Castro-Hermida, and E. Ares-Mazas.** 2005. *Giardia* in shellfish-farming areas: detection in mussels, river water and waste waters. *Vet. Parasitol.* **133:**13–18.

63. **Goyal, S. M., C. P. Gerba, and J. L. Melnick.** 1979. Human enteroviruses in oysters and their overlying waters. *Appl. Environ. Microbiol.* **37:**572–581.

64. **Grabow, W. O., V. Gauss-Muller, O. W. Prozesky, and F. Deinhardt.** 1983. Inactivation of hepatitis A virus and indicator organisms in water by free chlorine residuals. *Appl. Environ. Microbiol.* **46:**619–624.

65. **Graczyk, T. K., R. Fayer, E. J. Lewis, J. M. Trout, and C. A. Farley.** 1999. *Cryptosporidium* oocysts in Bent mussels (*Ischadium recurvum*) in the Chesapeake Bay. *Parasitol. Res.* **85:**518–521.

66. **Graczyk, T. K., R. C. Thompson, R. Fayer, P. Adams, U. M. Morgan, and E. J. Lewis.** 1999. *Giardia duodenalis* cysts of genotype A recovered from clams in the Chesapeake Bay subestuary, Rhode River. *Am. J. Trop. Med. Hyg.* **61:**526–529.

67. **Greenwood, M., G. Winnard, and B. Bagot.** 1998. An outbreak of *Salmonella enteritidis* phage type 19 infection associated with cockles. *Commun. Dis. Public Health* **1:**35–37.

68. **Griffin, D. W., K. A. Donaldson, J. H. Paul, and J. B. Rose.** 2003. Pathogenic human viruses in coastal waters. *Clin. Microbiol. Rev.* **16:**129–143.

69. **Halliday, M. L., L. Y. Kang, T. K. Zhou, M. D. Hu, Q. C. Pan, T. Y. Fu, Y. S. Huang, and S. L. Hu.** 1991. An epidemic of hepatitis A attributable to the ingestion of raw clams in Shanghai, China. *J. Infect. Dis.* **164:**852–859.

70. **Harrington, P. R., L. Lindesmith, B. Yount, C. L. Moe, and R. S. Baric.** 2002. Binding of Norwalk virus-like particles to ABH histo-blood group antigens is blocked by antisera from infected human volunteers or experimentally vaccinated mice. *J. Virol.* **76:**12335–12343.

71. **Hlady, W. G., and K. C. Klontz.** 1996. The epidemiology of *Vibrio* infections in Florida, 1981–1993. *J. Infect. Dis.* **173:**1176–1183.

72. **Honda, T., M. Arita, T. Takeda, M. Yoh, and T. Miwatani.** 1985. Non-O1 *Vibrio cholerae* produces two newly identified toxins related to *Vibrio parahaemolyticus* haemolysin and *Escherichia coli* heat-stable enterotoxin. *Lancet* **ii:**163–164.

73. **Hsu, F. C., Y. S. Shieh, J. van Duin, M. J. Beekwilder, and M. D. Sobsey.** 1995. Genotyping male-specific RNA coliphages by hybridization with oligonucleotide probes. *Appl. Environ. Microbiol.* **61**:3960–3966.

74. **Huang, P., T. Farkas, S. Marionneau, W. Zhong, N. Ruvoen-Clouet, A. L. Morrow, M. Altaye, L. K. Pickering, D. S. Newburg, J. LePendu, and X. Jiang.** 2003. Noroviruses bind to human ABO, Lewis, and secretor histo-blood group antigens: identification of 4 distinct strain-specific patterns. *J. Infect. Dis.* **188**:19–31.

75. **Hutson, A. M., R. L. Atmar, D. M. Marcus, and M. K. Estes.** 2003. Norwalk virus-like particle hemagglutination by binding to H histo-blood group antigens. *J. Virol.* **77**:405–415.

76. **Isaacson, M., J. Frean, J. He, J. Seriwatana, and B. L. Innis.** 2000. An outbreak of hepatitis E in Northern Namibia, 1983. *Am. J. Trop. Med. Hyg.* **62**:619–625.

77. **Ishii, S., W. B. Ksoll, R. E. Hicks, and M. J. Sadowsky.** 2006. Presence and growth of naturalized *Escherichia coli* in temperate soils from Lake Superior watersheds. *Appl. Environ. Microbiol.* **72**:612–621.

78. **Jeffery, B., T. Barlow, K. Moizer, S. Paul, and C. Boyle.** 2004. Amnesic shellfish poison. *Food Chem. Toxicol.* **42**:545–557.

79. **Jehl-Pietri, C., J. Dupont, C. Herve, D. Menard, and J. Munro.** 1991. Occurrence of faecal bacteria, *Salmonella* and antigens associated with hepatitis A virus in shellfish. *Zentralbl. Hyg. Umweltmed.* **192**:230–237.

80. **Jimenez-Clavero, M. A., E. Escribano-Romero, C. Mansilla, N. Gomez, L. Cordoba, N. Roblas, F. Ponz, V. Ley, and J. C. Saiz.** 2005. Survey of bovine enterovirus in biological and environmental samples by a highly sensitive real-time reverse transcription-PCR. *Appl. Environ. Microbiol.* **71**:3536–3543.

81. **Jones, S.** 2003. Tracking fecal pollution sources in New Hampshire shellfish waters using *Escherichia coli* ribotyping, p. 20–21. *In Workshop Report: Approaches to the Detection and Identification of Faecal Sewage Contamination in Coastal Waters.* Dunstaffnage Marine Laboratory, Oban, Scotland, 17 September 2003. Department for Environment, Food and Rural Affairs, London, United Kingdom.

82. **Joseph, S. W., and A. M. Carnahan.** 2000. Update on the genus *Aeromonas. ASM News* **66**:218–223.

83. **Kaneko, T., and R. R. Colwell.** 1973. Ecology of *Vibrio parahaemolyticus* in Chesapeake Bay. *J. Bacteriol.* **113**:24–32.

84. **Karamoko, Y., K. Ibenyassine, R. Aitmhand, M. Idaomar, and M. M. Ennaji.** 2005. Adenovirus detection in shellfish and urban sewage in Morocco (Casablanca region) by the polymerase chain reaction. *J. Virol. Methods* **126**:135–137.

85. **Kautter, D. A., A. J. Leblanc, A. L. A. LeBlanc, and R. K. Lynt.** 1974. Incidence of *Clostridium botulinum* in crabmeat from the blue crab. *Appl. Microbiol.* **28**:722.

86. **Kelsey, H., D. E. Porter, G. Scott, M. Neet, and D. White.** 2004. Using geographic information systems and regression analysis to evaluate relationships between land use and fecal coliform bacterial pollution. *J. Exp. Mar. Biol. Ecol.* **298**:197–209.

87. **Kelsey, R. H., G. I. Scott, D. E. Porter, B. Thompson, and L. Webster.** 2003. Using multiple antibiotic resistance and land use characteristics to determine sources of fecal coliform bacterial pollution. *Environ. Monit. Assess.* **81**:337–348.

88. **Kingsley, D. H., G. K. Meade, and G. P. Richards.** 2002. Detection of both hepatitis A virus and Norwalk-like virus in imported clams associated with food-borne illness. *Appl. Environ. Microbiol.* **68:**3914–3918.
89. **Kirkland, K. B., R. A. Meriwether, J. K. Leiss, and W. R. Mac Kenzie.** 1996. Steaming oysters does not prevent Norwalk-like gastroenteritis. *Public Health Rep.* **111:**527–530.
90. **Kohn, M. A., T. A. Farley, T. Ando, M. Curtis, S. A. Wilson, Q. Jin, S. S. Monroe, R. C. Baron, L. M. McFarland, and R. I. Glass.** 1995. An outbreak of Norwalk virus gastroenteritis associated with eating raw oysters. Implications for maintaining safe oyster beds. *JAMA* **273:**466–471.
91. **Koopmans, M., and E. Duizer.** 2004. Foodborne viruses: an emerging problem. *Int. J. Food Microbiol.* **90:**23–41.
92. **Koopmans, M., C. H. von Bonsdorff, J. Vinje, D. de Medici, and S. Monroe.** 2002. Foodborne viruses. *FEMS Microbiol. Rev.* **26:**187–205.
93. **Kopecka, H., S. Dubrou, J. Prevot, J. Marechal, and J. M. Lopez-Pila.** 1993. Detection of naturally occurring enteroviruses in waters by reverse transcription, polymerase chain reaction, and hybridization. *Appl. Environ. Microbiol.* **59:**1213–1219.
94. **Leber, A. L., and S. M. Novak.** 2003. Intestinal and urogenital amebae, flagellates, and ciliates, p. 1990–2007. *In* P. R. Murray, E. J. Baron, M. A. Pfaller, F. C. Tenover, and R. H. Yolken (ed.), *Manual of Clinical Microbiology*, 8th ed. ASM Press, Washington, D.C.
95. **Lee, T., W. C. Yam, T.-Y. Tam, B. S. W. Ho, M. H. Ng, and Malcolm J. Broom.** 1999. Occurrence of hepatitis A virus in green-lipped mussels (*Perna viridis*). *Water Res.* **33:**885–889.
96. **Lees, D.** 2000. Viruses and bivalve shellfish. *Int. J. Food Microbiol.* **59:**81–116.
97. **Legnani, P., E. Leoni, F. Soppelsa, and R. Burigo.** 1998. The occurrence of *Aeromonas* species in drinking water supplies of an area of the Dolomite Mountains, Italy. *J. Appl. Microbiol.* **85:**271–276.
98. **Le Guyader, F., E. Dubois, D. Menard, and M. Pommepuy.** 1994. Detection of hepatitis A virus, rotavirus, and enterovirus in naturally contaminated shellfish and sediment by reverse transcription-seminested PCR. *Appl. Environ. Microbiol.* **60:**3665–3671.
99. **Le Guyader, F., L. Haugarreau, L. Miossec, E. Dubois, and M. Pommepuy.** 2000. Three-year study to assess human enteric viruses in shellfish. *Appl. Environ. Microbiol.* **66:**3241–3248.
100. **Leon, P., E. Venegas, L. Bengoechea, E. Rojas, J. A. Lopez, C. Elola, and J. M. Echevarria.** 1999. Prevalence of infections by hepatitis B, C, D and E viruses in Bolivia. *Rev. Panam. Salud Publica* **5:**144–151. (In Spanish.)
101. **Leoni, E., C. Bevini, S. Degli Esposti, and A. Graziano.** 1998. An outbreak of intrafamiliar hepatitis A associated with clam consumption: epidemic transmission to a school community. *Eur. J. Epidemiol.* **14:**187–192.
102. **Ley, V., J. Higgins, and R. Fayer.** 2002. Bovine enteroviruses as indicators of fecal contamination. *Appl. Environ. Microbiol.* **68:**3455–3461.
103. **Lindesmith, L., C. Moe, S. Marionneau, N. Ruvoen, X. Jiang, L. Lindblad, P. Stewart, J. LePendu, and R. Baric.** 2003. Human susceptibility and resistance to Norwalk virus infection. *Nat. Med.* **9:**548–553.
104. **Lowry, P. W., A. T. Pavia, L. M. McFarland, B. H. Peltier, T. J. Barrett, H. B. Bradford, J. M. Quan, J. Lynch, J. B. Mathison, R. A. Gunn, et al.** 1989. Cholera in Louisiana. Widening spectrum of seafood vehicles. *Arch. Intern. Med.* **149:**2079–2084.

105. **Lucena, F., A. E. Duran, A. Moron, E. Calderon, C. Campos, C. Gantzer, S. Skraber, and J. Jofre.** 2004. Reduction of bacterial indicators and bacteriophages infecting faecal bacteria in primary and secondary wastewater treatments. *J. Appl. Microbiol.* **97**:1069–1076.

106. **Luo, K. X., L. Zhang, S. S. Wang, J. Nie, S. C. Yang, D. X. Liu, W. F. Liang, H. T. He, and Q. Lu.** 1999. An outbreak of enterically transmitted non-A, non-E viral hepatitis. *J. Viral Hepat.* **6**:59–64.

107. **Luthi, T. M., P. G. Wall, H. S. Evans, G. K. Adak, and E. O. Caul.** 1996. Outbreaks of foodborne viral gastroenteritis in England and Wales: 1992 to 1994. *Commun. Dis. Rep. CDR Rev.* **6**:R131–R136.

108. **Marionneau, S., N. Ruvoen, B. Le Moullac-Vaidye, M. Clement, A. Cailleau-Thomas, G. Ruiz-Palacois, P. Huang, X. Jiang, and J. Le Pendu.** 2002. Norwalk virus binds to histo-blood group antigens present on gastroduodenal epithelial cells of secretor individuals. *Gastroenterology* **122**:1967–1977.

109. **Martinez-Urtaza, J., M. Saco, G. Hernandez-Cordova, A. Lozano, O. Garcia-Martin, and J. Espinosa.** 2003. Identification of *Salmonella* serovars isolated from live molluscan shellfish and their significance in the marine environment. *J. Food Prot.* **66**:226–232.

110. **Mast, E. E., and M. J. Alter.** 1993. Epidemiology of viral hepatitis: an overview. *Semin. Virol.* **4**:273–283.

111. **McCarthy, S. A., and F. M. Khambaty.** 1994. International dissemination of epidemic *Vibrio cholerae* by cargo ship ballast and other nonpotable waters. *Appl. Environ. Microbiol.* **60**:2597–2601.

112. **McDonnell, S., K. B. Kirkland, W. G. Hlady, C. Aristeguieta, R. S. Hopkins, S. S. Monroe, and R. I. Glass.** 1997. Failure of cooking to prevent shellfish-associated viral gastroenteritis. *Arch. Intern. Med.* **157**:111–116.

113. **Mead, P. S., L. Slutsker, V. Dietz, L. F. McCaig, J. S. Bresee, C. Shapiro, P. M. Griffin, and R. V. Tauxe.** 1999. Food-related illness and death in the United States. *Emerg. Infect. Dis.* **5**:607–625.

114. **Merino, S., X. Rubires, S. Knochel, and J. Tomas.** 1995. Emerging pathogens: *Aeromonas* spp. *Int. J. Food Microbiol.* **28**:157–168.

115. **Middleton, P. J.** 1996. Viruses that multiply in the gut and cause endemic and epidemic gastroenteritis. *Clin. Diagn. Virol.* **6**:93–101.

116. **Miossec, L., F. Le Guyader, L. Haugarreau, and M. Pommepuy.** 2000. Magnitude of rainfall on viral contamination of the marine environment during gastroenteritis epidemics in human coastal population. *Rev. Epidemiol. Sante Pub.* **48**(Suppl. 2):2S62–2S71.

117. **Monfort, P., J. Minet, J. Rocourt, G. Piclet, and M. Cormier.** 1998. Incidence of *Listeria* spp. in Breton live shellfish. *Lett. Appl. Microbiol.* **26**:205–208.

118. **Morris, J. G., Jr.** 2003. Cholera and other types of vibriosis: a story of human pandemics and oysters on the half shell. *Clin. Infect. Dis.* **37**:272–280.

119. **Morse, D. L., J. J. Guzewich, J. P. Hanrahan, R. Stricof, M. Shayegani, R. Deibel, J. C. Grabau, N. A. Nowak, J. E. Herrmann, G. Cukor, et al.** 1986. Widespread outbreaks of clam- and oyster-associated gastroenteritis. Role of Norwalk virus. *N. Engl. J. Med.* **314**:678–681.

120. **Muniain-Mujika, I., M. Calvo, F. Lucena, and R. Girones.** 2003. Comparative analysis of viral pathogens and potential indicators in shellfish. *Int. J. Food Microbiol.* **83**:75–85.

121. **Nachamkin, I.** 2003. *Campylobacter* and *Arcobacter*, p. 716–726. *In* P. R. Murray, E. J. Baron, M. A. Pfaller, F. C. Tenover, and R. H. Yolken (ed.), *Manual of Clinical Microbiology*, 8th ed. ASM Press, Washington, D.C.

122. **Nasser, A. M., N. Zaruk, L. Tenenbaum, and Y. Netzan.** 2003. Comparative survival of *Cryptosporidium*, coxsackievirus A9 and *Escherichia coli* in stream, brackish and sea waters. *Water Sci. Technol.* **47**:91–96.

123. **National Marine Fisheries Service.** 1997. *Fisheries of the United States, 1996.* Fisheries Statistics and Economics Division, National Marine Fisheries Service, NOAA, U.S. Department of Commerce, Silver Spring, Md.

124. **National Marine Fisheries Service.** 1996. *Our Living Oceans: the Economic Status of U.S. Fisheries, 1996.* Office of Science and Technology, National Marine Fisheries Service, NOAA, Silver Spring, Md.

125. **Noble, R. T.** 2003. Using molecular methods for understanding impacts of non-point source runoff on economically important coastal recreational and shellfish harvesting waters. *In* M. Bruen (ed.), *Diffuse Pollution and Basin Management.* Proceedings of the 7th International Specialised IWA Conference, Dublin, Ireland.

126. **Noble, R. T., S. M. Allen, A. D. Blackwood, W. Chu, S. C. Jiang, G. L. Lovelace, M. D. Sobsey, J. R. Stewart, and D. A. Wait.** 2003. Use of viral pathogens and indicators to differentiate between human and non-human contamination in a microbial source tracking comparison study. *J. Water Health* **1**:195–207.

127. **Noble, R. T., D. F. Moore, M. K. Leecaster, C. D. McGee, and S. B. Weisberg.** 2003. Comparison of total coliform, fecal coliform, and *Enterococcus* bacterial indicator response for ocean recreational water quality testing. *Water Res.* **37**:1637–1643.

128. **Olsen, J. E., D. J. Brown, D. L. Baggesen, and M. Bisgaard.** 1992. Biochemical and molecular characterization of *Salmonella enterica* serovar Berta, and comparison of methods for typing. *Epidemiol. Infect.* **108**:243–260.

129. **Parveen, S., R. L. Murphree, L. Edmiston, C. W. Kaspar, K. M. Portier, and M. L. Tamplin.** 1997. Association of multiple-antibiotic-resistance profiles with point and nonpoint sources of *Escherichia coli* in Apalachicola Bay. *Appl. Environ. Microbiol.* **63**:2607–2612.

130. **Parveen, S., K. M. Portier, K. Robinson, L. Edmiston, and M. L. Tamplin.** 1999. Discriminant analysis of ribotype profiles of *Escherichia coli* for differentiating human and nonhuman sources of fecal pollution. *Appl. Environ. Microbiol.* **65**:3142–3147.

131. **Petric, M., and R. Tellier.** 2003. Rotaviruses, caliciviruses, astroviruses, and other diarrheic viruses, p. 1439–1451. *In* P. R. Murray, E. J. Baron, M. A. Pfaller, F. C. Tenover, and R. H. Yolken (ed.), *Manual of Clinical Microbiology*, 8th ed., ASM Press, Washington, D.C.

132. **Piergentili, P., M. Castellani-Pastoris, R. D. Fellini, G. Farisano, C. Bonello, E. Rigoli, and A. Zampieri.** 1984. Transmission of non O group 1 *Vibrio cholerae* by raw oyster consumption. *Int. J. Epidemiol.* **13**:340–343.

133. **Popovic, T., C. Bopp, O. Olsvik, and K. Wachsmuth.** 1993. Epidemiologic application of a standardized ribotype scheme for *Vibrio cholerae* O1. *J. Clin. Microbiol.* **31**:2474–2482.

134. **Potasman, I., A. Paz, and M. Odeh.** 2002. Infectious outbreaks associated with bivalve shellfish consumption: a worldwide perspective. *Clin. Infect. Dis.* **35**:921–928.

135. **Pringle, C. R.** 1998. Virus taxonomy—San Diego 1998. *Arch. Virol.* **143**:1449–1459.

136. **Reeve, G., D. L. Martin, J. Pappas, R. E. Thompson, and K. D. Greene.** 1989. An outbreak of shigellosis associated with the consumption of raw oysters. *N. Engl. J. Med.* **321**:224–227.

137. **Reyes, G. R., M. A. Purdy, J. P. Kim, K. C. Luk, L. M. Young, K. E. Fry, and D. W. Bradley.** 1990. Isolation of a cDNA from the virus responsible for enterically transmitted non-A, non-B hepatitis. *Science* **247**:1335–1339.

138. **Reyes, G. R., P. O. Yarbough, A. W. Tam, M. A. Purdy, C. C. Huang, J. S. Kim, D. W. Bradley, and K. E. Fry.** 1991. Hepatitis E virus (HEV): the novel agent responsible for enterically transmitted non-A, non-B hepatitis. *Gastroenterol. Jpn.* **26**(Suppl. 3):142–147.

139. **Richards, G.** 1988. Microbial purification of shellfish: a review of depuration and relaying. *J. Food Prot.* **51**:218–251.

140. **Richards, P. D., G. C. Fletcher, D. H. Buisson, and S. Fredericksen.** 1983. Virus depuration of the Pacific oyster (*Crassostrea gigas*) in New Zealand. *N. Z. J. Sci.* **26**:9–13.

141. **Richardson, W. H., S. S. Frei, and S. R. Williams.** 2004. A case of type F botulism in southern California. *J. Toxicol. Clin. Toxicol.* **42**:383–387.

142. **Ripabelli, G., M. L. Sammarco, I. Fanelli, and G. M. Grasso.** 2004. Detection of *Salmonella*, *Listeria* spp., *Vibrio* spp., and *Yersinia enterocolitica* in frozen seafood and comparison with enumeration for faecal indicators: implication for public health. *Ann. Ig.* **16**:531–539. (In Italian.)

143. **Rippey, S. R.** 1994. Infectious diseases associated with molluscan shellfish consumption. *Clin. Microbiol. Rev.* **7**:419–425.

144. **Rivera, S. C., T. C. Hazen, and G. A. Toranzos.** 1988. Isolation of fecal coliforms from pristine sites in a tropical rain forest. *Appl. Environ. Microbiol.* **54**:513–517.

145. **Rockx, B., M. De Wit, H. Vennema, J. Vinje, E. De Bruin, Y. Van Duynhoven, and M. Koopmans.** 2002. Natural history of human calicivirus infection: a prospective cohort study. *Clin. Infect. Dis.* **35**:246–253.

146. **Romalde, J. L., E. Area, G. Sanchez, C. Ribao, I. Torrado, X. Abad, R. M. Pinto, J. L. Barja, and A. Bosch.** 2002. Prevalence of enterovirus and hepatitis A virus in bivalve molluscs from Galicia (NW Spain): inadequacy of the EU standards of microbiological quality. *Int. J. Food Microbiol.* **74**:119–130.

147. **Rutala, W. A., F. A. Sarubi, Jr., C. S. Finch, J. N. McCormack, and G. E. Steinkraus.** 1982. Oyster-associated outbreak of diarrhoeal disease possibly caused by *Plesiomonas shigelloides*. *Lancet* **i**:739.

148. **Samadpour, M.** 2003. Identification and tracking of the sources of microbial pollution in shellfish growing areas. *In Workshop Report: Approaches to the Detection and Identification of Faecal Sewage Contamination in Coastal Waters*. Dunstaffnage Marine Laboratory, Oban, Scotland, 17 September 2003. Department for Environment, Food and Rural Affairs, London, United Kingdom.

149. **Schlech, W. F., III.** 2000. Foodborne listeriosis. *Clin. Infect. Dis.* **31**:770–775.

150. **Schlech, W. F., III, P. M. Lavigne, R. A. Bortolussi, A. C. Allen, E. V. Haldane, A. J. Wort, A. W. Hightower, S. E. Johnson, S. H. King, E. S. Nicholls, and C. V. Broome.** 1983. Epidemic listeriosis—evidence for transmission by food. *N. Engl. J. Med.* **308**:203–206.

151. **Schwab, K. J., F. H. Neill, M. K. Estes, T. G. Metcalf, and R. L. Atmar.** 1998. Distribution of Norwalk virus within shellfish following bioaccumulation and subsequent depuration by detection using RT-PCR. *J. Food Prot.* **61**:1674–1680.

152. **Scott, T. M., J. Caren, G. R. Nelson, T. M. Jenkins, and J. Lukasik.** 2004. Tracking sources of fecal pollution in a South Carolina watershed by ribotyping *Escherichia coli*: a case study. *Environ. Forensics* **5**:15–19.

153. **Scott, T. M., S. Parveen, K. M. Portier, J. B. Rose, M. L. Tamplin, S. R. Farrah, A. Koo, and J. Lukasik.** 2003. Geographical variation in ribotype profiles of *Escherichia coli* isolates from humans, swine, poultry, beef, and dairy cattle in Florida. *Appl. Environ. Microbiol.* **69**:1089–1092.

154. **Sekine, S., S. Okada, Y. Hayashi, T. Ando, T. Terayama, K. Yabuuchi, T. Miki, and M. Ohashi.** 1989. Prevalence of small round structured virus infections in acute gastroenteritis outbreaks in Tokyo. *Microbiol. Immunol.* **33**:207–217.

155. **Sha, J., L. Pillai, A. A. Fadl, C. L. Galindo, T. E. Erova, and A. K. Chopra.** 2005. The type III secretion system and cytotoxic enterotoxin alter the virulence of *Aeromonas hydrophila*. *Infect. Immun.* **73**:6446–6457.

156. **Shieh, Y., S. S. Monroe, R. L. Fankhauser, G. W. Langlois, W. Burkhardt III, and R. S. Baric.** 2000. Detection of Norwalk-like virus in shellfish implicated in illness. *J. Infect. Dis.* **181**(Suppl. 2):S360–S366.

157. **Shieh, Y. C., R. S. Baric, J. W. Woods, and K. R. Calci.** 2003. Molecular surveillance of enterovirus and Norwalk-like virus in oysters relocated to a municipal-sewage-impacted gulf estuary. *Appl. Environ. Microbiol.* **69**:7130–7136.

158. **Simpson, J. M., J. W. Santo Domingo, and D. J. Reasoner.** 2002. Microbial source tracking: state of the science. *Environ. Sci. Technol.* **36**:5279–5288.

159. **Solo-Gabriele, H. M., M. A. Wolfert, T. R. Desmarais, and C. J. Palmer.** 2000. Sources of *Escherichia coli* in a coastal subtropical environment. *Appl. Environ. Microbiol.* **66**:230–237.

160. **Stanley, J., N. Chowdry, N. Powell, and E. J. Threlfall.** 1992. Chromosomal genotypes (evolutionary lines) of *Salmonella berta*. *FEMS Microbiol. Lett.* **74**:247–252.

161. **Stull, T. L., J. J. LiPuma, and T. D. Edlind.** 1988. A broad-spectrum probe for molecular epidemiology of bacteria: ribosomal RNA. *J. Infect. Dis.* **157**:280–286.

162. **Sugieda, M., K. Nakajima, and S. Nakajima.** 1996. Outbreaks of Norwalk-like virus-associated gastroenteritis traced to shellfish: coexistence of two genotypes in one specimen. *Epidemiol. Infect.* **116**:339–346.

163. **Takeda, Y.** 1983. Thermostable direct hemolysin of *Vibrio parahaemolyticus*. *Pharmacol. Ther.* **19**:123–146.

164. **Talal, A. H., C. L. Moe, A. A. Lima, K. A. Weigle, L. Barrett, S. I. Bangdiwala, M. K. Estes, and R. L. Guerrant.** 2000. Seroprevalence and seroincidence of Norwalk-like virus infection among Brazilian infants and children. *J. Med. Virol.* **61**:117–124.

165. **Tamplin, M. L., J. K. Jackson, C. Buchrieser, R. L. Murphree, K. M. Portier, V. Gangar, L. G. Miller, and C. W. Kaspar.** 1996. Pulsed-field gel electrophoresis and ribotype profiles of clinical and environmental *Vibrio vulnificus* isolates. *Appl. Environ. Microbiol.* **62**:3572–3580.

166. **Terajima, J., K. Tamura, K. Hirose, H. Izumiya, M. Miyahara, H. Konuma, and H. Watanabe.** 2004. A multi-prefectural outbreak of *Shigella sonnei* infections associated with eating oysters in Japan. *Microbiol. Immunol.* **48**:49–52.

167. **Thompson, F. L., T. Iida, and J. Swings.** 2004. Biodiversity of vibrios. *Microbiol. Mol. Biol. Rev.* **68**:403–431.

168. **Tree, J. A., M. R. Adams, and D. N. Lees.** 2003. Chlorination of indicator bacteria and viruses in primary sewage effluent. *Appl. Environ. Microbiol.* **69:**2038–2043.

169. **U.S. Environmental Protection Agency.** 2002. *Implementation Guidance for Ambient Water Quality Criteria for Bacteria, May 2002 Draft.* EPA-823-B-02-003. Office of Water, U.S. Environmental Protection Agency, Washington, D.C.

170. **U.S. Environmental Protection Agency.** 2005. *Microbial Source Tracking Guide Document.* EPA-600-R-05-064. Office of Research and Development, U.S. Environmental Protection Agency, Cincinnati, Ohio.

171. **Vidal, R., V. Solari, N. Mamani, X. Jiang, J. Vollaire, P. Roessler, V. Prado, D. O. Matson, and M. L. O'Ryan.** 2005. Caliciviruses and foodborne gastroenteritis, Chile. *Emerg. Infect. Dis.* **11:**1134–1137.

172. **Vilde, J. L., A. Huchon, M. Mignon, H. Scherrer, E. Bergogne-Berezin, and J. Pierre.** 1980. Typically appearing infection due to *Listeria monocytogenes* after ingestion of oysters. *Nouv. Presse Med.* **9:**3281. (In French.)

173. **Wallace, B. J., J. J. Guzewich, M. Cambridge, S. Altekruse, and D. L. Morse.** 1999. Seafood-associated disease outbreaks in New York, 1980–1994. *Am. J. Prev. Med.* **17:**48–54.

174. **Ward, B. Q., B. J. Carroll, E. S. Garrett, and G. B. Reese.** 1967. Survey of the U.S. Gulf Coast for the presence of *Clostridium botulinum. Appl. Microbiol.* **15:**629–636.

175. **Webster, L. F., B. C. Thompson, M. H. Fulton, D. E. Chestnut, R. F. Van Dolah, A. K. Leight, and G. I. Scotta.** 2004. Identification of sources of *Escherichia coli* in South Carolina estuaries using antibiotic resistance analysis. *J. Exp. Mar. Biol. Ecol.* **298:**179–195.

176. **Wetz, J. J., E. K. Lipp, D. W. Griffin, J. Lukasik, D. Wait, M. D. Sobsey, T. M. Scott, and J. B. Rose.** 2004. Presence, infectivity, and stability of enteric viruses in seawater: relationship to marine water quality in the Florida Keys. *Mar. Pollut. Bull.* **48:**698–704.

177. **Wilhelmi, I., E. Roman, and A. Sanchez-Fauquier.** 2003. Viruses causing gastroenteritis. *Clin. Microbiol. Infect.* **9:**247–262.

178. **Wilson, I. G., and J. E. Moore.** 1996. Presence of *Salmonella* spp. and *Campylobacter* spp. in shellfish. *Epidemiol. Infect.* **116:**147–153.

179. **Xu, X.-J., M. R. Ferguson, V. L. Popov, C. W. Houston, J. W. Peterson, and A. K. Chopra.** 1998. Role of a cytotoxic enterotoxin in *Aeromonas*-mediated infections: development of transposon and isogenic mutants. *Infect. Immun.* **66:**3501–3509.

180. **Yao, G. B.** 1989. A brief account of the 1988 seminar in Shanghai on viral hepatitis A. *Zhonghua Yi Xue Za Zhi* **69:**113–114. (In Chinese.)

Microbial Source Tracking
Edited by Jorge W. Santo Domingo and Michael J. Sadowsky

Statistical Issues in Microbial Source Identification

6

Jayson D. Wilbur and John E. Whitlock

In recent years, a number of different approaches to microbial source tracking (MST) have been proposed (51, 60, 64). Some approaches require cultivation of a target organism as a preliminary step before traits of interest can be measured. These methods are commonly referred to as "cultivation-dependent" approaches. By contrast, "cultivation-independent" approaches do not require cultivation, such as those that employ PCR to amplify a limited number of genetic markers which indicate the presence of a particular organism or trait of interest (5, 6).

Another distinction often made among the various approaches to source tracking is based on whether or not the method explicitly employs a statistical rule for source identification. Approaches which explicitly use statistical rules are commonly referred to as being "library dependent" because they identify the most likely source for a sample of unknown origin by comparing its traits to those of a collection, or library, of samples from each potential source category. Likewise, approaches to source tracking that do not require the use of an explicit statistical rule for source identification are considered to be "library independent."

It is worth noting that, from a statistical perspective, the distinction between library-dependent and library-independent approaches seems to be somewhat artificial. This is because even the so-called library-independent methods require the implicit use of a statistical rule for source identification. That is to say, even with approaches based on the measurement of a single trait that is highly specific for a particular source category of interest, there

JAYSON D. WILBUR, Department of Mathematical Sciences, Worcester Polytechnic Institute, Worcester, MA 01609-2280. JOHN E. WHITLOCK, Division of Mathematics and Sciences, Hillsborough Community College, Tampa, FL 33614.

is some probability of error. Further, as "library-independent" markers for more source categories are discovered, explicit statistical classification rules will be required to associate a sample of unknown origin to its most likely source using the multiple markers and to quantify the uncertainty in that identification.

Therefore, while the focus of this chapter is almost exclusively on statistical issues and procedures commonly employed in cultivation-dependent approaches to microbial source tracking, the material in this chapter is also relevant more broadly. However, due to the current focus on epidemiological methods by those employing cultivation-independent approaches to source tracking, there are relatively few references to these approaches and their applications found in this chapter.

Measuring Method Performance

Performance measures

While approaches to MST vary widely, the underlying problem of source identification is the same. The performance of these various methods should be evaluated and compared based on their relative sensitivity and specificity with respect to each potential contaminant source. The "sensitivity" of a method with respect to a particular source is the probability that a sample from that source will be correctly identified as originating from that source by the method. Likewise, the "specificity" of a method with respect to a particular source is the probability that a sample which is not from that source is correctly identified as not originating from that source. Thus, if we denote the relative frequency for the four possible results of a source identification using the notation in Table 1 then we can express sensitivity as the ratio $Se = TP/(TP + FN)$ and specificity as the ratio $Sp = TN/(TN + FP)$ (77, 78).

Methods with higher specificity and sensitivity for each potential contaminant source are preferred to methods with lower specificity and sensitivity. However, in practice, efforts to increase specificity of a particular method with respect to a given source will usually decrease its sensitivity with respect to that source. Conversely, efforts to increase the sensitivity of a method will decrease its specificity. For example, one can imagine a method for source identification which always identified the contaminant source as most likely being of human origin. This method would have 100% sensitivity with

Table 1 Four possible outcomes of source identification

Actual status	Test positive	Test negative
Positive	True positive (TP)	False negative (FN)
Negative	False positive (FP)	True negative (TN)

respect to human sources because all samples of human origin would be correctly identified. However, this method would have 0% specificity because no samples that were not of human origin would be correctly identified.

Thus, while admittedly an extreme case, this example emphasizes the fact that a balance of specificity and sensitivity is usually necessary. Furthermore, it should be noted that when there is more than one potential contaminant source, one method may not be highly specific with respect to every source; the same is true with sensitivity (57, 67). Therefore, if one is not constrained to using a single method for source identification, an approach which incorporates several methods would be preferable in order to take advantage of their relative specificities and sensitivities to various sources.

Two other useful measures of method performance are the positive and negative predictive value. The "positive predictive value" of a method with respect to a particular source is the probability that a positive test is a true positive. Likewise, the "negative predictive value" of a method with respect to a particular source is the probability that a negative test is a true negative. Thus, using the notation of Table 1, we can express the positive predictive value (PPV) by the ratio PPV = TP/(TP + FP) and the negative predictive value (NPV) by the ratio NPV = TN/(TN + FN). Unlike sensitivity and specificity, which together measure the accuracy of the test, the PPV and NPV are, in some sense, measures of confidence in source identification. However, it should be noted that because of the inherent relatedness of these four measures, there is no valid single-number summary for the performance of an approach to source tracking.

Estimating performance measures

In order to estimate the sensitivity, specificity, PPV, and NPV of a source-identification method with respect to each of several potential contaminant sources, a random sample of test isolates must be collected from each source category proportional to their prevalence in the environment. The importance of sampling test isolates proportional to prevalence cannot be emphasized enough, since any departure from this assumption will result in biased estimates of sensitivity, specificity, PPV, and NPV. Further, it is important to emphasize that these test isolates must not have been used at any point during the development of the source-identification method or, again, the estimators for these performance measures will be biased. Once the test isolates have been appropriately sampled, the source-identification method is employed and a table of cross-classifications based on the results is constructed. Table 2 displays some notation for cross-classification in the context of a study with four potential source categories. This notation will be used throughout this chapter.

Table 2 Notation for cross-classification for four source categories

Origin source	Classification source				
	1	2	3	4	Total
1	n_{11}	n_{12}	n_{13}	n_{14}	$n_1 = \sum_{j=1}^{4} n_{1j}$
2	n_{21}	n_{22}	n_{23}	n_{24}	$n_2 = \sum_{j=1}^{4} n_{2j}$
3	n_{31}	n_{32}	n_{33}	n_{34}	$n_3 = \sum_{j=1}^{4} n_{3j}$
4	n_{41}	n_{42}	n_{43}	n_{44}	$n_4 = \sum_{j=1}^{4} n_{4j}$
Total	$n_1^* = \sum_{i=1}^{4} n_{i1}$	$n_2^* = \sum_{i=1}^{4} n_{i2}$	$n_3^* = \sum_{i=1}^{4} n_{i3}$	$n_4^* = \sum_{i=1}^{4} n_{i4}$	$n = \sum_{i=1}^{4}\sum_{j=1}^{4} n_{ij}$

If we let n_{ij} denote the number of isolates from source i identified as coming from source j, then

$$n_i = \sum_{j=1}^{G} n_{ij}$$

is the total number of test isolates from source i,

$$n_j^* = \sum_{i=1}^{G} n_{ij}$$

is the total number of test isolates classified as coming from source j, and

$$n = \sum_{i=1}^{G}\sum_{j=1}^{G} n_{ij}$$

is the total number of test isolates, where G is the total number of potential source categories under consideration. Using this notation we can estimate Se_i, the sensitivity of the method with respect to source i, by the corresponding sample proportion:

$$\hat{Se}_i = \frac{\text{number of true positive tests for source } i}{\text{number of total samples from source } i} = \frac{n_{ii}}{n_i}$$

Furthermore, if we can assume that the collection of test isolates is a representative random sample from the environment, then the following 100(1 − α)% confidence interval for Se_i can be constructed in order to quantify the margin of error due to sampling variation:

$$\tilde{\text{Se}}_i - z_{1-\alpha/2}\sqrt{\frac{\tilde{\text{Se}}_i\left(1-\tilde{\text{Se}}_i\right)}{n_i}} \le \text{Se}_i \le \tilde{\text{Se}}_i + z_{1-\alpha/2}\sqrt{\frac{\tilde{\text{Se}}_i\left(1-\tilde{\text{Se}}_i\right)}{n_i}}$$

In the above expression, $z_{1-\alpha/2}$ denotes the 100(1 − α/2)th percentile of the standard normal distribution and

$$\tilde{\text{Se}}_i = \frac{n_{ii} + 0.5z_{1-\alpha/2}^2}{n_i + z_{1-\alpha/2}^2}$$

is a modified estimator for Se_i which results in improved interval estimates, particularly when n_i is small or Se_i is close to 1. Note that for a 95% confidence interval, (i.e., $\alpha = 0.05$), $z_{1-\alpha/2} = 1.96$ and

$$\tilde{\text{Se}}_i = \frac{n_{ii} + 0.5z_{1-\alpha/2}^2}{n_i + z_{1-\alpha/2}^2} \approx \frac{n_{ii} + 2}{n_i + 4}$$

(see reference 59).

Similarly, we can estimate Sp_i, the specificity of the method with respect to source i, by the corresponding sample proportion:

$$\hat{\text{Sp}}_i = \frac{\text{number of true negative tests for source } i}{\text{number of total samples not from source } i} = \frac{\left(n - n_i^*\right) - \left(n_i - n_{ii}\right)}{n - n_i}$$

Again, a 100(1 − α)% confidence interval for Sp_i can be constructed:

$$\tilde{\text{Sp}}_i - z_{1-\alpha/2}\sqrt{\frac{\tilde{\text{Sp}}_i\left(1-\tilde{\text{Sp}}_i\right)}{n - n_j}} \le \text{Sp}_i \le \tilde{\text{Sp}}_i + z_{1-\alpha/2}\sqrt{\frac{\tilde{\text{Sp}}_i\left(1-\tilde{\text{Sp}}_i\right)}{n - n_i}}$$

This interval, like the confidence interval for specificity described above, uses the biased estimator

$$\tilde{\text{Sp}}_i = \frac{\left(n - n_i^*\right) - \left(n_i - n_{ii}\right) + 0.5z_{1-\alpha/2}^2}{n - n_i + z_{1-\alpha/2}^2}$$

for the proportion Sp_i in order to decrease the width of the interval, while maintaining a 100(1 − α)% confidence level.

Positive and negative predictive values can also be estimated similarly by

$$\hat{P}pv_i = \frac{\text{number of true positive tests for source } i}{\text{number of total positive tests for source } i} = \frac{n_{ii}}{n_i^*}$$

$$\hat{N}pv_i = \frac{\text{number of true negative tests for source } i}{\text{number of total negative tests for source } i} = \frac{(n - n_i^*) - (n_i - n_{ii})}{n - n_i^*}$$

Additionally, the following confidence intervals can be constructed:

$$\tilde{P}pv_i - z_{1-\alpha/2}\sqrt{\frac{\tilde{P}pv_i(1 - \tilde{P}pv_i)}{n_i^*}} \le \tilde{P}pv_i \le \tilde{P}pv_i + z_{1-\alpha/2}\sqrt{\frac{\tilde{P}pv_i(1 - \tilde{P}pv_i)}{n_i^*}}$$

$$\tilde{N}pv_i - z_{1-\alpha/2}\sqrt{\frac{\tilde{N}pv_i(1 - \tilde{N}pv_i)}{n - n_i^*}} \le \tilde{N}pv_i \le \tilde{N}pv_i + z_{1-\alpha/2}\sqrt{\frac{\tilde{N}pv_i(1 - \tilde{N}pv_i)}{n - n_i^*}}$$

where

$$\tilde{P}pv_i = \frac{n_{ii} + 0.5z_{1-\alpha/2}^2}{n_i^* + z_{1-\alpha/2}^2}$$

and

$$\tilde{N}pv_i = \frac{(n - n_i^*) - (n_i - n_{ii}) + 0.5z_{1-\alpha/2}^2}{n - n_i^* + z_{1-\alpha/2}^2}$$

are modified estimators as before.

As an example of these calculations, consider the results of cross-classification obtained by Whitlock et al. (71) which are reproduced in Table 3. The original study used antibiotic resistance profiles to discriminate between four known sources of fecal contamination in the watershed of Stevenson Creek, Clearwater, Florida: cow, dog, human, and wildlife. It should be noted that this cross-classification was performed on the same isolates used to construct the source-identification method and no special effort was made to sample proportionally to the expected prevalence in the environment. Therefore, as mentioned previously, the estimates of the performance mea-

Table 3 Results of cross-classification for Whitlock data[a]

Origin	Classification of isolates				
	Cow	Dog	Human	Wildlife	Total
Cow	321	54	9	0	$n_1 = 384$
Dog	38	333	57	52	$n_2 = 480$
Human	102	195	702	169	$n_3 = 1{,}168$
Wildlife	6	48	169	231	$n_4 = 366$
Total	$n_1^* = 467$	$n_2^* = 630$	$n_3^* = 849$	$n_4^* = 452$	$n = 2{,}398$

[a]Data are from Whitlock et al. (71).

sures presented in Table 4 are somewhat biased. However, they suffice as a demonstration of the numerical calculations discussed in this section.

Classification rates

Unfortunately, relatively few source-tracking studies report more than one of these measures of "method performance" (33). Instead, measures of "classifier performance" are commonly reported, such as the average rate of correct classification (ARCC). As its name might imply, the ARCC is the average of the marginal rates of correct classification (RCCs) for each group of known origin. The RCC for source i is the proportion of source i isolates correctly identified as coming from source i. Using the notation defined previously and displayed in Table 2, the RCC for source i can be expressed as follows:

$$\text{RCC}_i = \frac{\text{number of source } i \text{ isolates correctly classified}}{\text{number of source } i \text{ isolates}} = \frac{n_{ii}}{n_i}$$

It should be emphasized that while the mathematical expression for RCC_i is identical to that of $\hat{\text{Se}}_i$ stated previously, the two estimators measure different quantities because of the collection of samples upon which they are based. $\hat{\text{Se}}_i$ is an estimator for the probability that the method will correctly identify a future environmental sample from source i and thus is computed based on a new random sample of isolates not previously used in constructing the classifier. By comparison, the RCC_i simply measures the performance of the classifier for the data at hand and is most useful for comparison with other classification rules under consideration.

The ARCC is simply the average of the RCC_is across the G different source categories.

Table 4 Estimation of specificity and sensitivity based on data in Whitlock et al. (71)[a]

	Sensitivity			**Specificity**			**Positive predictive value**			**Negative predictive value**		
Source	Estimate	95% LCL	95% UCL	Estimate	95% LCL	95% UCL	Estimate	95% LCL	95% UCL	Estimate	95% LCL	95% UCL
Cow	0.8359	0.7951	0.8698	0.9275	0.9153	0.9380	0.6874	0.6437	0.7295	0.9674	0.9584	0.9744
Dog	0.6938	0.6509	0.7334	0.8452	0.8282	0.8607	0.5286	0.4894	0.5676	0.9169	0.9030	0.9289
Human	0.6010	0.5726	0.6288	0.8805	0.8610	0.8975	0.8269	0.7998	0.8524	0.6992	0.6758	0.7215
Wildlife	0.6311	0.5803	0.6792	0.8912	0.8769	0.9041	0.5111	0.4649	0.5571	0.9306	0.9184	0.9411

[a]LCL, lower confidence limit; UCL, upper confidence limit.

$$\text{ARCC} = \frac{1}{G}\sum_{i=1}^{G} \text{RCC}_i = \frac{1}{G}\sum_{i=1}^{G} \frac{n_{ii}}{n_i}$$

ARCC is an unbiased estimator for the probability that an isolate from one of these four source categories is correctly classified under the assumption that it is equally likely to have come from any one of the source categories a priori.

A similar measure of classifier performance found in the source tracking literature is the expected rate of correct classification (ECC) (1). This measure that can be expressed by

$$\text{ECC} = \sum_{i=1}^{G} \frac{n_{ii}}{n}$$

is simply the proportion of all test isolates for which the method correctly identified the source. The ECC is an unbiased estimator for the probability of correct classification under the assumption that either (i) the proportional representation of each source category in the collection of test isolates roughly corresponds to the relative likelihood that each is the true source, a priori, or (ii) each isolate has the same probability of being correctly identified, regardless of its true origin.

It may be useful to note that when the number of test isolates from each of the G source categories is the same (i.e., $n_i = n/G$), the ARCC and ECC are equivalent.

$$\text{ARCC} = \frac{1}{G}\sum_{i=1}^{G} \frac{n_{ii}}{n_i} = \frac{1}{G}\sum_{i=1}^{G} \frac{n_{ii}}{n/G} = \frac{1}{G}\sum_{i=1}^{G} \frac{Gn_{ii}}{n} = \sum_{i=1}^{G} \frac{n_{ii}}{n} = \text{ECC}$$

In practice, the cross-classifications used to estimate either the ARCC or the ECC should be based on cross-validation (1, 75) rather than simple resubstitution (11). Further discussion of this and other topics related to discriminant analysis are addressed in detail later in the chapter.

One other performance measure used in some source-tracking studies (71, 75) is the average rate of misclassification (ARMC), which, analogous to the ARCC, is the average of the marginal rates of misclassification (RMCs) as defined by

$$\text{RMC}_i = \frac{\text{number of isolates incorrectly classified as source } i}{\text{number of isolates not classified as source } i} = \frac{n_i^* - n_{ii}}{n - n_i}$$

for source category i.

$$\text{ARMC} = \frac{1}{G}\sum_{i=1}^{G}\text{MC}_i = \frac{1}{G}\sum_{i=1}^{G}\frac{n_i^* - n_{ii}}{n - n_i}$$

Whitlock et al. (71) also introduce a rather conservative confidence interval ARMC ± 4SD_{ARMC}, where

$$\text{SD}_{\text{ARMC}} = \frac{1}{G-1}\sum_{i=1}^{G}(\text{MC}_i - \text{ARMC})^2 .$$

The ARMC was intended as a threshold for considering isolates of unknown origin as true positives and is loosely based on the average rate at which isolates are misclassified into various source categories and is buffered by the SD for a more conservative limit to prevent false positives. However, the PPV described previously is a preferred measure and could be modified to adjust for disproportionate representation from various sources. As a numerical example of these measures we apply the ARCC, ECC, and ARMC to the Whitlock et al. (71) data.

$$\text{ARRC} = \frac{1}{4}\sum_{i=1}^{4}\text{RRC}_i = \frac{1}{4}\sum_{i=1}^{4}\frac{n_{ii}}{n_i} = \frac{1}{4}\left(\frac{321}{384} + \frac{333}{480} + \frac{702}{1{,}168} + \frac{231}{366}\right) = 0.6905$$

$$\text{ECC} = \sum_{i=1}^{4}\frac{n_i}{n} = \frac{321 + 333 + 702 + 231}{2{,}398} = 0.6618$$

$$\text{ARMC} \pm 4\text{SD}_{\text{ARMC}} = \frac{1}{4}\sum_{i=1}^{4}\text{MC}_i \pm \frac{4}{3}\sum_{i=1}^{4}(\text{MC}_i - \text{ARMC})^2$$

$$= 0.1139 \pm 4(0.0339)$$

These coincide with the numerical results reported in Whitlock et al. (71).

Sampling Considerations

Regardless of the approach to be used for source identification, a representative random sample of isolates must be collected from each potential source category. For some methods, such as those which target a single host-specific genetic marker, these samples are only needed in order to determine expected levels of method performance (i.e., sensitivity, specificity, etc.) with respect to each of the potential contaminant sources. For other methods, an initial collection of representative samples from each potential source category, the so-called library, is also required so that statistical rules for source identification can be established.

This task is challenging, as one must not only construct a library that is large and diverse enough to be representative of the population being measured but must also consider the ecology of a diverse group of indicator organisms. In this section, we discuss several sources of variation in populations of indicator organisms as they relate to sampling.

Temporal and spatial variation

When constructing a library of indicators, the temporal and spatial (including geographic) variation in indicator populations in the area from which samples of unknown isolates will eventually be taken is perhaps one of the most important considerations (68). There is no evidence that the most commonly used indicators such as *Escherichia coli* and the enterococci are species-specific over the entire geographic range of the host species, regardless of whether the method used is genotypic or phenotypic. Some concerns of this nature, especially regarding *E. coli*, have been addressed by Gordon (26).

Many source tracking studies use libraries constructed from isolates proximate to the areas sampled (32, 71, 75). Wiggins et al. (75) found that small libraries tended to result in high rates of correct classification but have limited generalizability to a broad geographic range. Larger libraries, on the other hand, may become more representative of larger geographic populations, even though the reported rate of correct classification may be lower.

The results of Wiggins et al. (75) also suggested that the temporal variation of bacterial populations is also problematic for library-based methods since libraries may need to be continually updated to reflect changing populations. Representative testing, such as that employed by Wiggins et al. (75), is an essential step in ensuring that the library being constructed will reflect natural populations of indicators as much as possible. Studies on the phenotypic and/or genetic structure of the indicator populations should precede the construction of libraries or at least should be conducted during the sampling phase to aid in assessing the size of the samples collected from hosts.

Differential survival

Correct application of MST techniques requires knowledge of biological aspects of the indicator organisms including longevity, diversity, and distribution in natural ecosystems. For example, differential survival or culturability of various *Enterococcus* species (47) may influence the species composition of indicator bacteria in natural waters, complicating the use of the enterococci as predictors of particular host sources of pollution. The potential for prolonged survival, resuspension, and even growth of indicators in sediments and soils (24, 52, 58, 62, 66) may also cause difficulties in distinguishing between recent and preexisting fecal contamination.

Environmental waters are not the primary habitat for fecal indicator organisms and, thus, present challenges to survival (7). For example, survival of *E. coli* for weeks to months outside of host animals has been demonstrated repeatedly, and reported survival rates vary a great deal depending upon conditions and strains examined (13, 21, 43, 55, 62). The passage of the organisms from the host gastrointestinal tract to a different external environment may alter the composition of the surviving population (26, 72). Moreover, because some strains of *E. coli* may be better suited to survive out of the host in various microhabitats, they would be sampled more frequently over time compared to strains that die off quickly outside the host. The bacteria that are analyzed by source tracking methods will, in many instances, have been in the environment long enough for such selection to take place, having been exposed to several microhabitats (i.e., feces, wastewater collection system, soil, sediment, or water column) before being sampled. Populations of indicator bacteria in the environment may therefore differ from the original populations within the host (27, 72). If source tracking methods are to be reliable, this selection process should be reflected in libraries of isolates from known sources that are used to classify isolates from water.

Process variation

Both genotypic and phenotypic MST methods may rely on processing groups of samples together in batches in the laboratory. While ideally there will be no added variation due to a batch effect, many source tracking methods that have been reported in the literature have a considerable amount of error associated with correctly classifying samples. Processing the samples along with positive controls or specimens with a known and consistent fingerprint is recommended to identify batch variation. These controls may also be useful for indicating that the procedure is working properly, as well as for allowing for normalization of digital images used in band-matching processing. It is also prudent to process samples in randomized blocks to minimize the effect of batches on fingerprints. For example, when specimens are initially fingerprinted to build a database, the gels should contain some specimens from each library source (e.g., cow, dog, human). Consequently, the batch effect will not be confounded with the source-to-source variation.

Measuring representativeness

Diversity measurements of indicator populations in various hosts may also be used to test representativeness (3). Each host species from which samples are taken may support significantly different diversity indicators. Consequently, the number of indicators required to be representative of each host may be different. Anderson et al. (3) found that individual horses have significantly diverse populations of *E. coli* bacteria based on ribotype profiles. Accumulation curves (8, 46) were created that reflect the number of ribotype patterns observed related to sample size. As sample sizes of *E. coli* increased, new patterns continued to be observed in some host species (e.g., horse), while other hosts' populations reached an early plateau. This implies that the number of samples required from individual hosts may vary depending on the host.

Some studies have constructed libraries using composite samples (i.e., using a mixture of fecal material) of several individuals (75). This is an attempt to randomly select isolates from a large number of host individuals with the rationale that it is more congruent with real sources of contamination. A more statistically rigorous procedure would determine the sample size required from each individual using an accumulation curve and collect that many isolates from each of a sufficient number of individuals, again using an accumulation curve.

Collecting water samples to isolate indicators from unknown sources requires true representation in order to maximize the sensitivity of the analysis. However, the samples must be collected properly to help yield specificity of the analysis. Ideally, samples should be taken in replicate across the area from which contamination is to be inferred. A single sample of water taken from a single point is statistically representative of that point and not the water body. Replication should include samples randomly taken throughout the water body. If possible, these samples should be processed separately to yield truly replicated results. It may be found that various locations throughout the water body are contaminated with different sources of indicators. Composite samples are undesirable as they constitute sacrificial pseudoreplication and, while increasing sensitivity, may diminish specificity.

Discriminant Analysis

This section presents a critical review of statistical techniques for discriminant analysis in microbial source tracking. Due to the frequent misuse of the term "discriminant analysis" in the source tracking literature (31, 57), it is first necessary to provide a proper definition of the term. Discriminant analysis is a term used to refer to the broad collection of all techniques for the definition and description of group separation and for the construction and evaluation of rules for assigning samples of unknown origin to predefined groups (40). Thus, by this definition, discriminant analysis includes several tasks commonly performed in the course of a source tracking study, including definition of source categories, description of group separation, and construction and assessment of rules for source identification.

In the context of source tracking, initial definition of source categories is generally not a statistical problem but rather one of land use. However, the description of this group separation either graphically (16, 34) or through the use of hypothesis testing (1, 18, 32, 71) is an often-neglected element of discriminant analysis in source tracking. Conducting an initial evaluation of category separation based on the observed data is essential, since it is unlikely that an effective rule for source identification can be constructed if there is not sufficient group separation. Statistical techniques for describing group separation can generally be categorized as being either exploratory or confirmatory. Exploratory analyses aim to identify patterns of variation in the data relevant to assumptions and hypotheses, as well as any striking deviations from this general pattern (e.g., outliers). By contrast, confirmatory analyses consist of more-formal tests of the validity of these assumptions and hypotheses.

Common techniques for exploratory discriminant analysis include Fisher's linear discriminant analysis (22), multidimensional scaling (69), and other techniques for cluster analysis (42). Confirmatory discriminant analyses generally involve techniques for testing of hypotheses of group difference, such as a multivariate analysis of variance (40, 76).

Exploratory discriminant analysis

The goal of Fisher's linear discriminant analysis (FLDA), also called "canonical discriminant analysis," is to construct a small number of new variables which contain most of the information relevant to discrimination between predefined groups in the full set of original variables. That is to say, if the original variables, such as binary indicator profiles of antibiotic resistance, are represented by $X_1, X_2, \ldots, X_p$, FLDA constructs linear combinations of the form $Y = a_1X_1 + a_2X_2 + \ldots + a_pX_p$ such that the variable weights $a_1, a_2, \ldots, a_p$ are chosen to maximize the discriminatory power of Y as measured by the ratio of its between-group to within-group sums of squares (i.e., the typical "F ratio"), subject to certain constraints on the weights (40). The resulting variables can then be plotted in two or three dimensions for graphical assessment of the degree of group separation. This procedure is useful in the context of source tracking to evaluate the degree of separation between groups of samples from different source categories (16) or groups of samples from the same source category collected at different times or under different conditions. It should be noted that while FLDA is most commonly used as a graphical tool in exploratory analyses, procedures for source identification based on FLDA do exist.

It is important to note that while this procedure makes no explicit distributional assumption for the data, it does characterize the data for each group and, consequently, characterizes group separation in terms of means and a matrix of pairwise covariances which is assumed to be the same for all groups, which would be an appropriate assumption for multivariate normal data. This is not necessarily a good assumption for many types of source-tracking data. However, its performance as a method for graphically exploring the degree of group separation is not strongly affected by mild departures from this assumption.

An alternative technique for exploratory discriminant analysis which does not make as many assumptions about the data is multidimensional scaling (MDS). MDS is a technique for graphically representing a matrix of dissimilarities (or distances) in relatively few dimensions.

In the context of microbial source tracking, MDS plots are based on a matrix of numerical interisolate dissimilarity measures. Patterns of interisolate variation can be represented in a two- or three-dimensional plot in which distances between points are roughly proportional to the dissimilarity between the isolates they represent. It should be noted that, unlike FLDA, MDS does not take into account any assumed or hypothesized groupings, such as source categories. Thus, an MDS plot which exhibits strong grouping by source category is a good indication that source identification will be possible, since no group structure is imposed a priori as with FLDA. Figure 1 displays FLDA and MDS plots of the same data for comparison. The FLDA plot shows that the linear discriminant functions can separate the three groups when they are known and the MDS plot shows considerable natural separation between the three groups, even when they are not taken into account in the analysis.

The underlying assumption of MDS is that the measure of dissimilarity is a valid measure of interobject distance in the context of the problem. Common choices for similarity measurements among binary profiles include the similarity coefficients of Jaccard (38) or Dice (15) or the matching coefficient (65). Similarities between quantitative profiles, such as densitometric curves, are quantified in terms of correlation (54), Mahalanobis distance (48), or simple Euclidean distance. The properties and appropriateness of these measures for MDS are discussed further in the following section.

Measures of similarity and distance. Most similarity coefficients for the binary profiles can be expressed in terms of four quantities, namely, the number of 1-1 matches, denoted by a, the number of 1-0 mismatches, denoted by b, the number of 0-1 mismatches, denoted by c, and the number of 0-0 matches,

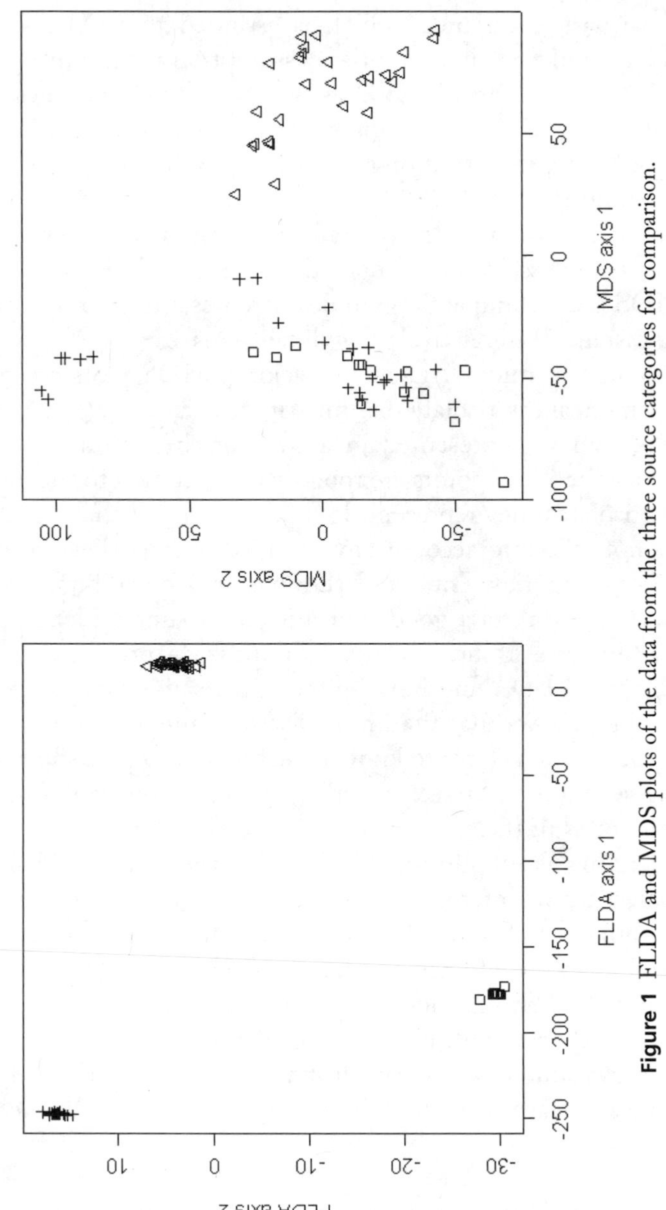

Figure 1 FLDA and MDS plots of the data from the three source categories for comparison.

Table 5 Common notation for similarity coefficients between binary profiles

Profile 1	Profile 2	
	1	0
1	a	b
0	c	d

denoted by d. Using this notation, which is summarized in Table 5, the three coefficients mentioned previously are given the following expressions:

$$\text{Jaccard} = \frac{a}{a+b+c} \qquad \text{Dice} = \frac{a}{a+b+c} \qquad \text{Matching} = \frac{a+d}{a+b+c+d}$$

Each of these coefficients ranges in value from 0 to 1. Values near 1 indicate extreme similarity and values near 0 indicate extreme dissimilarity. In order to evaluate the validity of each coefficient for a particular collection of indicators, it is helpful to consider certain extreme situations. For example, when there are no mismatches (i.e., $b = c = 0$), each of these coefficients equals 0. However, when there are no 1-1 matches (i.e., $a = 0$), the matching coefficient is not necessarily equal to 0. This may indicate that the matching coefficient is an invalid measure of similarity in the context of some analyses. Another consideration is whether or not the corresponding dissimilarity measure (generally, 1 minus the similarity measure) satisfies the properties of a distance metric (i.e., symmetry, triangle inequality, etc.). In particular, it should be noted that the coefficient of Dice does not satisfy these conditions. Therefore, while the Dice coefficient might be a valid measure of similarity in certain situations, its use for MDS may result in some minor distortion. Shi (63) and Duarte et al. (17) make a much more extensive comparison of these coefficients and several alternative measures of similarity for binary profiles. Likewise, the common similarity and distance measures for quantitative profiles can be expressed in terms of the profile of one sample $X = (X_1, X_2, \ldots, X_p)$ and another $Y = (Y_1, Y_2, \ldots, Y_p)$ and certain summary measures like the sample mean vectors, denoted $\overline{X}$ and $\overline{Y}$, and the sample covariance matrix denoted by S.

$$\text{Correlation} = \frac{\sum_{k=1}^{p}\left(X_k - \overline{X}\right)\left(Y_k - \overline{Y}\right)}{\sqrt{\sum_{k=1}^{p}\left(X_k - \overline{X}\right)^2 \sum_{k=1}^{p}\left(Y_k - \overline{Y}\right)^2}}$$

$$\text{Mahalanobis} = \sqrt{(X-Y)' S^{-1} (X-Y)}$$

$$\text{Euclidean} = \sqrt{\sum_{k=1}^{p} (X_k - Y_k)^2}$$

Confirmatory discriminant analysis

An exploratory analysis should usually be sufficient to establish whether or not the source categories can be distinguished based on the observed variables. However, it is also important to confirm whether or not there is significant separation between groups or whether or not patterns of spatial or temporal dependence are significant. Multivariate analysis of variance (MANOVA) can be used to test for significant spatial or temporal effects across multiple groups. Confirmatory analyses such as these have been applied in the context of source tracking to determine whether or not there is a significant difference between each pair of groups (71).

However, in order for valid conclusions to be drawn from a typical MANOVA, data must satisfy the assumption of multivariate normality. This assumption is rarely met by MST data and therefore resampling-based methods for MANOVA (2) or a nonparametric approach (1) which do not assume multivariate normality are preferable. Unfortunately, this methodology is not currently available in standard software packages. Therefore, it should be emphasized that results of standard MANOVA must be interpreted with caution.

Source identification

Statistical methods for predictive discriminant analysis are commonly referred to as classification rules. This section describes the general process of constructing and evaluating classification rules in the context of microbial source tracking. Certain types of classification rules and their corresponding assumptions are more appropriate for different types of MST data, but the same general process should be followed in the construction and assessment of classification rules regardless of the type of data (35):

(i) Randomly divide data into training isolates (~50%), validation isolates (~25%), and test isolates (~25%).
(ii) Use the training isolates to construct various classification rules.
(iii) Estimate the accuracy of each rule by attempting to classify the validation isolates.

(iv) Select the most accurate rules and refine them. Refinement techniques include variable selection and the adjustment of tuning parameters.

(v) Once a (single) best rule has been selected, use the test isolates to estimate generalizability of the rule. This step is important for estimating how well the classification rule will work in application in the environment.

It should be noted that appropriate selection and application of statistical analyses can have a significant impact on study conclusions (57, 67). Therefore, it is important to provide a reasonable justification and detailed description of any statistical methods employed, in order to provide a basis for evaluation of the significance and validity of results and conclusions. In some published source-tracking studies, statistical analyses are described in sufficient detail to enable others to follow a similar approach and, here, to permit a critical review. However, in other cases, crucial details of statistical methods are not stated in sufficient detail to provide a context for the interpretation of results or the evaluation of conclusions. Furthermore, most authors justify their choice of statistical methodology by simply citing earlier application of the same methodology to a similar problem (e.g., references 11 and 31). This practice, while common, is improper, since it allows the use of inappropriate statistical methodology to persist.

In reading this review it is important to keep in mind that, over time, there has been a general trend of improvement in the selection of appropriate methods for statistical analyses. However, because selection of statistical methods is often improperly justified by citation of an earlier inappropriate application, some older articles frequently cited as justification for statistical methods are included in this review.

Model-Based Classification Rules

The two most common phenotypic approaches to microbial source tracking are based on traits of antibiotic resistance (45, 73) and carbon source utilization (31, 37). In most studies using phenotypic approaches for source identification, linear discriminant functions are used to identify the most likely source for an isolate of unknown origin (11, 18, 28, 30–33, 70, 71, 73–75). Linear discriminant analysis is a discriminant procedure based on the assumption of multivariate normality. That is to say, it assumes the phenotypic profiles from each source follow a multivariate normal distribution. Implicit in this assumption is the notion that the dependence structure among the variables is fully characterized by a matrix of pairwise covariances. Linear discriminant analysis makes the additional assumption that this matrix of pairwise covariances is the same for data from each source.

The most common approach to classification using linear discriminant functions measures the distance from the profile of the isolate of unknown origin to the average profile for each source. This distance can be expressed in mathematical notation as

$$D_i^2\left(x_{\text{unknown}}\right)=\left(x_{\text{unknown}}-\overline{x_i}\right)' S_i^{-1}\left(x_{\text{unknown}}-\overline{x_i}\right)$$

where x_{unknown} represents the profile of an isolate of unknown origin, $\overline{x_i}$ is the mean profile for source i, and S is an estimate of the common covariance matrix. The procedure then identifies the most likely source of the unknown isolate as the one for which $D_i^2\ (x_{\text{unknown}})$ is smallest.

It should be noted that this procedure is commonly referred to in the source tracking literature simply as "discriminant analysis" (DA) (31). This use of terminology is incorrect because it does not fully communicate the nature of the procedure performed. As mentioned previously, the statistical term "discriminant analysis" refers more generally to the collection of all statistical methods for distinguishing between groups. The more descriptive name "linear discriminant analysis" (LDA) should be used to describe this method. It should also be noted that LDA has been confused with Fisher's linear discriminant analysis (FLDA) in the source tracking literature (11, 18). The source of this confusion may be the equivalence of LDA and FLDA when there are only two source categories. However, when there are more than two categories, the results of the two methods can differ (56).

The assumptions of linear discriminant analysis are rarely met by source tracking data. First, the assumption of a common covariance matrix for different categories is unlikely to hold. For example, resistance to two antibiotics might be positively correlated for one source category and negatively correlated for another source category. Additionally, because the variance of a binary variable can be expressed as a function of its mean (i.e., variance = mean [1 − mean]), when binary profiles are used for antibiotic resistance analysis, the assumption that resistance profiles differ between source categories contradicts the assumption of a common covariance matrix. For example, if 90% of isolates from human sources are resistant to a certain antibiotic and only 40% of isolates from wildlife sources are resistant to the same antibiotic, then this trait will help to distinguish human and wildlife sources. However, this difference implies that the variance differs between the two sources:

$$\text{Variance}_{\text{Human}} = \text{Mean}_{\text{Human}}\ (1 - \text{Mean}_{\text{Human}}) = 0.90\ (1 - 0.90) = 0.09$$

and

$$\text{Variance}_{\text{Wildlife}} = \text{Mean}_{\text{Wildlife}} (1 - \text{Mean}_{\text{Wildlife}}) = 0.40 (1 - 0.40) = 0.24$$

which violates the assumption of a common covariance matrix across sources.

Quadratic discriminant analysis (QDA) is a discriminant procedure similar to linear discriminant analysis in that it assumes profiles from each source follow a multivariate normal distribution but does not assume a common covariance matrix for each source category. As in linear discriminant analysis, the distance is measured from the profile of the isolate of unknown origin to the average profile for each source. This distance can be expressed in mathematical notation as

$$D_i^2\left(x_{\text{unknown}}\right) = \left(x_{\text{unknown}} - \overline{x_i}\right)' S_i^{-1} \left(x_{\text{unknown}} - \overline{x_i}\right)$$

where x_{unknown} again represents the profile of an isolate of unknown origin, $\overline{x_i}$ is the mean profile for source i, but S_i is an estimator for the covariance matrix for profiles from source i. As before, the most likely source is the one for which $D_i^2\,(x_{\text{unknown}})$ is smallest.

While linear and quadratic discriminant analysis are both fairly robust to mild departures from the multivariate normal distribution, it should be noted that the assumption of multivariate normality cannot hold for binary profiles commonly used for antibiotic resistance analysis (73). Additionally, LDA and QDA both use estimates of covariance matrices that tend to be poor unless sample sizes are very large and there are no extreme outlying observations. Thus, classification rules based on linear discriminant analysis and quadratic discriminant analysis can often perform poorly when sample sizes are not very large.

Logistic discrimination is an alternative to linear and quadratic discriminant analyses which does not assume multivariate normality or a common covariance matrix and does not require an estimate of the covariance matrix. Instead, logistic discriminant analysis models the log posterior odds that an isolate comes from source category i as a linear combination of the variables that make up the phenotypic profile:

$$\log\left(\frac{\pi_i}{1-\pi_i}\right) = \sum_{k=1}^{p} \beta_{ik} x_{\text{unknown}[k]}$$

The procedure then identifies the most likely source as the one for which the estimated posterior probability of group membership

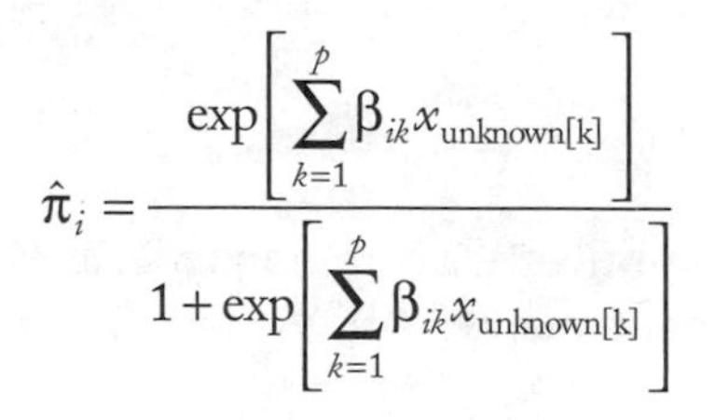

$$\hat{\pi}_i = \frac{\exp\left[\sum_{k=1}^{p} \beta_{ik} x_{\text{unknown}[k]}\right]}{1 + \exp\left[\sum_{k=1}^{p} \beta_{ik} x_{\text{unknown}[k]}\right]}$$

is largest. Logistic discrimination has been used successfully in the context of source tracking with carbon source utilization profiles (33) and antibiotic resistance profiles (57). Generally, logistic discrimination and LDA will produce similar results. However, logistic discrimination is preferred when there are extreme outlying observations. This is because outliers have a strong influence on the covariance estimates used in LDA but are naturally downweighted by the logistic model, which gives more weight to points near the decision boundary (19, 35).

Several other model-based statistical procedures for discriminant analysis and classification exist in the statistical literature (35, 49). Some of these methods, such as penalized discriminant analysis (1) and support vector machines (79), have been employed with success in the analysis of source-tracking data. The relative success of different procedures depends primarily on the shape of the optimal boundary between groups, with some of these methods outperforming methods like LDA when boundary shapes are more complex than the separating hyperplane implied by LDA. However, algorithms for these are not readily available in the software commonly used for microbial source tracking but rather in more mathematically and statistically oriented software packages.

Feature Selection

It is often the case that the performance of a statistical discriminant procedure can be improved by excluding some information which does not aid in discrimination (i.e., noise). This phenomenon has been observed for phenotypic methods in the context of antibiotic resistance analysis (28) and carbon source utilization analysis (31), and for genotypic methods, where Albert et al. (1) utilized a subset of the coefficients from a wavelet decomposition of repetitive extragenic palindromic-PCR profiles. This process of feature selection can lead to improved classifier performance by reducing the chance of overfitting the statistical discriminant model to the noise in the observed data (35, 57). Additionally, it has been pointed out that reducing the number of antibiotics and concentrations in an antibiotic resistance analysis can reduce costs significantly (18).

Several algorithms for variable selection in discriminant analysis exist, and some are available in commonly used statistical software packages such as SAS. Use of such algorithms reduces the chance of user error and reduces the time required for analysis as compared to a trial-and-error approach (28, 31). Stepwise procedures are essentially well-defined, automated versions of a trial-and-error approach (e.g., SAS Proc STEPDISC). However, procedures which examine all possible subsets of variables are preferred (e.g., "Selection" in SAS Proc LOGISTIC), because stepwise procedures are not guaranteed to even consider the optimal subset of variables.

Similarity-Based Classification Rules

The most common genotypic approaches to microbial source tracking are based on DNA fingerprinting techniques such as repetitive extragenic palindromic-PCR (10, 16, 29, 39), ribotyping (9, 53), and pulsed-field gel electrophoresis profiles (41). Most studies using these genotypic approaches for source identification employ similarity-based classification rules to identify the most likely source for an isolate of unknown origin (1, 34, 36, 39, 41, 50). These classification rules do not assume any explicit form for the data distribution, such as multivariate normality. Instead they simply assume that (in some particular sense) isolates with similar profiles are likely to come from the same source.

There are several varieties of similarity-based rules, but in general they share the characteristic of classifying isolates of unknown origin based on the source(s) of those most similar for which the origin is known. The three most common types of similarity-based rules in the source-tracking literature are the maximum-similarity, average-similarity, and *k*-nearest-neighbor rules (1, 34, 40, 57). The maximum-similarity classification rule simply assigns isolates of unknown origin to the source of the most similar isolate in the library. The average-similarity classification rule measures the similarity between the isolate of unknown origin and all isolates of known origin and then assigns the unknown isolate to the source with which the unknown isolate profile has the highest average similarity.

The *k*-nearest-neighbor classification rule (12, 23) is somewhat of a compromise between these two methods. For some specified value of k, the k most similar objects are identified and the isolate of unknown origin is assigned to the source with the largest representation among the k nearest neighbors. It should be noted that the maximum-similarity classification rule is generally equivalent to the $k = 1$ nearest-neighbor classification rule. Surprisingly little research has been conducted regarding the choice of the value of k. However, for the simple case of two multivariate normal populations of comparable group sizes, Enas and Choi (20) recommend selecting k

to be approximately between $n^{2/8}$ and $n^{3/8}$, depending on whether there are small or large differences between the group covariance matrices. So, even for sample sizes of $n = 1{,}000$, the recommended value of k is somewhere between 5 and 13. Thus, a large number of neighbors is not advisable. Devroye et al. (14) also provide some results on the optimal selection of k.

It is also important to note the biological assumptions implied by each of these similarity-based classification rules. First, the average-similarity rule corresponds to a population genetics approach, much like LDA and QDA. The k-nearest-neighbor approach instead assumes that isolates with similar profiles are likely to have originated from the same source. Using values of k greater than 1 implies that isolates with a slightly different genotype than the isolate of unknown origin are likely to be from the same source category. Thus, the maximum-similarity rule corresponds to the assumption that only the isolate with the most similar genotypic profile contains any information about the source category of the isolate of unknown origin.

Thresholding

In certain situations, however, the closest match is not similar enough to provide reliable source identification. Thus, several methods for evaluating the quality and reliability of source identifications have been proposed. The software package BioNumerics (Applied Maths, Belgium) offers its users two ad hoc procedures to evaluate the quality of source identifications, which are described in detail by Ritter et al. (57). Its "bootstrap ID" and "quality factors" have been employed in source-tracking studies to reduce the number of false identifications (34, 57, 67). However, the statistical properties of these measures have not been fully investigated. Therefore, no critical evaluation of them is given in this chapter, but it is left as an opportunity for further statistical research.

The use of a threshold, or reject option, in classification is a better-studied method for improving the reliability of source identifications (24). In a general sense, this technique simply requires a certain "threshold" of evidence before a source identification is permitted. Otherwise, no identification is made, in order to reduce the likelihood of a false-positive identification. The practice of "epidemiological matching" is an application of thresholding to maximum-similarity classification. Specifically, if the most similar isolate of known origin is less similar than the threshold, then no identification is made. The threshold is usually based on the level of internal similarity among replicate isolates in the library.

Software for Discriminant Analysis

Most of the techniques and tools for discriminant analysis discussed in this chapter require software for implementation. Several different software packages have been used in source-tracking studies, including MATLAB (1), R (79), SAS-JMP (28), and MINITAB (18). In addition, the BioNumerics system (9, 16, 39, 50) and DiversiLab (36) are particularly attractive because they include tools for image analysis, data management, and discriminant analysis. However, the most commonly used software package for discriminant analysis is SAS (11, 53, 60, 61, 71, 73–75). It is likely that SAS is used more often than other packages because of its wide availability and flexibility. For example, SAS procedures can be used to perform nearly all analyses described in this chapter, including LDA and QDA (Proc DISCRIM), logistic discrimination (Proc LOGISTIC), FLDA (Proc CANDISC), MDS (Proc MDS), and hierarchical clustering (Proc CLUSTER) (44).

SUMMARY AND FUTURE DIRECTIONS

This chapter reviews current statistical practice in microbial source tracking with particular attention to evaluation of method performance, sampling considerations, and discriminant analysis. In the area of evaluating method performance, the terminology and methodology commonly used to evaluate diagnostic medical tests were introduced as a common framework for evaluating the performance of methods with respect to source categories. Measures currently used to evaluate the performance of methods for source identification were also presented, but there was little correspondence between these measures and quantities commonly of interest in the evaluation of diagnostic testing.

In the area of sampling, considerable work has been done to understand various sources and patterns of variation in populations of indicator organisms. Continuing work is needed to more fully understand these sources of variation so that appropriate sampling plans can be developed to obtain representative random samples from each potential source category for use in the evaluation of method performance and discriminant analysis.

Lastly, methods for addressing exploratory, confirmatory, and predictive discriminant analysis were presented. A wide variety of classification rules have been employed for source identification, and recent comparison studies (34, 57, 67) have demonstrated that method selection can have a substantial impact on source identification. Recent ad hoc procedures to evaluate and improve the reliability of source identifications were also discussed. However, the development of valid measures of reliability and confidence in source identifications is an important problem requiring further statistical research.

ACKNOWLEDGMENTS

We thank the book editors as well as Luis Tenorio of the Colorado School of Mines and Kerry J. Ritter of the Southern California Coastal Water Research Project for their valuable comments and suggestions.

REFERENCES

1. **Albert, J. M., J. Munakata-Marr, L. Tenorio, and R. L. Siegrist.** 2003. Statistical evaluation of bacterial source tracking data obtained by rep-PCR DNA fingerprinting of *Escherichia coli*. *Environ. Sci. Technol.* **37:**4554–4560.
2. Reference deleted.
3. **Anderson, M. A., J. E. Whitlock, and V. J. Harwood.** 2005. Frequency distributions of *Escherichia coli* subtypes in various fecal sources: application to bacterial source tracking methods. *Appl. Environ. Microbiol.* **71:**3041–3048.
4. Reference deleted.
5. **Bernhard, A. E., and K. G. Field.** 2000. A PCR assay to discriminate human and ruminant feces on the basis of host differences in *Bacteroides-Prevotella* genes encoding 16S rRNA. *Appl. Environ. Microbiol.* **66:**4571–4574.
6. **Bernhard, A. E., T. Goyard, M. T. Simonich, and K. G. Field.** 2003. Application of a rapid method for identifying fecal pollution sources in a multi-use estuary. *Water Res.* **37:**909–913.
7. **Bissonnette, G. K., J. J. Jezeski, G. A. McFeters, and D. G. Stuart.** 1975. Influence of environmental stress on enumeration of indicator bacteria from natural waters. *Appl. Microbiol.* **29:**186–194.
8. **Bunge, J., and M. Fitzpatrick.** 1993. Estimating the number of species: a review. *J. Am. Stat. Assoc.* **88:**364–373.
9. **Carson, C. A., B. L. Shear, M. R. Ellersieck, and A. Asfaw.** 2001. Identification of fecal *Escherichia coli* from humans and animals by ribotyping. *Appl. Environ. Microbiol.* **67:**1503–1507.
10. **Carson, C. A., B. L. Shear, M. R. Ellersieck, and J. D. Schnell.** 2003. Comparison of ribotyping and repetitive extragenic palindromic-PCR for identification of fecal *Escherichia coli* from humans and animals. *Appl. Environ. Microbiol.* **69:**1836–1839.
11. **Choi, S., W. Chu, J. Brown, S. J. Becker, V. J. Harwood, and S. C. Jiang.** 2003. Application of enterococci antibiotic resistance patterns for contamination source identification at Huntington Beach, California. *Mar. Pollut. Bull.* **46:**748–755.
12. **Dasarathy, B. V.** 1991. *Nearest Neighbor: Pattern Classification Techniques.* IEEE Computer Society Press, Los Alamitos, Calif.
13. **Davies, C. M., J. A. Long, M. Donald, and N. J. Ashbolt.** 1995. Survival of fecal microorganisms in marine and freshwater sediments. *Appl. Environ. Microbiol.* **61:**1888–1896.
14. **Devroye, L., L. Gyorfi, and G. Lugosi.** 1996. *A Probabilistic Theory of Pattern Recognition.* Springer-Verlag, New York, N.Y.
15. **Dice, L. R.** 1945. Measures of the amount of ecological association between species. *Ecology* **26:**297–302.
16. **Dombek, P. E., L. K. Johnson, S. T. Zimmerley, and M. J. Sadowsky.** 2000. Use of repetitive DNA sequences and the PCR to differentiate *Escherichia coli* isolates from human and animal sources. *Appl. Environ. Microbiol.* **66:**2572–2577.7.
17. **Duarte, J. M., J. B. dos Santos, and L. C. Melo.** 1999. Comparison of similarity coefficients based on RAPD markers in the common bean. *Genet. Mol. Biol.* **22:**427–432.
18. **Ebdon, J. E., J. L. Wallism, and H. D. Taylor.** 2004. A simplified low-cost approach to antibiotic resistance profiling for faecal source tracking. *Water Sci. Technol.* **50:**185–191.
19. **Efron, B.** 1975. The efficiency of logistic regression compared to discrimination analysis. *J. Am. Stat. Assoc.* **70:**892–898.

20. **Enas, G. G., and S. C. Choi.** 1986. Choice of the smoothing parameter and efficiency of k-nearest neighbor classification. *Comput. Math. Appl.* **12**:235–244.
21. **Fish, J. T., and G. W. Pettibone.** 1995. Influence of freshwater sediment on the survival of Escherichia coli and Salmonella sp. as measured by three methods of enumeration. *Lett. Appl. Microbiol.* **20**:277–281.
22. **Fisher, R. A.** 1936. The use of multiple measures in taxonomic problems. *Ann. Eugen.* **7**:179–188.
23. **Fix, E., and J. L. Hodges.** 1952. *Discriminatory Analysis: Nonparametric Discrimination: Small Sample Performance.* Technical Report No. 11. Project No. 21-49-004. USAF School of Aviation Medicine, Randolph Field, Tex.
24. **Fujioka, R. S., C. Sian-Denton, M. Borja, J. Castro, and K. Morphew.** 1999. Soil: the environmental source of *Escherichia coli* and enterococci in Guam's streams. *J. Appl. Microbiol.* **85**:83S–89S.
25. Reference deleted.
26. **Gordon, D. M.** 2001. Geographical structure and host specificity in bacteria and the implications for tracing the source of coliform contamination. *Microbiology* **147**:1079–1085.
27. **Gordon, D. M., S. Bauer, and J. R. Johnson.** 2002. The genetic structure of *Escherichia coli* populations in primary and secondary habitats. *Microbiology* **148**:1513–1522.
28. **Graves, A. K., C. Hagedorn, A. Teetor, M. Mahal, A. M. Booth, and R. B. Reneau.** 2002. Antibiotic resistance profiles to determine sources of fecal contamination in a rural Virginia watershed. *J. Environ. Qual.* **31**:1300–1308.
29. **Haack, S. K., L. R. Fogarty, and C. Wright.** 2003. *Escherichia coli* and enterococci at beaches in the Grand Traverse Bay, Lake Michigan: sources, characteristics, and environmental pathways. *Environ. Sci. Technol.* **37**:3275–3282.
30. **Hagedorn, C., S. L. Robinson, J. R. Filtz, S. M. Grubbs, T. A. Angier, and R. B. Beneau.** 1999. Determining sources of fecal pollution in a rural Virginia watershed with antibiotic resistance patterns in fecal streptococci. *Appl. Environ. Microbiol.* **65**:5522–5531.
31. **Hagedorn, C., J. B. Crozier, K. A. Mentz, A. M. Booth, A. K. Graves, N. J. Nelson, and R. B. Reneau.** 2003. Carbon source utilization profiles as a method to identify sources of fecal pollution in water. *J. Appl. Microbiol.* **94**:792–799.
32. **Harwood, V. J., J. Whitlock, and V. Withington.** 2000. Classification of antibiotic resistance patterns of indicator bacteria by discriminant analysis: use in predicting the source of fecal contamination in subtropical waters. *Appl. Environ. Microbiol.* **66**:3698–3704.
33. **Harwood, V. J., B. Wiggins, C. Hagedorn, R. D. Ellender, J. Gooch, J. Kern, M. Samadpour, A. C. H. Chapman, B. J. Robinson, and B. C. Thompson.** 2003. Phenotypic library-based microbial source tracking methods: efficacy in the California collaborative study. *J. Water Health* **1**:153–166.
34. **Hassan, W. M., S. Y. Wang, and R. D. Ellender.** 2005. Methods to increase fidelity of repetitive extragenic palindromic PCR fingerprint-based bacterial source tracking efforts. *Appl. Environ. Microbiol.* **71**:512–518.
35. **Hastie, T., R. Tibshirani, and J. Friedman.** 2001. *The Elements of Statistical Learning.* Springer, New York, N.Y.
36. **Healy, M., K. Reece, D. Walton, J. Huong, K. Shah, and D. P. Kontoyiannis.** 2004. Identification to the species level and differentiation between strains of *Aspergillus* clinical isolates by automated repetitive-sequence-based PCR. *J. Clin. Microbiol.* **42**:4016–4024.

37. **Holmes, B., M. M. Costa, S. L. W. Ganner, and O. M. Stevens.** 1994. Evaluation of Biolog system for identification of some gram-negative bacteria of clinical importance. *J. Clin. Microbiol.* **32:**1970–1975.

38. **Jaccard, P.** 1901. Distribution de la flore alpine dans le Bassin des Dranes et dans quelques régions voisines. *Bull. Soc. Vaud. Sci. Nat.* **37:**241–272.

39. **Johnson, L. K., M. B. Brown, E. A. Carruthers, J. A. Ferguson, P. E. Dombek, and M. J. Sadowsky.** 2004. Sample size, library composition, and genotypic diversity among natural populations of *Escherichia coli* from different animals influence accuracy of determining sources of fecal pollution. *Appl. Environ. Microbiol.* **70:**4478–4485.

40. **Johnson, R. A., and D. W. Wichern.** 2002. *Applied Multivariate Statistical Analysis.* Prentice Hall, Upper Saddle River, N.J.

41. **Kariuki, S., C. Gilks, J. Kimari, A. Obanda, J. Muyodi, P. Waiyaki, and C. A. Hart.** 1999. Genotype analysis of *Escherichia coli* strains isolated from children and chickens living in close contact. *Appl. Environ. Microbiol.* **65:**472–476.

42. **Kaufman, L., and P. J. Rousseeuw.** 1990. *Finding Groups in Data: an Introduction to Cluster Analysis.* John Wiley & Sons, Inc., Hoboken, N.J.

43. **Kerr, M., M. Fitzgerald, J. J. Sheridan, D. A. McDowell, and I. S. Blair.** 1999. Survival of *Escherichia coli* O157:H7 in bottled natural mineral water. *J. Appl. Microbiol.* **87:**833–841.

44. **Khattree, R., and D. N. Naik.** 2000. *Multivariate Data Reduction and Discrimination with SAS Software.* Wiley, Hoboken, N.J.

45. **Kibbey, H. J., C. Hagedorn, and E. L. McCoy.** 1978. Use of fecal streptococci as indicators of pollution in soil. *Appl. Environ. Microbiol.* **35:**711–717.

46. **Koellner, T., A. M. Hersperger, and T. Wohlgemuth.** 2004. Rarefaction method for assessing plant species diversity on a regional scale. *Ecography* **27:**532–544.

47. **Lleo, M. M., B. Bonato, M. C. Tafi, C. Signoretto, M. Boaretti, and P. Canepari.** 2001. Resuscitation rate in different enterococcal species in the viable but non-culturable state. *J. Appl. Microbiol.* **91:**1095–1102.

48. **Mahalanobis, P. C.** 1936. On the generalized distance in statistics. *Proc. Natl. Inst. Sci. Calcutta India* **12:**49–55.

49. **McLachlan, G. J.** 1992. *Discriminant Analysis and Statistical Pattern Recognition.* Wiley, Hoboken, N.J.

50. **McLellan, S. L., A. D. Daniels, and A. K. Salmore.** 2003. Genetic characterization of *Escherichia coli* populations from host sources of fecal pollution using DNA fingerprinting. *Appl. Environ. Microbiol.* **69:**2587–2594.

51. **Meays, C. L., K. Broersma, R. Nordin, and A. Mazumder.** 2004. Source tracking fecal bacteria in water: a critical review of current methods. *J. Environ. Manage.* **73:**71–79.

52. **Obiri-Danso, K., and K. Jones.** 2000. Intertidal sediments as reservoirs for hippurate negative campylobacters, salmonellae, and faecal indicators in three EU recognized bathing waters in northwest England. *Water Res.* **23:**519–527.

53. **Parveen, S., K. M. Portier, K. Robinson, L. Edmiston, and M. L. Tamplin.** 1999. Discriminant analysis of ribotype profiles of *Escherichia coli* for differentiating human and nonhuman sources of fecal pollution. *Appl. Environ. Microbiol.* **65:**3142–3147.

54. **Pearson, K.** 1896. Mathematical contributions to the theory of evolution. III. Regression, heredity and panmixia. *Philos. Trans. R. Soc. Lond. B* **187:**253–318.

55. **Personne, J. C., F. Poty, L. Vaute, and C. Drogue.** 1998. Survival, transport and dissemination of *Escherichia coli* and enterococci in a fissured environment. Study of a flood in a karstic aquifer. *J. Appl. Microbiol.* **84:**431–438.

56. **Ripley, B. D.** 1996. *Pattern Recognition and Neural Networks.* Cambridge University Press, New York, N.Y.

57. **Ritter, K. J., E. Carruthers, C. A. Carson, R. D. Ellender, V. J. Harwood, K. Kingsley, C. Nakatsu, M. Sadowsky, B. Shear, B. West, J. E. Whitlock, B. A. Wiggins, and J. D. Wilbur.** 2003. Assessment of statistical methods used in library-based approaches to microbial source tracking. *J. Water Health* **1:**209–223.

58. **Roll, B. M., and R. S. Fujioka.** 1997. Sources of faecal indicator bacteria in a brackish, tropical stream and their impact on recreational water quality. *Water Sci. Technol.* **35:**179–186.

59. **Samuels, M. L., and J. A. Witmer.** 2003. *Statistics for the Life Sciences*, 3rd ed. Prentice Hall, Upper Saddle River, N.J.

60. **Scott, T. M., J. B. Rose, T. M. Jenkins, S. R. Farrah, and J. Lukasik.** 2002. Microbial source tracking: current methodology and future directions. *Appl. Environ. Microbiol.* **68:**5796–5803.

61. **Scott, T. M., S. Parveen, K. M. Portier, J. B. Rose, M. L. Tamplin, S. R. Farrah, A. Koo, and J. Lukasik.** 2003. Geographical variation in ribotype profiles of *Escherichia coli* isolates from humans, swine, poultry, beef, and dairy cattle in Florida. *Appl. Environ. Microbiol.* **69:**1089–1092.

62. **Sherer, B. M., J. R. Miner, J. A. Moore, and J. D. Buckhouse.** 1992. Indicator bacterial survival in stream sediments. *J. Environ. Qual.* **21:**591–595.

63. **Shi, G. R.** 1993. Multivariate data analysis in palaeoecology and palaeobiogeography—a review. *Palaeogeogr. Palaeoclimatol. Palaeoecol.* **105:**199–234.

64. **Simpson, J. M., J. W. Santo Domingo, and D. J. Reasoner.** 2002. Microbial source tracking: state of the science. *Environ. Sci. Technol.* **36:**5279–5288.

65. **Sokal, R. R., and C. D. Michener.** 1958. A statistical method for evaluating systematic relationships. *Univ. Kansas Sci. Bull.* **38:**1409–1438.

66. **Solo-Gabriele, H. M., M. A. Wolfert, T. R. Desmarais, and C. J. Palmer.** 2000. Sources of *Escherichia coli* in a coastal subtropical environment. *Appl. Environ. Microbiol.* **66:**230–237.

67. **Stoeckel, D. M., M. V. Mathes, K. E. Hyer, C. Hagedorn, H. Kator, J. Lukasik, T. L. O'Brien, T. W. Fenger, M. Samadpour, K. M. Strickler, and B. A. Wiggins.** 2004. Comparison of seven protocols to identify fecal contamination sources using *Escherichia coli*. *Environ. Sci. Technol.* **38:**6109–6117.

68. **Tian, Y. Q., P. Gong, J. D. Radke, and J. Scarborough.** 2002. Spatial and temporal modeling of microbial containments on grazing farmlands. *J. Environ. Qual.* **31:**860–869.

69. **Torgerson, W. S.** 1952. Multidimensional scaling: I. Theory and method. *Psychometrika* **17:**401–419.

70. **Webster, L. F., B. C. Thompson, M. H. Fulton, D. E. Chestnut, R. F. Van Dolah, A. K. Leight, and G. I. Scott.** 2004. Identification of sources of *Escherichia coli* in South Carolina estuaries using antibiotic resistance analysis. *J. Exp. Mar. Biol. Ecol.* **298:**179–195.

71. **Whitlock, J. E., D. T. Jones, and V. J. Harwood.** 2002. Identification of the sources of fecal coliforms in an urban watershed using antibiotic resistance analysis. *Water Res.* **36:**4273–4282.

72. **Whittam, T. S.** 1989. Clonal dynamics of *Escherichia coli* in its natural habitat. *Antonie Leeuwenhoek* **55**:23–32.

73. **Wiggins, B. A.** 1996. Discriminant analysis of antibiotic resistance patterns in fecal streptococci, a method to differentiate human and animal sources of fecal pollution in natural waters. *Appl. Environ. Microbiol.* **62**:3997–4002.

74. **Wiggins, B. A., R. W. Andrews, R. A. Conway, C. L. Corr, E. J. Dobratz, D. P. Dougherty, J. R. Eppard, S. R. Knupp, M. C. Limjoco, J. M. Mettenburg, J. M. Rinehardt, J. Sonsino, R. L. Torrijos, and M. E. Zimmerman.** 1999. Use of antibiotic resistance analysis to identify nonpoint sources of fecal pollution. *Appl. Environ. Microbiol.* **65**:3483–3486.

75. **Wiggins, B. A., P. W. Cash, W. S. Creamer, S. E. Dart, P. P. Garcia, T. M. Gerecke, J. Han, B. L. Henry, K. B. Hoover, E. L. Johnson, K. C. Jones, J. G. McCarthy, J. A. McDonough, S. A. Mercer, M. J. Noto, H. Park, M. S. Phillips, S. M. Purner, B. M. Smith, E. N. Stevens, and A. K. Varner.** 2003. Representativeness testing of multiwatershed libraries using antibiotic resistance analysis. *Appl. Environ. Microbiol.* **69**:3399–3405.

76. **Wilks, S. S.** 1932. Certain generalization in the analysis of variance. *Biometrika* **24**:471–494.

77. **Woodward, M.** 1999. *Epidemiology: Study Design and Data Analysis.* Chapman and Hall/CRC Press, Boca Raton, Fla.

78. **Yerushalmy, J.** 1947. Statistical problems in assessing methods of medical diagnosis, with special reference to X-ray techniques. *Pub. Health Rep.* **62**:1432–1449.

79. **Zhong, X.** 2004. *A Study of Several Statistical Methods for Classification with Application to Microbial Source Tracking.* M.S. thesis. Worcester Polytechnic Institute, Worcester, Mass.

Microbial Source Tracking
Edited by Jorge W. Santo Domingo and Michael J. Sadowsky

Application of Microbial Source Tracking to Human Health and National Security

7

Cindy H. Nakatsu, Peter T. Pesenti, and Albert Rhodes

Microbial source tracking (MST), the determination of a source of microorganisms into an environment, is a phrase that has been used by various microbiology disciplines. However, the targeted microorganisms, analytical approaches, and environment often differ between disciplines. Nevertheless, there are commonalities between the disciplines, since the overall desired outcome, the identification of sources, is similar. Up to this point, chapters in this book have outlined the research and application of microbial source tracking for determining sources of pathogens into food and water resources. It focused on microbial source tracking by two different disciplines, water quality and food sciences, which rarely interact. Water quality scientists, usually trained as aquatic and soil scientists, track fecal contamination into aquatic ecosystems. Alternatively, food scientists attempt to directly monitor the presence of common pathogens found in food products. This chapter concentrates on a third even more diverse group that was not addressed in previous chapters, the biological threat, environmental monitoring, and forensics community. This group is composed of a variety of disciplines, including scientists from many fields, public health officials, and law enforcement, who are investigating the means to track the deliberate release of biological threat agents (BTAs). Their ultimate goals are similar to the other MST communities, namely, ensuring the quality and safety of our major resources.

Threats to U.S. homeland security have resulted in the deployment of biosurveillance programs to detect the deliberate release of BTAs. The

CINDY H. NAKATSU, Department of Agronomy, Purdue University, West Lafayette, IN 47907-2054. PETER T. PESENTI, Environmental Microbiology, Department of Homeland Security, Directorate for Science and Technology, Biological Countermeasures, Washington, DC 20528. ALBERT RHODES, Homeland Security Institute, 2900 S. Quincy St., Ste. 800, Arlington, VA 22206.

BioWatch program, a cooperative network between the Department of Homeland Security (DHS), U.S. Environmental Protection Agency, and the Centers for Disease Control and Prevention (CDC) (http://www.ostp.gov/html/10-20-03%20jhm%20BioSecurity%202003.pdf), has been collecting and analyzing air samples from major urban centers across the United States. The surveillance outcome from all these groups is essential for public health, law enforcement, and national security. Methods being developed and currently used by the homeland security community to detect, identify, and differentiate microorganisms can be used in other source-tracking efforts. At the same time, past experiences and challenges that have been faced by those using microbial source tracking for fecal contamination and pathogen detection in foods can be used by the threat-agent-recognition community.

This chapter discusses some of the research and development efforts in bioforensic technologies underway in the DHS Science and Technology Directorate. Comments on some of the common fundamental research challenges still required in microbial ecology to aid in tracking sources of microbial contamination are also made. The greatest challenge is likely the characterization of diversity, abundance, and distribution of microbes in their natural reservoirs. Without knowing the naturally occurring microorganisms, it is impossible to detect the introduction of a BTA. For example, under the assumption that the natural habitat of enteric bacteria (e.g., *Escherichia coli* or *Enterococcus* species) was limited to mammalian intestines, health department personnel have monitored them as indicators of fecal contamination to assess potential risk. However, research has now shown that these indicator bacteria can survive in the environment and, in some cases, have become a part of the natural microflora. The chapter concludes with examples of some outbreaks that have occurred and the responses by various relevant groups.

DHS Research and Development Programs for Bioforensics

The Bioforensics Program, within the Science and Technology Directorate, is a capabilities-driven initiative focused on providing advanced evidentiary analytical capabilities to federal law enforcement investigators and other agencies with an interest in attribution. These capabilities provide a tool set required to conduct the comprehensive analysis, characterization, and evaluation of a diverse set of biological threat agents that may be associated with a biocrime or bioterror event. This new forensics science will be a supportive element in a comprehensive criminal-investigative process with the ultimate goal of attribution, apprehension, and prosecution of the perpetrator(s). Figure 1 illustrates the analytical process employed for biological evidence. For the development of a robust analytical tool kit for bioforensics,

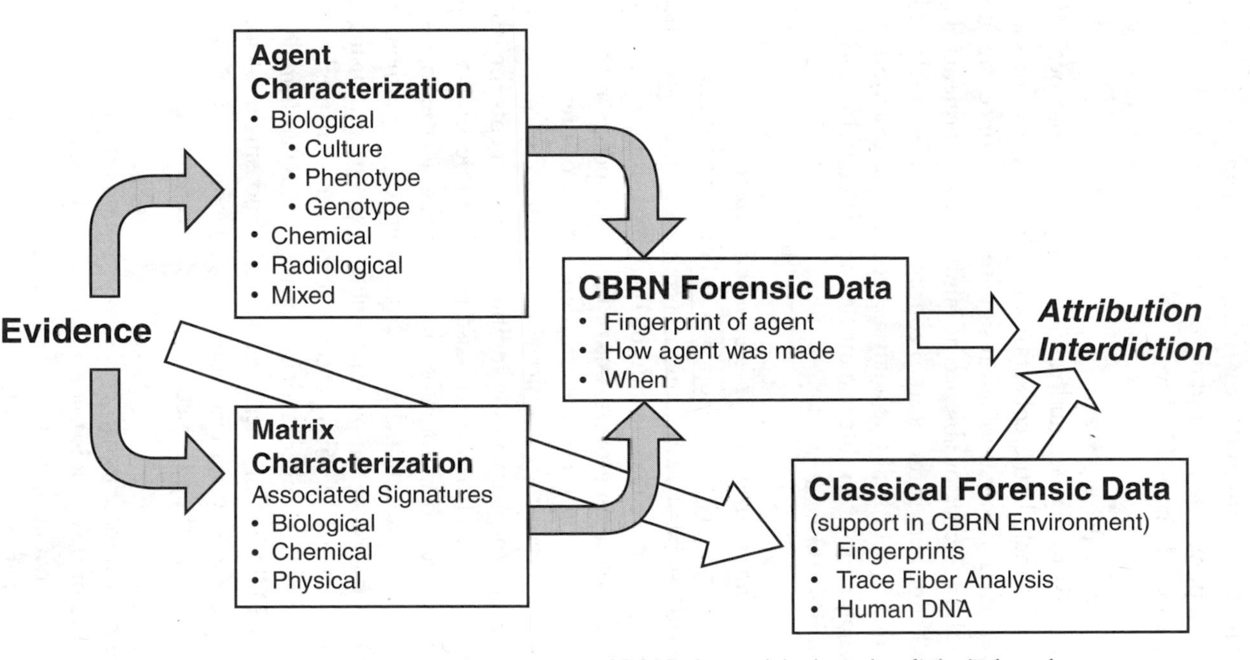

Figure 1 Analytical process employed for biological evidence. CBRN, chemical, biological, radiological, nuclear.

the program supports a comprehensive research and development program to fuel discovery of next-generation methods and techniques. This program addresses each of the steps in the analytical process and supports both intramural and extramural research and development in three focus areas. These areas include those described below.

Bioforensic Sample Management

The objective is to develop and validate operational protocols for sample management that include the collection of viable threat agents and nondenatured molecular and immunological signatures and concentration of these signatures in the presence or absence of inhibitory components. The critical technical challenge here is the need to concentrate nucleic acids and proteins in sufficient quantity to facilitate analysis protocols. Extraction and concentration of nanoliter-to-microliter quantities of nucleic acids and proteins are major goals of this thrust area.

Molecular Signature Analysis of Bioforensic Samples

The objective here is to develop and validate operational protocols for molecular-based comparative genomic assays to assist in the identification and ultimate phylogenetic characterization of BTA organisms with respect to the degree of relatedness among organisms of the same species at the strain level. These activities focus on the near-term deployment of previously developed molecular forensic assays to the National Bioforensics Analysis Center, as well as efforts to develop new assays for additional threat agents for which signatures do not yet exist. This area will also pursue the feasibility of proteomic analysis for application to bioforensic analysis. In the near term, research continues to focus on improved nucleic acid-based genotyping schemes to better answer questions regarding sample relatedness. One very promising area is the integration of leading-edge mass spectrometry detection methods with the application of variable-number tandem repeat (VNTR) and single-nucleotide polymorphism DNA typing methods. VNTRs and single-nucleotide polymorphisms represent sources of variation in the genome and are recognized as powerful tools for examining genetic relationships among species and strains.

Physical and Chemical Analysis of Bioforensic Samples

The objective of this area is to develop and validate operational protocols for physical and chemical analysis of evidence containing biothreat agents, including threat agent matrices. This program will focus on the development and validation of standard procedures and methods for identifying inorganic and organic signatures associated with a biological threat agent's growth, harvest, and processing conditions. The ultimate goal is to provide information on the geographic relatedness of BTAs used in geographically or temporally different biocrimes or bioterrorism events. In addition, to the extent technically possible, this area will provide unique sample characterization tools to determine the date of BTA production and potential geographic origin of the materials used in the growth of the agents.

Biological Agents of Concern to Homeland Security

The Select Agents Program at the CDC provides current lists of agents and toxins provided by the Department of Health and Human Services (HHS) and the U.S. Department of Agriculture (Tables 1 and 2). The CDC regulate the possession, use, and transfer of potential biologic agents, and distribution under the rules of the Select Agent Program and the CDC Office of Emergency Preparedness and Response (http://www.bt.cdc.gov/agent/agentlist-category.asp) places these agents into three categories, A, B, and C (53). These categories are based from highest to lowest with respect to ease of dissemination or transmission, potential for major public-health impact (e.g., high mortality), potential for public panic and social disruption, and requirements for special public-health preparedness. Currently the bacterial agents in category A are as follows: *Bacillus anthracis* (anthrax), *Clostridium botulinum* (botulism), *Francisella tularensis* (tularemia), and *Yersinia pestis* (plague). The viral agents are as follows: variola major (smallpox), filoviruses (Ebola and Marburg hemorrhagic fevers), and arenaviruses (Lassa fever, Argentine hemorrhagic fever). Category B contains *Brucella* species (brucellosis), *Burkholderia mallei* (Glander's disease), *Burkholderia pseudomallei* (melioidosis), *Chlamydia psittaci* (psittacosis), *Clostridium perfringens* (gas gangrene), *Coxiella burnetii* (Q fever), *Staphylococcus* species (staphylococcal enterotoxin B), *Rickettsia prowazekii* (typhus fever), known food-borne (e.g., *Salmonella* spp., *E. coli* O157:H7, *Shigella* spp.) and waterborne (e.g., *Vibrio cholerae*, *C. parvum*) illness agents, and agents of viral encephalitis (alphaviruses) including Venezuelan equine encephalitis, eastern equine encephalitis, and western equine encephalitis. Toxins also are covered by the Laboratory Registration/Select Agents Transfer rules. An example is ricin toxin, which can be extracted from *Ricinus communis* (castor beans). Category C is comprised of emerging infectious disease agents such as Nipah virus and hantavirus. Reservoirs for many of these potential agents are in soils or animal hosts (Tables 1 and 2). There is potential for human exposure to these agents by air inhalation or water and/or food consumption.

Challenges for Microbial Ecology

To differentiate indigenous from introduced microbial populations in a community, it is essential to have an understanding of the natural microflora

Table 1 CDC list of select agents and toxins

Microorganism[a]	Disease(s)	Natural reservoir(s)	Natural mode(s) of infection
Bacteria			
*Bacillus anthracis**	Anthrax	Soil long-term reservoir infected domestic, zoo, and wild animals	Inhalation and direct contact with infected animal
*Brucella abortus,** *B. melitensis,** *B. suis**	Mediterranean fever, undulant fever, Malta fever	Cattle, sheep, goats, swine, dogs	Ingestion, usually milk products, and possibly direct contact or inhalation
*Burkholderia mallei**	Glanders	Horses, also donkeys and mules	Direct contact with infected animals through the skin and mucosal surfaces
*Burkholderia pseudomallei**	Melioidosis or Whitmore's disease	Water and soil	Inhalation of dust, ingestion of contaminated water, and contact with contaminated soil especially through skin abrasions or wounds
*Clostridium botulinum**	Botulism	Water, soil, and home-canned foods	Ingestion of toxin; germinate and produce toxin in wounds or the gastrointestinal tract of infants
*Clostridium perfringens**	Gas gangrene	Cattle and sheep or soil	Ingestion
*Francisella tularensis**	Tularemia	True reservoir unknown; can also infect rodents, rabbits, muskrats, ticks	Direct contact with infected animals, ingestion, inhalation of bacteria
Yersinia pestis	Plague	Rodents and their fleas	Tick bites, aerosol or direct contact with infected animals
Rickettsiae			
*Coxiella burnetii**	Q fever	Cattle, sheep, and goats; has also been found in other animals	Aerosol is major means of transmission; also contact with infected tissues or soiled laundry or ingestion
Rickettsia prowazekii	Typhus fever	Rodents	Bites from fleas, ticks, lice, mites
Rickettsia rickettsii	Rocky Mountain spotted fever	Ticks	Ticks

Fungi			
*Coccidioides immitis**	Coccidioidomycosis	Soil in semiarid areas of California San Joaquin Valley	Inhalation of spores
Coccidioides posadasii	Coccidioidomycosis	Soil in semiarid areas of southwest U.S., Mexico, South America	Inhalation of spores
Viruses			
Ceropithecine herpesvirus 1 (herpes B virus)	Encephalomyelitis	Macaque monkeys	Animal bites, scratches, or exposure to the tissues or secretions of macaques
Crimean Congo hemorrhagic fever virus	Hemorrhagic fever	Domestic and wild animals	Direct contact with blood or other infected tissues from livestock or tick bite
Eastern equine encephalitis virus*	Encephalitis	Birds	Mosquitoes
Ebola virus	Hemorrhagic fever	Unknown but possibly bats	Person to person, direct contact with infected blood, secretions, organs or semen; route from primary source unknown
Hendra virus*	Respiratory and neurological disease	Horses	Direct contact with respiratory secretions of infected animals
Lass fever virus	Fever	Rodents	Person to person, inhalation, direct contact with broken skin, or ingestion
Marburg virus	Hemorrhagic fever	Unknown	Person to person, direct contact with infected blood, secretions, organs or semen; route from primary source unknown
Monkeypox virus	Monkeypox	African squirrel, rats, mice, rabbits, monkeys	Direct contact of fluids from infected animals or humans or objects they have infected
Nipah virus*	Encephalitis	Unknown but possibly bats	Direct contact with infected animals
Rift Valley fever virus*	Fever	Domestic animals	Mosquitoes

(continued)

Table 1 CDC list of select agents and toxins *(continued)*

Microorganism	Disease(s)	Natural reservoir(s)	Natural mode(s) of infection
Viruses (continued)			
South American hemorrhagic fever virus group	Hemorrhagic fever	Rodents	Inhalation, direct contact with broken skin, or ingestion
Tick-borne encephalitis complex virus	Encephalitis, meningitis, or meningoencephalitis	Mainly small rodents but also possibly cattle, bats, and sheep	Tick bites or consumption of milk from infected animals
Variola major and Variola minor viruses	Smallpox	Humans	Person to person, direct contact with infected bodily fluids or contaminated objects
Venezuelan equine encephalitis virus	Encephalitis	Horses, birds, and mosquitoes	Mosquito bites
Toxins			
Abrin	Poison	Rosary or jequirity pea seeds	Inhalation or ingestion
Botulinum neurotoxins*	Neurological poison	*Clostridium botulinum, C. butyricum,* and *C. baratii*	Ingestion
Conotoxins	Paralytic poison	Pacific cone snails	Ingestion
Diacetoxyscirpenol	Mycotoxin poison	*Fusarium* species	Inhalation or ingestion
Ricin	Poison	Castor bean seeds	Inhalation or ingestion
Saxitoxin	Poison	Dinoflagellates	Ingestion of contaminated seafood
Shiga toxins*	Shigellosis	*Shigella dysenteriae*; also *S. flexneri, S. boydii*, and *S. sonnei*	Fecal-oral route
Staphylococcal enterotoxins*	Food poisoning	*Staphylococcus aureus*	Ingestion of food handled by infected human, milk contaminated by cow, or eggs
T-2 toxin*	Mycotoxin poison	*Fusarium* species	Inhalation or ingestion
Tetrodotoxin	Poison	Pufferfish (*Tetraodon* spp.)	Ingestion

[a]*, agents on both HHS and USDA select agents and toxin list. Source, HHS and USDA select agents and toxins (http://www.cdc.gov/od/sap/docs/salist.pdf).

Table 2 USDA select agents and toxins list[a]

Agent	Disease target
Bacteria	
Mycoplasma capricolum/M.F38/ *M. mycoides capri*	Goats
Mycoplasma mycoides subsp. *mycoides*	Cattle and buffalo
Rickettsiae	
Cowdria ruminantium (heartwater)	Ruminants
Viruses	
African horse sickness virus	Horses, mules, donkeys
African swine fever virus	Swine, wild and domestic
Akabane virus	Cattle, sheep, and goats
Avian influenza virus (highly pathogenic)[b]	Birds, wild and domestic
Bluetongue virus (exotic)	Sheep
Bovine spongiform encephalopathy agent	Cattle
Camel pox virus	Camels
Classical swine fever virus	Pigs and wild boars
Foot-and-mouth disease virus	Cattle, sheep, swine, and goats
Goat pox virus	Goat
Japanese encephalitis virus	Domestic pigs and wild birds
Lumpy skin disease virus	Cattle and water buffalo
Malignant catarrhal fever virus (alcelaphine herpesvirus type 1)	Domestic cattle and buffaloes and many species of wild ruminants
Menangle virus	Pigs
Newcastle disease virus (velogenic)	Poultry
Peste des petits ruminants virus	Sheep and goats
Rinderpest virus	Cattle, domestic buffalo, and some species of wildlife
Sheep pox virus	Sheep
Swine vesicular disease virus	Pigs
Vesicular stomatitis virus (exotic)	Cattle, horses, pigs, and some species of wildlife

[a]Sources, HHS and USDA select agents and toxins (http://www.cdc.gov/od/sap/docs/salist.pdf).
[b]Known cases of human infection.

inhabiting an environment. The long existence of microorganisms on this planet has enabled them to inhabit almost every ecosystem on Earth, even the most extreme environments (5). While it may seem like a relatively easy task to identify and understand this vast array of microbial communities, in reality it has been challenging, as demonstrated by a few key studies. Over 20 years ago Staley and Konopka reported the "great plate-count anomaly" (63), in which only 1 to 5% of total bacterial cells counted by microscope in environmental samples are cultivated in the laboratory. Then, using DNA reassociation kinetics, Torsvik et al. (69) demonstrated that cultivation represented only a small fraction of the total microbial diversity. They estimated that there are 4,000 to 10,000 different bacterial genomes in a gram of soil. In the last 2 decades, with the advent of molecular biological methods, an understanding and characterization of microbial diversity and function associated with natural and engineered systems has been gleaned. Recently, sequence analysis of DNA extracted from the Sargasso Sea revealed the presence of 1,800 genome species and a total of 1,412 distinct small subunit rRNA genes (73). Non-cultivation-based methods have been used extensively to study microbial diversity in marine systems, yet these numbers again demonstrate our limited understanding of natural microbial communities (24).

In most ecosystems, our current knowledge is limited to the dominant cultivable microorganisms studied under select conditions. This has restricted understanding of microorganisms across ecosystems, even in common environments such as the soil (70) and the human gut (18). Although it is not essential to understand all the ecological, physiological, and phylogenetic diversity of microbes on this planet, there are some fundamental issues that must be understood regardless of the context to conduct microbial source tracking. The natural history (e.g., distribution and preferred habitat) of the target organism used in any source-tracking study must be known to determine if it is applicable to the study objectives. Secondly, the variability of the monitored target trait must be understood so its stability over time and the level of achieved distinction can be determined. The method used must be able to differentiate the target organism among all of the naturally occurring populations in that environment. To accomplish this, the microbial ecology of the pathogen, its reservoirs, and the vehicle for transmission, typically soil, water, air, and animals, must be understood (27, 30). In all these environments, the abundance of naturally occurring target organisms must first be determined. Then, conditions for the survival and persistence of newly introduced organisms must be understood.

Sources of Potential Human Pathogens of Concern to Biosecurity

In 2001, Taylor and colleagues (67) reviewed the literature and found 1,415 species of infectious human pathogens, comprised of 217 viruses and prions, 538 bacteria and rickettsia, 307 fungi, 66 protozoa, and 287 helminths. They found that the greatest challenge facing scientists is the detection and identification of newly emerged pathogens (19). These new pathogens may come from a number of sources, including the following: (i) known pathogens detected in a new location (19); (ii) zoonotic pathogens that are traditionally present at a location but typically infect animals and not humans (79); and

(iii) new diseases that have not been previously documented in any locations (79). In some cases it is possible to quickly identify the organism regardless of the source, but in other cases it may take weeks or months to determine the causative agent. Epidemiology can be used to track disease transmission to target the source, but without some previous knowledge of the pathogen it may take considerable time to determine the original source (reservoir) to undertake control tactics. Even with knowledge of the source and the means and rate at which the disease spreads, it may take considerable time and money to remedy or control the disease.

A factor that complicates the detection and attribution of a disease outbreak to its source is whether or not the pathogen is endemic to the sampling site (8). In the case of microbial indicators of fecal contamination, these organisms are widely distributed in the environment and can be attributed to a number of wild and domestic sources (60). Under certain conditions they can be found in high enough numbers to constitute a public health concern. For example, it is common to see a rise in the numbers of fecal organisms in estuary waters following a large rainfall event. In certain cases these organisms can be traced to domestic sources, but generally they are from environmental sources and are natural inhabitants of the watershed. The technical challenge to microbial source tracking is one of discerning the source of contamination within a rich microbial background. Although the risk of eating contaminated shellfish that arises as a result of these episodic rainfall events is real, it is controllable with proper monitoring and harvest restrictions. Thus, there is a need to understand the potential sources or reservoirs of potential pathogens in the environment, namely, in soils, water, and air.

Soil

Soil is a complex ecosystem with undoubtedly the greatest diversity of microorganisms (71). Not only are pathogens introduced into soil from intentional treatments such as manure and sewage applications, but soil also may be the preferred habitat for a large reservoir of pathogens. Thus, these pathogens can contaminate drinking water and food (66). Many soil pathogens are opportunistic, causing human disease under conditions when the host is immunocompromised. Quandaries to understanding zoonotic diseases, such as "why and when does a pathogen make a transition from its natural habitat to infect a human host and how can it be controlled?" also need to be addressed in soil. For example, melioidosis is an infection caused by the saprophytic gram-negative soil bacterium *Burkholderia pseudomallei* (2, 9). Bacteria of the genus *Burkholderia* are common soil inhabitants that have been studied extensively, not only to better understand their pathogenicity mechanism but also for their beneficial role in bioremediation and plant growth promotion (13, 44, 55). Melioidosis is endemic to Southeast Asia and northern Australia but often is misdiagnosed, confounding its true distribution. It can be highly fatal to humans, causing death within 48 hours of infection, which has made it a microbe of interest to those studying potential biological threat agents from a security perspective. Currently there are no rapid diagnostic tests available, and most clinicians rely on cultivation. Recent completion of sequencing of the genome of *B. pseudomallei* revealed genomic islands that comprise 6.1% of the genome (32). These islands contribute to genotypic variability between isolates, providing a potential means to differentiate the most virulent strains. Various genotyping methods such as pulsed-field gel electrophoresis (10, 38), multilocus sequence typing (11), and randomly amplified polymorphic DNA analysis (42) are being explored for strain differentiation. Comparative genomics between isolates and with the closely related recently sequenced *B. mallei* genome (52) (the agent of Glander's disease) may provide discriminatory genetic markers that can be used for rapid diagnostics, biosurveillance, detection, and forensics.

Water

Water in many cases is not the reservoir for pathogens but often may be the vector. Water, a valuable essential resource for our survival, also has potential to be a major source of infection. Drinking water and surface waters are monitored on a regular basis throughout the country to test their quality. Surveys are typically conducted using indicator organisms and not pathogens. There are a number of diseases that can be contracted by water consumption, often after pathogen deposition from an animal or human host reservoir. In the past, people often contracted cholera, typhoid fever, and dysentery, but incidences of these diseases have almost been eliminated in countries where modern water treatment is conducted (68). Recent outbreaks of infection by *Cryptosporidium*, *E. coli* O157:H7, hepatitis A, Norwalk virus, *Giardia*, and *Shigella* have occurred from poorly treated water (51, 59). People have become infected from organisms resistant to water treatment technology or by ingesting untreated drinking water, recreational water, food irrigated with contaminated waters, and contaminated shellfish. Usually gastroenteritis is contracted, but also dermatitis and meningoencephalitis are evident. Many of these organisms are of concern not only because of their natural routes of entry into water resources but also because of potential deliberate release.

An indigenous aquatic organism, a dinoflagellate named *Pfiesteria piscicida*, has been reported to produce toxins that can result in major fish kills (6, 7). There are some questions regarding the different levels of exotoxin production by *Pfiesteria* (15, 64), but these varying observations may be caused by differences between test strains used (7). Regardless of the mode of infection, the occurrence of this disease can cause economic havoc in the fishing industry and impact human health by impairing learning and higher cognitive functions (26). This is an excellent example of the importance of understanding ecological factors that contribute to a disease and its control as well as the need for methods to differentiate between virulent strains. *Pfiesteria*-related fish kills have been documented mainly in eutrophic, poorly flushed estuaries (25). Toxicity of *Pfiesteria* appears to be enhanced in the presence of bacteria and fish (7). Early detection and understanding of the ecology of this organism may aid in controlling or preventing the occurrence of this disease. The occurrence and abundance of this pathogen are being tracked using a number of molecular genetic approaches, including real-time quantitative PCR (4, 54, 81). The genotypic approaches used in fecal-source identification may be useful in tracking specific strains or genotypes of *Pfiesteria*.

Air

Air is viewed more as vector rather than a reservoir for pathogens. Diseases can be contracted by inhalation of microorganisms from infected people, animals, or even inanimate objects (75). Examples of some viral diseases contracted from infected humans are chickenpox (varicella), flu (influenza), measles (rubeola), German measles (rubella), mumps (mumps), and smallpox (variola). The bacterial diseases whooping cough (*Bordetella pertussis*), meningitis (*Neisseria* species), diphtheria (*Corynebacterium diphtheriae*), pneumonia (*Mycoplasma pneumoniae, Streptococcus* species), and tuberculosis (*Mycobacterium tuberculosis*) also are contracted by inhalation. *Legionella pneumophila,* which can grow in fresh water, air-conditioning systems, or water storage tanks (20, 65), is the causative agent of Legionnaire's disease, which is contracted from inhaling of contaminated aerosols. Spores from various fungi and actinomycetes growing on organic matter can also be inhaled causing respiratory disorders. The spores of the fungus *Histoplasma capsulatum*, found in soil contaminated with bat or bird droppings, are the causative agent of the pulmonary infection histoplasmosis (77). Another fungal disease contracted by spore inhalation is coccidioidomycosis, which is caused by *Coccidioides immitis*, a fungus that grows in desert soils of Central, South, and North America (14). Some of these organisms are included on lists as potential intentional threat agents because air dispersal can be widespread and simultaneously affect large numbers of the population if administered correctly.

Preparedness for Release of Biological Threat Agents

The detection of a clandestine release of a bioterrorism agent presents two related challenges. In the first case, the organism of concern is not endemic to the sampling site. Examples for this situation would be the detection of smallpox or Ebola virus in one of the BioWatch cities in the continental United States (45). As these organisms are not endemic to the United States, their detection would be investigated as a result of an intentional release. Confirmation of the presence of either of these biothreat agents in an environmental sample would result in an immediate and expensive response by the public health and law enforcement community. The second case involves the detection of biothreat pathogens that are endemic to the sampling area but are generally present at low levels and a limited incidence of infection (e.g., *B. anthracis, Y. pestis, F. tularensis*) (8, 40). In this case, monitoring is similar to the tracking of fecal contamination using microbial sources described above. Detection within a region where a threat agent is endemic requires both identification and quantification of the organisms. An endemic organism at background levels is not a bioterrorism threat but is one of public health concern. On the other hand, the same organism at levels that significantly exceed the background could be the indication of an intentional release. A graded response thus requires public health officials to weigh the magnitude of the threat (i.e., abundance) against the health impact (i.e., virulence, lethality). This sounds straightforward; however, it is complicated by the limited amount of information on the distribution and abundance of endemic biothreat organisms in the United States.

Ribosomal RNA gene sequences can be used to identify species of potential pathogens or indicator organisms. However, differentiation at a strain level is often necessary to assess their virulence and/or their original source. Species producing clonal populations simplify the task of microbial source tracking and bioterrorism detection. Any of the various genotyping and phenotyping methods currently in use are able to distinguish between genetically identical members of a community. However, genetic plasticity appears to be a common trait in many bacteria, especially pathogens (74). This can aid in the differentiation of some pathogens but can also hinder identification unless traits can be found that allow differentiation of the strain of interest. For example, pathogenicity islands that carry genes responsible for toxicity are often a source of genetic variability (28). After the acquisition of Shiga-toxin genes, *E. coli* became the virulent strain O157:H7. A comparison of a number of genotyping methods to differentiate *E. coli* O157:H7 revealed that fingerprinting methods such as repetitive extragenic palindromic PCR did not differentiate strains from different food outbreaks, whereas pulsed-field gel electrophoresis using a restriction enzyme cutting within the pathogenicity island could differentiate strains from different sources (29). This genetic plasticity also is likely an important factor in understanding new and reemerging diseases and their treatment (57). Viral pathogens also are known for their genetic instability (49), which can lead to zoonotic pathogens infecting humans. Abiotic and biotic interactions with these microbial populations are likely important contributors to these genetic changes (12, 79).

Examples of Recent Outbreaks

A number of natural disease outbreaks have tested the readiness of the health care and political communities to respond and quickly and effectively manage them. A few recent outbreaks that can serve as examples are those of Sin Nombre hantavirus pulmonary syndrome (17, 56), West Nile disease (50), severe acute respiratory syndrome (SARS) (61), United Kingdom foot-and-mouth disease (31, 78), Asian soybean rust (37, 80), and avian influenza (33, 43). The ideal scenario is the accurate identification of a disease threat ahead of time or when relatively few cases have occurred. Information from these outbreaks can be used in predictive mathematical models that can capture the variable dynamics of disease reproduction and transmission to predict disease spread and progression in the future (47). For the sake of discussion, three cases are examined below. Each illustrates how the variability between the introduction and spread of different types of diseases impacts efforts to track a pathogen to its source.

Severe Acute Respiratory Syndrome—Emerging Human Infectious Disease

The SARS virus came to public attention quickly because it could be lethal not only to young, old, and immunocompromised individuals but also to those that were apparently very healthy prior to infection (61, 62). Quarantining of all patients and potential carriers was used to control the spread of disease, but this did not occur until the disease was reported globally. Ultimately it was lethal to 774 of the 8,098 patients that had contracted the disease. Epidemiologists rapidly determined that the disease started in southern China's Guangzhou province (83). Using many available modern molecular biology tools, reverse transcriptase PCR, and nucleotide sequencing, the disease was identified as a new virus belonging to the coronavirus family (16, 39, 46). However, the actual source is still under dispute, although many believe it is a zoonotic virus that crossed from civet cats, prevalent in the open food markets, to humans (41, 82). The SARS pandemic was a good example of scientists, epidemiologists, and the medical community working together and utilizing the various tools now available to quickly identify a new disease (21). Investigation has revealed that the first cases of illness occurred in November of 2002, but the outbreak was not reported until February of 2003. The World Health Organization (WHO) issued a global alert in March 2003. However, the scientific and medical community worked expeditiously and the genome sequence of this new virus was completed, annotated, and published in May of the same year

(46). The SARS epidemic occurred naturally but it illustrated the rapid spread of an infectious deadly disease that was controlled by global cooperation. This event also has aided in preparations for other potential pandemics, such as avian influenza (1, 3, 76).

Soybean Rust—Emerging Fungal Crop Disease

Soybean rust (*Phakopsora pachyrhizi*) is a fungal disease that attacks the foliage of soybean plants, resulting in the reduction of pod set, pod fill, and seed quality and severely reducing grain yield. This disease was first reported in Japan in 1902 and has been endemic to Asia and Australia for decades. It has been spreading around the world, reaching Africa in the 1990s, Hawaii in 1994 (37), Brazil in 2001, and then on to other South American countries (80). Rapid spread of the disease has been facilitated by the transport of soybean rust spores in air currents. Most recently, in November of 2004, tropical storms transported the disease to North America (http://www.usda.gov/soybeanrust/). Investigators were able to rapidly identify the agent as *P. pachyrhizi*, differentiating it from the closely related species *P. meibomiae* (58) because a PCR method had been developed several years earlier (22). This prompted U.S. federal, state, university, and industry groups to work together to survey, report, predict, and manage soybean rust for the 2005 growth season. By August of 2005, the disease was confirmed in five southern states (Alabama, Florida, Georgia, Mississippi, and South Carolina) by the U.S. Department of Agriculture (http://www.usda.gov/soybeanrust/). Soybean is a major economic crop and drastic measures such as wide-scale culling used against animal-borne pathogens have not been used for the control of plant diseases. Instead, crop surveillance is being used to track the spread, and when the disease is identified, immediate control measures such as fungicide application and use of agronomic treatments that reduce disease are used. Long-term objectives are the development of rust-resistant cultivars of soybean. The soybean rust outbreak is an example of successful management of a disease threat to a major agricultural crop (72).

Mail Anthrax Attacks—Deliberate Release of a Pathogen

Anthrax is caused by the gram-positive endospore-forming bacterium *Bacillus anthracis*. It has a long history as a human pathogen, and several authorities speculate that it was the cause of the sixth Egyptian plague recorded in the book of Genesis (23). In its endospore form, *B. anthracis* is resistant to heat and desiccation, is stable for long periods of time, can be ground to a fine powder, and produces a high number of casualties when properly disseminated. These properties make *B. anthracis* an ideal biological warfare agent, and it was extensively studied by both the United States and the Soviet Union in their biological warfare programs. Its potential to produce mass casualties was tragically demonstrated in 1979 (48). An accidental release of an aerosol from a secret biological weapons facility in Sverdlovsk (Ekaterinburg), Russia, caused 96 human cases of cutaneous and pulmonary anthrax, and 64 eventually died. Acquisition by or proliferation of this capability to hostile nations and terrorist groups has been of great international concern.

The United States experienced its first *B. anthracis* attack on September 18, 2001 (34). Letters containing a fine powder of spores were mailed to several news media organizations and the offices of two senators. Eleven people who handled these letters contracted pulmonary anthrax as a result of inhaling aerosolized spores. Of these, five would ultimately die from anthrax-related causes. In addition, the letters caused 11 cases of cutaneous anthrax resulting in no fatalities.

During the subsequent investigation, it was determined that the letters contained two different qualities of *B. anthracis*-laced powder. Genotyping of strains isolated from the powders and clinical specimens revealed that they all were genetically identical to the Ames strain, which is studied at the U.S. Army Medical Research Institute of Infectious Diseases (USAMRIID) in Fort Detrick, Maryland. Unlike many other bacteria, the genome of *B. anthracis* is genetically stable, making strain differentiation challenging (35, 36). Few genetic polymorphisms were found using amplified fragment length polymorphism (35), but it is possible that finer-scale genotyping methods, such as multiple-locus VNTR analysis (36), might enable strain differentiation. For source tracking and attribution purposes, sequence analysis might be the only approach that can resolve subtle genotypic variations. Sequence analysis requires knowledge of both where these genetic elements reside in the genome and an understanding of their rates of mutation. For certain pathogens for which detailed understanding of the genome is lacking, the only approach may be total genome sequencing to reveal variable regions.

POTENTIAL FUTURE DIRECTIONS

Approaches for studying microorganisms are changing due to the availability of whole genome sequences and bioinformatics approaches that have spurred the "-omics" era, comparative genomics, proteomics, and "metabolomics." In the various fields of MST, new methods are being investigated to rapidly identify microorganisms, to understand host-microbe interactions, and to develop new treatment strategies. Electronic media have made much of this technological knowledge readily available to scientists across disciplines. Essential to rapid progress is the combining of new approaches with conventional methods used to understand the ecology, physiology, and biochemistry of organisms. By understanding features unique to a group of organisms, such as the nature of infection and molecular signatures, strategies can be developed to identify and treat infectious organisms using new antibiotics and vaccines. High-throughput methods such as DNA microarrays are being developed to identify either the infectious organism or the gene essential for pathogenicity. In either case it is essential to identify specific target genes to be used on these microarrays. Comparative genomics of microbial genome sequences can aid in the identification of genes both common and unique to infectious organisms. Advances in proteomics and metabolomics combined with new imaging technology can also aid in identification of molecular mechanisms of infection and pathogenesis.

There is opportunity for all those interested ultimately in human health safety, whether they are directly involved in issues related to human disease, plant disease, animal disease, water quality, air quality, soil quality, or food safety, to join their resources. It is the combined knowledge of these fields that will allow the most efficient approach for national security. Although each field is unique, many standardized approaches or essential factors for evaluation of technologies are common to all. Targets, whether biotic or abiotic, may differ between groups, but many of the same questions must be addressed. Currently there are many issues of importance requiring scientific attention, but financial resources often limit research that can be done. These issues are often looming threats in the general public's minds, pressing politicians to focus on the superficially obvious, while often missing the most fundamentally important research because rewards are not always immediate.

REFERENCES

1. **Ahmad, K.** 2005. Increased Asian collaboration in fight against avian flu. *Lancet Infect. Dis.* **5**:9.
2. **Aldhous, P.** 2005. Tropical medicine: melioidosis? Never heard of it. *Nature* **434**:692–693.
3. **Aldhous, P., and S. Tomlin.** 2005. Avian flu special: avian flu: are we ready? *Nature* **435**:399.
4. **Bowers, H. A., T. Tengs, H. B. Glasgow, J. M. Burkholder, P. A. Rublee, and D. W. Oldach.** 2000. Development of real-time PCR assays for rapid detection of *Pfiesteria piscicida* and related dinoflagellates. *Appl. Environ. Microbiol.* **66**:4641–4648.
5. **Bull, A. T.** 2003. *Microbial Diversity and Bioprospecting.* ASM Press, Washington, D.C.
6. **Burkholder, J. M., and H. B. Glasgow.** 1997. *Pfiesteria piscicida* and other *Pfiesteria*-like dinoflagellates: behavior, impacts, and environmental controls. *Limnol. Oceanogr.* **42**:1052–1075.
7. **Burkholder, J. M., A. S. Gordon, P. D. Moeller, J. M. Law, K. J. Coyne, A. J. Lewitus, J. S. Ramsdell, H. G. Marshall, N. J. Deamer, S. C. Cary, J. W. Kempton, S. L. Morton,**

and P. A. Rublee. 2005. Demonstration of toxicity to fish and to mammalian cells by *Pfiesteria* species: comparison of assay methods and strains. *Proc. Natl. Acad. Sci. USA* **102:**3471–3476.

8. **Chang, M. H., M. K. Glynn, and S. L. Groseclose.** 2003. Endemic, notifiable bioterrorism-related diseases, United States, 1992–1999. *Emerg. Infect. Dis.* **9:**556–564.
9. **Cheng, A. C., and B. J. Currie.** 2005. Melioidosis: epidemiology, pathophysiology, and management. *Clin. Microbiol. Rev.* **18:**383–416.
10. **Cheng, A. C., N. P. Day, M. J. Mayo, D. Gal, and B. J. Currie.** 2005. Burkholderia pseudomallei strain type, based on pulsed-field gel electrophoresis, does not determine disease presentation in melioidosis. *Microbes Infect.* **7:**104–109.
11. **Cheng, A. C., D. Godoy, M. Mayo, D. Gal, B. G. Spratt, and B. J. Currie.** 2004. Isolates of *Burkholderia pseudomallei* from northern Australia are distinct by multilocus sequence typing, but strain types do not correlate with clinical presentation. *J. Clin. Microbiol.* **42:**5477–5483.
12. **Cleaveland, S., M. K. Laurenson, and L. H. Taylor.** 2001. Diseases of humans and their domestic mammals: pathogen characteristics, host range and the risk of emergence. *Philos. Trans. R. Soc. Lond. B* **356:**991–999.
13. **Coenye, T., and P. Vandamme.** 2003. Diversity and significance of *Burkholderia* species occupying diverse ecological niches. *Environ. Microbiol.* **5:**719–729.
14. **Cox, R. A., and D. M. Magee.** 2004. Coccidioidomycosis: host response and vaccine development. *Clin. Microbiol. Rev.* **17:**804–839.
15. **Drgon, T., K. Saito, P. M. Gillevet, M. Sikaroodi, B. Whitaker, D. N. Krupatkina, F. Argemi, and G. R. Vasta.** 2005. Characterization of ichthyocidal activity of *Pfiesteria piscicida*: dependence on the dinospore cell density. *Appl. Environ. Microbiol.* **71:**519–529.
16. **Drosten, C., S. Gunther, W. Preiser, S. van der Werf, H. R. Brodt, S. Becker, H. Rabenau, M. Panning, L. Kolesnikova, R. A. M. Fouchier, A. Berger, A. M. Burguiere, J. Cinatl, M. Eickmann, N. Escriou, K. Grywna, S. Kramme, J. C. Manuguerra, S. Muller, V. Rickerts, M. Sturmer, S. Vieth, H. D. Klenk, A. Osterhaus, H. Schmitz, and H. W. Doerr.** 2003. Identification of a novel coronavirus in patients with severe acute respiratory syndrome. *N. Engl. J. Med.* **348:**1967–1976.
17. **Duchin, J. S., F. T. Koster, C. J. Peters, G. L. Simpson, B. Tempest, S. R. Zaki, T. G. Ksiazek, P. E. Rollin, S. Nichol, E. T. Umland, R. L. Moolenaar, S. E. Reef, K. B. Nolte, M. M. Gallaher, J. C. Butler, R. F. Breiman, M. Burkhart, N. Kalishman, R. Voorhees, J. Voorhees, M. Samuel, M. Tanuz, L. Hughes, S. Wictor, G. Oty, L. Nims, S. Castle, B. Bryt, C. M. Sewell, P. Reynolds, T. Brown, L. Sands, K. Komatsu, C. Kioski, K. Fleming, J. Doll, C. Levy, T. M. Fink, P. Murphy, B. England, M. Smolinski, B. Erickson, W. Slanta, G. Gellert, P. Schillam, R. E. Hoffman, S. Lanser, C. Nichols, L. Hubbardpourier, J. Cheek, A. Craig, R. Haskins, B. Muneta, S. John, J. Kitzes, J. Hubbard, M. Carroll, R. Wood, C. North, P. Bohan, N. Cobb, R. Zumwalt, P. McFeely, H. Levy, G. Mertz, S. Young, K. Foucar, B. Hjelle, J. McLaughlin, S. Allen, S. Simpson, T. Merlin, M. Schmidt, L. Simonsen, C. Vitek, C. Dalton, R. Helfand, P. Ettestadt, J. Tappero, A. Khan, L. Chapman, R. Pinner, K. Wachsmuth, A. Kaufmann, J. Wenger, and J. McDade.** 1994. Hantavirus pulmonary syndrome—a clinical description of 17 patients with a newly recognized disease. *N. Engl. J. Med.* **330:**949–955.
18. **Eckburg, P. B., E. M. Bik, C. N. Bernstein, E. Purdom, L. Dethlefsen, M. Sargent, S. R. Gill, K. E. Nelson, and D. A. Relman.** 2005. Diversity of the human intestinal microbial flora. *Science* **308:**1635–1638.

19. **Fauci, A. S., N. A. Touchette, and G. K. Folkers.** 2005. Emerging infectious diseases: a 10-year perspective from the National Institute of Allergy and Infectious Diseases. *Emerg. Infect. Dis.* **11:**519–525.
20. **Fields, B. S., R. F. Benson, and R. E. Besser.** 2002. *Legionella* and Legionnaires' disease: 25 years of investigation. *Clin. Microbiol. Rev.* **15:**506–526.
21. **Finlay, B. B., R. H. See, and R. C. Brunham.** 2004. Rapid response research to emerging infectious diseases: lessons from SARS. *Nat. Rev. Microbiol.* **2:**602–607.
22. **Frederick, R. D., C. L. Snyder, G. L. Peterson, and M. R. Bonde.** 2002. Polymerase chain reaction assays for the detection and discrimination of the soybean rust pathogens *Phakopsora pachyrhizi* and *P. meibomiae*. *Phytopathology* **92:**217–227.
23. **Friedlander, A. M.** 1997. Anthrax, p. 467–478. *In* F. R. Sidell, E. T. Takafuji, and D. R. Franz (ed.), *Medical Aspects of Chemical and Biological Warfare*. Office of the Surgeon General, Borden Institute, Walter Reed Army Medical Center, Washington, D.C.
24. **Giovannoni, S.** 2004. Evolutionary biology: oceans of bacteria. *Nature* **430:**515–516.
25. **Glasgow, H. B., J. M. Burkholder, M. A. Mallin, N. J. Deamer-Melia, and R. E. Reed.** 2001. Field ecology of toxic *Pfiesteria* complex species and a conservative analysis of their role in estuarine fish kills. *Environ. Health Perspect.* **109:**715–730.
26. **Grattan, L. M., D. Oldach, T. M. Perl, M. H. Lowitt, D. L. Matuszak, C. Dickson, C. Parrott, R. C. Shoemaker, C. L. Kauffman, M. P. Wasserman, J. R. Hebel, P. Charache, and J. G. Morris.** 1998. Learning and memory difficulties after environmental exposure to waterways containing toxin-producing *Pfiesteria* or *Pfiesteria*-like dinoflagellates. *Lancet* **352:**532–539.
27. **Guernier, V., M. E. Hochberg, and J. F. O. Guegan.** 2004. Ecology drives the worldwide distribution of human diseases. *PLoS Biol.* **2:**740–746.
28. **Hacker, J., G. Blum-Oehler, I. Muhldorfer, and H. Tschape.** 1997. Pathogenicity islands of virulent bacteria: structure, function and impact on microbial evolution. *Mol. Microbiol.* **23:**1089–1097.
29. **Hahm, B.-K., Y. Maldonado, E. Schreiber, A. K. Bhunia, and C. H. Nakatsu.** 2003. Subtyping of foodborne and environmental isolates of Escherichia coli by multiplex-PCR, rep-PCR, PFGE, ribotyping and AFLP. *J. Microbiol. Methods* **53:**387–399.
30. **Haydon, D. T., S. Cleaveland, L. H. Taylor, and M. K. Laurenson.** 2002. Identifying reservoirs of infection: a conceptual and practical challenge. *Emerg. Infect. Dis.* **8:**1468–1473.
31. **Haydon, D. T., R. R. Kao, and R. P. Kitching.** 2004. The U.K. foot-and-mouth disease outbreak—the aftermath. *Nat. Rev. Microbiol.* **2:**675–681.
32. **Holden, M. T. G., R. W. Titball, S. J. Peacock, A. M. Cerdeno-Tarraga, T. Atkins, L. C. Crossman, T. Pitt, C. Churcher, K. Mungall, S. D. Bentley, M. Sebaihia, N. R. Thomson, N. Bason, I. R. Beacham, K. Brooks, K. A. Brown, N. F. Brown, G. L. Challis, I. Cherevach, T. Chillingworth, A. Cronin, B. Crossett, P. Davis, D. DeShazer, T. Feltwell, A. Fraser, Z. Hance, H. Hauser, S. Holroyd, K. Jagels, K. E. Keith, M. Maddison, S. Moule, C. Price, M. A. Quail, E. Rabbinowitsch, K. Rutherford, M. Sanders, M. Simmonds, S. Songsivilai, K. Stevens, S. Tumapa, M. Vesaratchavest, S. Whitehead, C. Yeats, B. G. Barrell, P. C. F. Oyston, and J. Parkhill.** 2004. Genomic plasticity of the causative agent of melioidosis, *Burkholderia pseudomallei*. *Proc. Natl. Acad. Sci. USA* **101:**14240–14245.
33. **Horimoto, T., and Y. Kawaoka.** 2001. Pandemic threat posed by avian influenza A viruses. *Clin. Microbiol. Rev.* **14:**129–149.

34. Jernigan, D. B., P. L. Raghunathan, B. P. Bell, R. Brechner, E. A. Bresnitz, J. C. Butler, M. Cetron, M. Cohen, T. Doyle, M. Fischer, C. Greene, K. S. Griffith, J. Guarner, J. L. Hadler, J. A. Hayslett, R. Meyer, L. R. Petersen, M. Phillips, R. Pinner, T. Popovic, C. P. Quinn, J. Reefhuis, D. Reissman, N. Rosenstein, A. Schuchat, W. J. Shieh, L. Siegal, D. L. Swerdlow, F. C. Tenover, M. Traeger, J. W. Ward, I. Weisfuse, S. Wiersma, K. Yeskey, S. Zaki, D. A. Ashford, B. A. Perkins, S. Ostroff, J. Hughes, D. Fleming, J. P. Koplan, and J. L. Gerberding. 2002. Investigation of bioterrorism-related anthrax, United States, 2001: epidemiologic findings. *Emerg. Infect. Dis.* 8:1019–1028.

35. Keim, P., A. Kalif, J. Schupp, K. Hill, S. E. Travis, K. Richmond, D. M. Adair, M. Hughjones, C. R. Kuske, and P. Jackson. 1997. Molecular evolution and diversity in *Bacillus anthracis* as detected by amplified fragment length polymorphism markers. *J. Bacteriol.* 179:818–824.

36. Keim, P., L. B. Price, A. M. Klevytska, K. L. Smith, J. M. Schupp, R. Okinaka, P. J. Jackson, and M. E. Hugh-Jones. 2000. Multiple-locus variable-number tandem repeat analysis reveals genetic relationships within *Bacillus anthracis*. *J. Bacteriol.* 182:2928–2936.

37. Killgore, E., R. Heu, and D. E. Gardner. 1994. First report of soybean rust in Hawaii. *Plant Dis.* 78:1216.

38. Koonpaew, S., M. N. Ubol, S. Sirisinha, N. J. White, and S. C. Chaiyaroj. 2000. Genome fingerprinting by pulsed-field gel electrophoresis of isolates of *Burkholderia pseudomallei* from patients with melioidosis in Thailand. *Acta Trop.* 74:187–191.

39. Kuiken, T., R. A. M. Fouchier, M. Schutten, G. F. Rimmelzwaan, G. van Amerongen, D. van Riel, J. D. Laman, T. de Jong, G. van Doornum, W. Lim, A. E. Ling, P. K. S. Chan, J. S. Tam, M. C. Zambon, R. Gopal, C. Drosten, S. van der Werf, N. Escriou, J. C. Manuguerra, K. Stohr, J. S. M. Peiris, and A. Osterhaus. 2003. Newly discovered coronavirus as the primary cause of severe acute respiratory syndrome. *Lancet* 362: 263–270.

40. Kuske, C. R., S. M. Barns, C. C. Grow, E. P. Williams, L. Merrill, J. Richardson, J. Buckingham, and J. Dunbar. Environmental survey for four pathogenic bacteria and closely related species using phylogenetic and functional genes. *Appl. Environ. Microbiol.*, in press.

41. Lai, M. M. C. 2003. SARS virus: the beginning of the unraveling of a new coronavirus. *J. Biomed. Sci.* 10:664–675.

42. Leelayuwat, C., A. Romphruk, A. Lulitanond, S. Trakulsomboon, and V. Thamlikitkul. 2000. Genotype analysis of *Burkholderia pseudomallei* using randomly amplified polymorphic DNA (RAPD): indicative of genetic differences amongst environmental and clinical isolates. *Acta Trop.* 77:229–237.

43. Li, K. S., Y. Guan, J. Wang, G. J. D. Smith, K. M. Xu, L. Duan, A. P. Rahardjo, P. Puthavathana, C. Buranathai, T. D. Nguyen, A. T. S. Estoepangestie, A. Chaisingh, P. Auewarakul, H. T. Long, N. T. H. Hanh, R. J. Webby, L. L. M. Poon, H. Chen, K. F. Shortridge, K. Y. Yuen, R. G. Webster, and J. S. M. Peiris. 2004. Genesis of a highly pathogenic and potentially pandemic H5N1 influenza virus in eastern Asia. *Nature* 430:209–213.

44. Mahenthiralingam, E., T. A. Urban, and J. B. Goldberg. 2005. The multifarious, multireplicon *Burkholderia cepacia* complex. *Nat. Rev. Microbiol.* 3:144–156.

45. Marburger, J. 2003. *Keynote Address on National Preparedness.* Biosecurity 2003. [Online.] http://www.ostp.gov/html/10-20-03%20jhm%20BioSecurity%202003.pdf.

46. **Marra, M. A., S. J. M. Jones, C. R. Astell, R. A. Holt, A. Brooks-Wilson, Y. S. N. Butterfield, J. Khattra, J. K. Asano, S. A. Barber, S. Y. Chan, A. Cloutier, S. M. Coughlin, D. Freeman, N. Girn, O. L. Griffin, S. R. Leach, M. Mayo, H. McDonald, S. B. Montgomery, P. K. Pandoh, A. S. Petrescu, A. G. Robertson, J. E. Schein, A. Siddiqui, D. E. Smailus, J. E. Stott, G. S. Yang, F. Plummer, A. Andonov, H. Artsob, N. Bastien, K. Bernard, T. F. Booth, D. Bowness, M. Czub, M. Drebot, L. Fernando, R. Flick, M. Garbutt, M. Gray, A. Grolla, S. Jones, H. Feldmann, A. Meyers, A. Kabani, Y. Li, S. Normand, U. Stroher, G. A. Tipples, S. Tyler, R. Vogrig, D. Ward, B. Watson, R. C. Brunham, M. Krajden, M. Petric, D. M. Skowronski, C. Upton, and R. L. Roper.** 2003. The genome sequence of the SARS-associated coronavirus. *Science* **300:**1399–1404.
47. **Matthews, L., and M. Woolhouse.** 2005. New approaches to quantifying the spread of infection. *Nat. Rev. Microbiol.* **3:**529.
48. **Meselson, M., J. Guillemin, M. Hughjones, A. Langmuir, I. Popova, A. Shelokov, and O. Yampolskaya.** 1994. The Sverdlovsk anthrax outbreak of 1979. *Science* **266:**1202–1208.
49. **Moya, A., E. C. Holmes, and F. Gonzalez-Candelas.** 2004. The population genetics and evolutionary epidemiology of RNA viruses. *Nat. Rev. Microbiol.* **2:**279–288.
50. **Nash, D., F. Mostashari, A. Fine, J. Miller, D. O'Leary, K. Murray, A. Huang, A. Rosenberg, A. Greenberg, M. Sherman, S. Wong, M. Layton, G. L. Campbell, J. T. Roehrig, D. J. Gubler, W. J. Shieh, S. Zaki, and P. Smith.** 2001. The outbreak of West Nile virus infection in the New York City area in 1999. *N. Engl. J. Med.* **344:**1807–1814.
51. **Neumann, N. F., D. W. Smith, and M. Belosevic.** 2005. Waterborne disease: an old foe re-emerging? *J. Environ. Eng. Sci.* **4:**155–171.
52. **Nierman, W. C., D. DeShazer, H. S. Kim, H. Tettelin, K. E. Nelson, T. Feldblyum, R. L. Ulrich, C. M. Ronning, L. M. Brinkac, S. C. Daugherty, T. D. Davidsen, R. T. Deboy, G. Dimitrov, R. J. Dodson, A. S. Durkin, M. L. Gwinn, D. H. Haft, H. Khouri, J. F. Kolonay, R. Madupu, Y. Mohammoud, W. C. Nelson, D. Radune, C. M. Romero, S. Sarria, J. Selengut, C. Shamblin, S. A. Sullivan, O. White, Y. Yu, N. Zafar, L. Zhou, and C. M. Fraser.** 2004. Structural flexibility in the *Burkholderia mallei* genome. *Proc. Natl. Acad. Sci. USA* **101:**14246–14251.
53. **Rotz, L. D., A. S. Khan, S. R. Lillibridge, S. M. Ostroff, and J. M. Hughes.** 2002. Public health assessment of potential biological terrorism agents. *Emerg. Infect. Dis.* **8:**225–230.
54. **Rublee, P. A., D. L. Remington, E. F. Schaefer, and M. M. Marshall.** 2005. Detection of the dinozoans *Pfiesteria piscicida* and *P. shumwayae*: a review of detection methods and geographic distribution. *J. Eukaryot. Microbiol.* **52:**83–89.
55. **Salles, J. F., J. A. van Veen, and J. D. van Elsas.** 2004. Multivariate analyses of *Burkholderia* species in soil: effect of crop and land use history. *Appl. Environ. Microbiol.* **70:**4012–4020.
56. **Schmaljohn, C., and B. Hjelle.** 1997. Hantaviruses: a global disease problem. *Emerg. Infect. Dis.* **3:**95–104.
57. **Schmidt, H., and M. Hensel.** 2004. Pathogenicity islands in bacterial pathogenesis. *Clin. Microbiol. Rev.* **17:**14–56.
58. **Schneider, R. W., C. A. Hollier, H. K. Whitam, M. E. Palm, J. M. McKemy, J. R. Hernandez, L. Levy, and R. DeVries-Paterson.** 2005. First report of soybean rust caused by *Phakopsora pachyrhizi* in the continental United States. *Plant Dis.* **89:**774.
59. **Sharma, S., P. Sachdeva, and J. S. Virdi.** 2003. Emerging water-borne pathogens. *Appl. Microbiol. Biotechnol.* **61:**424–428.

60. **Simpson, J. M., J. W. Santo Domingo, and D. J. Reasoner.** 2002. Microbial source tracking: state of the science. *Environ. Sci. Technol.* **36:**5279–5288.

61. **Skowronski, D. A., C. Astell, R. C. Brunham, D. E. Low, M. Petric, R. L. Roper, P. J. Talbot, T. Tam, and L. Babiuk.** 2005. Severe acute respiratory syndrome (SARS): a year in review. *Annu. Rev. Med.* **56:**357–381.

62. **Stadler, K., V. Masignani, M. Eickmann, S. Becker, S. Abrignani, H. D. Klenk, and R. Rappuoli.** 2003. SARS—beginning to understand a new virus. *Nat. Rev. Microbiol.* **1:**209–218.

63. **Staley, J. T., and A. Konopka.** 1985. Measurement of in situ activities of nonphotosynthetic microorganisms in aquatic and terrestrial habitats. *Annu. Rev. Microbiol.* **39:**321–346.

64. **Stoecker, D. K., M. W. Parrow, J. M. Burkholder, and H. B. Glasgow.** 2002. Grazing by microzooplankton on *Pfiesteria piscicida* cultures with different histories of toxicity. *Aquat. Microb. Ecol.* **28:**79–85.

65. **Swanson, M. S., and B. K. Hammer.** 2000. Legionella pneumophila pathogenesis: a fateful journey from amoebae to macrophages. *Annu. Rev. Microbiol.* **54:**567–613.

66. **Tauxe, R. V.** 1997. Emerging foodborne diseases: an evolving public health challenge. *Emerg. Infect. Dis.* **3:**425–434.

67. **Taylor, L. H., S. M. Latham, and M. E. J. Woolhouse.** 2001. Risk factors for human disease emergence. *Philos. Trans. R. Soc. Lond. B* **356:**983–989.

68. **Theron, J., and T. E. Cloete.** 2002. Emerging waterborne infections: contributing factors, agents, and detection tools. *Crit. Rev. Microbiol.* **28:**1–26.

69. **Torsvik, V., J. Goksøyr, and F. L. Daae.** 1990. High diversity in DNA of soil bacteria. *Appl. Environ. Microbiol.* **56:**782–787.

70. **Torsvik, V., and L. Ovreas.** 2002. Microbial diversity and function in soil: from genes to ecosystems. *Curr. Opin. Microbiol.* **5:**240–245.

71. **Torsvik, V., L. Ovreas, and T. F. Thingstad.** 2002. Prokaryotic diversity—magnitude, dynamics, and controlling factors. *Science* **296:**1064–1066.

72. **U.S. Department of Agriculture.** 2005. *A Coordinated Framework for Soybean Rust Surveillance, Reporting, Prediction, Management and Outreach.* [Online.] http://www.aphis.usda.gov/ppq/ep/soybean_rust/coordfram041405.pdf.

73. **Venter, J. C., K. Remington, J. F. Heidelberg, A. L. Halpern, D. Rusch, J. A. Eisen, D. Wu, I. Paulsen, K. E. Nelson, W. Nelson, D. E. Fouts, S. Levy, A. H. Knap, M. W. Lomas, K. Nealson, O. White, J. Peterson, J. Hoffman, R. Parsons, H. Baden-Tillson, C. Pfannkoch, Y.-H. Rogers, and H. O. Smith.** 2004. Environmental genome shotgun sequencing of the Sargasso Sea. *Science* **304:**66–74.

74. **Wain, J., D. House, D. Pickard, G. Dougan, and G. Frankel.** 2001. Acquisition of virulence-associated factors by the enteric pathogens *Escherichia coli* and *Salmonella enterica*. *Philos. Trans. R. Soc. Lond. B* **356:**1027–1034.

75. **Walther, B. A., and P. W. Ewald.** 2004. Pathogen survival in the external environment and the evolution of virulence. *Biol. Rev. Camb. Philos. Soc.* **79:**849–869.

76. **Webster, R., and D. Hulse.** 2005. Controlling avian flu at the source. *Nature* **435:**415–416.

77. **Woods, J. P.** 2002. *Histoplasma capsulatum* molecular genetics, pathogenesis, and responsiveness to its environment. *Fungal Genet. Biol.* **35:**81–97.

78. **Woolhouse, M., and A. Donaldson.** 2001. Managing foot-and-mouth—the science of controlling disease outbreaks. *Nature* **410:**515–516.

79. **Woolhouse, M. E. J.** 2002. Population biology of emerging and re-emerging pathogens. *Trends Microbiol.* **10:**S3–S7.

80. **Yorinori, J. T., W. M. Paiva, R. D. Frederick, L. M. Costamilan, P. F. Bertagnolli, G. E. Hartman, C. V. Godoy, and J. Nunes.** 2005. Epidemics of soybean rust (*Phakopsora pachyrhizi*) in Brazil and Paraguay from 2001 to 2003. *Plant Dis.* **89:**675–677.

81. **Zhang, H., and S. J. Lin.** 2002. Detection and quantification of *Pfiesteria piscicida* by using the mitochondrial cytochrome *b* gene. *Appl. Environ. Microbiol.* **68:**989–994.

82. **Zhong, N. S.** 2004. Management and prevention of SARS in China. *Philos. Trans. R. Soc. Lond. B* **359:**1115–1116.

83. **Zhong, N. S., B. J. Zheng, Y. M. Li, L. L. M. Poon, Z. H. Xie, K. H. Chan, P. H. Li, S. Y. Tan, Q. Chang, J. P. Xie, X. Q. Liu, J. Xu, D. X. Li, K. Y. Yuen, J. S. M. Peiris, and Y. Guan.** 2003. Epidemiology and cause of severe acute respiratory syndrome (SARS) in Guangdong, People's Republic of China, in February, 2003. *Lancet* **362:**1353–1358.

Microbial Source Tracking
Edited by Jorge W. Santo Domingo and Michael J. Sadowsky

The Future of Microbial Source Tracking Studies

8

Michael J. Sadowsky, Douglas R. Call, and Jorge W. Santo Domingo

Microbial source tracking (MST) is differentiated from traditional microbial water quality efforts by the need to identify the host species from which the bacteria originate, rather than identifying an individual point source (152, 174). As such, tools that are successful for subtyping individual bacterial isolates (e.g., pulsed-field gel electrophoresis) may be less useful for identifying sources at the level of the host animal. This is partly because most bacterial species contain considerable genetic variation and this diversity is often poorly correlated with host species. Despite recent advances in the development and application of phenotypic- and genotypic-based microbial source tracking methodologies, new technologies and approaches clearly need to be investigated and employed to satisfy the demand for rapid and accurate detection and identification systems. Currently, issues of plasticity of phenotypic markers, in part due to horizontal gene transfer (44), and the inherent variation in microbial phenotypes due to environmental and genomic variation within single microbial populations (177) negatively impact available methods. Thus, given these limitations, future MST technologies will most likely need to be non-library-based and have sufficient throughput to examine a large number of samples in order to overcome demands for speed, accuracy, and cost effectiveness.

Ideally, the direct detection of the pathogens could provide a better indication of the microbial quality of a water sample. Moreover, if pathogens

MICHAEL J. SADOWSKY, Department of Soil, Water & Climate and BioTechnology Institute, University of Minnesota, 1991 Upper Buford Circle, 439 BorH, St. Paul, MN 55108. DOUGLAS R. CALL, Department of Veterinary Microbiology & Pathology, Washington State University, Pullman, WA 99164-7040. JORGE W. SANTO DOMINGO, U.S. Environmental Protection Agency, ORD/NRMRL/WSWRD/MCCB, 26 West Martin Luther King Dr., MS 387, Cincinnati, OH 45268.

could be specifically assigned to fecal sources, it would then be possible to implement management practices that targeted the pathogen(s) of interest. However, there are a relatively large number of waterborne pathogens, and in most cases the analytic tools available lack sensitivity to detect low infectious doses of these microorganisms. Consequently, the enumeration of indicator bacteria continues to be the regulatory standard used to measure public health risks. In this chapter, we will summarize some of the challenges that people interested in source identification are currently facing, provide information on the microbial ecology of the gut and potential microorganisms to use in source-tracking studies, discuss research tools that will likely improve marker discovery, and offer some suggestions on the future needs for determining sources of microorganisms impacting waterways.

Challenges in Microbial Source Tracking

Current challenges in applying MST technologies to a variety of environments and situations are numerous (174). The obstacles depend on the type of method used for source identification. For example, in the case of library-dependent methods, some of the obstacles involve the use of large libraries of known-source bacteria, the inability to "type" many environmental isolates, the nontransportability of regional known-source libraries, the presence of a large number of cosmopolitan isolates, the requirement for complex statistical analyses, low rates of correct classification, excessive false-positive and false-negative reactions, and the lack of adequate quality control procedures. In library-independent methods, markers can be used in a binary, semiquantitative, or quantitative fashion, but there are a number of assumptions for each of these approaches. One major assumption is that representation of the source via marker detection is measurable, constant, and approximately equal (the latter is not necessary, but makes it a lot easier). Other issues relate to the lack of criteria for sampling design and geographic distribution of genetic markers. Based on this above list, the challenges appear to be numerous. However, it is hoped that the application of currently available, but not yet applied, technologies and future advances in detection and characterization methodologies will most likely alleviate many of these problems. Some suggested ways in which this might occur are discussed below.

As was discussed in Chapter 2, most of the currently used MST methodologies require the use of libraries of known-source organisms to identify unknown environmental isolates. However, it has become apparent that the population structure of most currently used indicator organisms is diverse (73, 90, 91, 119, 120), and thus, large numbers of microbial strains need to be included in host source libraries. The composition and size of these libraries is often empirically determined. However, the minimum size of a useful library is not currently known. It has been estimated, from rarefaction analysis and MST studies, that a library size of up to 2×10^5 to 5×10^5 isolates may be needed to capture all the genetic diversity present in *Escherichia coli* (91; M. Samadpour, personal communication). This is not surprising in light of the significant genetic variation that *E. coli* exhibits in natural environments (189). Moreover, in order to be useful in space and time, the library used must be representative of those microbes in the intestinal tracts of animals across wide geographical areas, have little food-induced change in structure, and be capable of determining sources of fecal contamination throughout the year.

The Organism(s) of Choice

There is considerable debate as to the choice of organism(s) to be used as an indicator of fecal contamination of waterways and for MST studies. The logical selection of the organism(s) to be used is often dictated by microbial survival and habitat issues. Enterococci are being assayed to identify putative markers for MST; these microbes are good targets because they are easy to cultivate on selective media and there is a well-established track record for using this group of organisms in water quality monitoring. However, other researchers argue that targeting *E. coli* is a more logical choice, since this bacterium is more frequently monitored than enterococci and is used by most states and local agencies as an indicator of fecal pollution (152). Others argue that equally good information can come from studies done using bacteriophages (161), enteric viruses (59), or other bacteria (64). As expected, there are both pros and cons for the use of each of these organisms in MST studies. However, the central question still remains as to which microbe should be targeted for MST studies. One strategy would be to focus on species/strains that comprise the largest proportion of the gut flora and for which there has been a very long commensal or mutualistic relationship. These relationships would allow for the evolution of the greatest degree of host specificity, yielding better correlation with the presence/absence of a particular microbe and host origin. For example, while bacterial communities in the human gut can achieve remarkably high densities (188), these organisms only represent 8 of 55 known bacterial divisions (9). More than half of the human gut microbiota belongs to the *Cytophaga-Flavobacterium-Bacteroides* and the *Firmicutes* (e.g., *Clostridium* and *Eubacterium*) divisions. While *E. coli* is perhaps the most widely used bacterium for MST studies in nonmarine environments, the γ-*Proteobacteria* typically do not dominate the mammalian gut community (148, 149). Indeed, the first library-independent markers described for microbial source tracking came from the *Cytophaga-Flavobacterium-Bacteroides* group (11, 12). Alternately, another strategy would be to choose an organism with the following attributes: (i) its survival best mimics pathogenic microorganisms in the environment, (ii) it displays limited population heterogeneity, (iii) it is frequently found in the intestinal tract in a host-specific manner, and (iv) it is easy to culture and quantify.

Microbial Ecology and Diversity

Bacteria

The identification of sources of fecal pollution depends on the assumption that particular fecal microorganisms preferentially colonize a specific gut environment over other potential habitats. Clearly, the evolution of host specificity or preferential host distribution of gut microorganisms could be explained as the net result of a series of dynamic and multitrophic cellular and biochemical interactions between the host and microbes. These interactions have been estimated to involve 300 to 400 different bacterial species (57, 123) in the mammalian gastrointestinal (GI) tract, although this might be a conservative estimate considering the fact that the numbers are based on culturing techniques and phenotypic characterization. In terms of genomic content, the mole percent G+C content of gut bacterial genomes varies from low (30% G+C in clostridia) to moderately high (≥60% in bifidobacteria). Eukaryotic and archaeal microorganisms (48), as well as bacteriophages (phages) (161) and enteric viruses (21), also inhabit the gut systems of most warm-blooded animals. Similar microbial groups might inhabit cold-blooded animal gut systems, with the invertebrate gut being among the most-studied gut environments (50). While the gut systems of cold-blooded animals (i.e., poikilotherms) are also reservoirs of fecal microbes, their relevance to public health is poorly understood. However, the fact that many aquatic organisms can harbor (15), and some (i.e., shellfish) in many cases can even concentrate, fecal microorganisms, including fecal indicators, suggests that they can also play a role in the microbial quality of natural water systems.

There are four basic types of digestive systems in animals: simple stomach or monogastric (man, pig), equine-modified simple stomach (horse), ruminant or polygastric (sheep, cow, goat, deer), and avian (chicken, turkey). In general terms, the types of food ingested are also different among the different groups, varying from primarily herbivorous diets (horse, sheep, cow, and goat) to an omnivorous diet (human). Gastrointestinal systems are predominantly anoxic and greatly populated by bacteria of the genera *Bacteroides*, *Clostridium*, *Bifidobacterium*, *Eubacterium*, *Lactobacillus*, *Fusobacterium*, *Peptococcus*, *Peptostreptococcus*, *Rhodococcus*, *Ruminococcus*, *Streptococcus*, and *Lactococcus* (5, 115, 123, 153). It should be noted that historical knowledge on the ecology of many gut microorganisms strictly relied on the use of inefficient cultivation techniques. The use of novel molecular biology tools is revealing new information on the taxonomy of gut microbes, which in turn is providing a new picture regarding the ecology, abundance, and diversity of the gut microorganisms, most of which have escaped cultivation (48).

While similar microbial groups are present in different gut systems, differences in the microbial composition of different animal guts have been documented using 16S rRNA gene-based methods. For example, based on analyses of 16S rRNA gene clone libraries, bacteria affiliated with the *Cytophaga-Flexibacter-Bacteroides* group may represent less than 5%, 11%, 27 to 49%, and 40% of gut microflora in chickens (108, 196), pigs (105), humans (184), and cows (166), respectively. In contrast, less than 0.1% of *Proteobacteria* sequences (including *Escherichia coli*) were reported to be in the human intestinal tract (48). Microbial composition is also different within the different parts of the GI tract (113, 183, 184). Facultative anaerobes are numerically dominant in the upper part of the colon, while strict anaerobic bacteria predominate in the lower part. A more in-depth description of the diversity and ecology of some selected gut microbial groups will follow, with emphasis on those microorganisms that are relevant to environmental monitoring and fecal source identification.

Low-G + C gram-positive bacteria are the predominant members of the gut systems of most warm-blooded animals, representing up to 70 and 81% of 16S rRNA clone sequences found in horses (36) and pigs (105), respectively. Most of the low-G + C gram-positive bacteria in the gut are anaerobic spore-forming rods of the genus *Clostridium,* although *Selenomonas* and *Ruminococcus* may predominate in animals consuming high-starch-containing diets (99). Many clostridial species are involved in fermentation processes. The densities of *Clostridium coccoides* and *Clostridium leptum* subgroups in human feces have been reported to be $\log_{10}$ 10.3 and 9.9 cells per g (wet weight), respectively (116). In a recent European survey, each of these clostridial subgroups, on average, represented more than 25% of the total human colonic microbiota. Molecular studies have demonstrated the presence of 16S rRNA genes of clostridial groups of less than 97% sequence similarity to previously cultured bacteria, suggesting that novel phylotypes might be significant in gut systems (23). Similarly, Eckburg et al. (48) reported that 62% of the bacterial phylotypes found in the human gut were novel, and most were members of the phyla *Firmicutes* and *Bacteroidetes*. Several clostridrial species are considered to be important human pathogens as well as potential agents of bioterrorism (95). In addition, *Clostridium perfringens* has been suggested as a potential indicator of human wastewater fecal contamination. Like many fecal bacteria, this species has been isolated from the feces of other animals (24, 74). Moreover, strains with an identical ribotype to *Clostridium difficile* might be shared by humans and multiple animals (8), suggesting a wide distribution of similar clostridial genotypes in warm-blooded animals. Since many clostridial species are present in soil and survive outside the host, they should not solely be considered of fecal origin. Consequently, members of this genus might not be an ideal target in source identification studies.

Bacteroides species are common inhabitants of gut systems, and, as a result, it has been suggested that this group may be a useful alternate indicator of fecal pollution (56). Taxonomically, this genus belongs to the highly diverse *Bacteroidetes* phylum. The survival of *Bacteroides* outside of the gut is believed to be poor, which is a relevant trait when tracing recent fecal pollution events. However, a subcluster of this phylum was recently shown to be globally distributed in aquatic environments (35, 71, 130, 171), suggesting that some *Bacteroidetes* have developed efficient survival mechanisms. In addition, sediments are known to be an ecological, but perhaps transient, habitat for some *Bacteroidetes* (125, 155, 169). Within the genus *Bacteroides*, *Bacteroides vulgatus, Bacteroides thetaiotaomicron*, *Bacteroides uniformis*, and *Bacteroides fragilis* have normally been associated with human feces, although close relatives to these species have been documented in pigs, pets, and waterfowl (43). *Bacteroides* species are known to complement the host physiology by hydrolyzing polysaccharides (193). Sequence analysis of *Bacteroides*-like clone libraries has shown that many members of this group have escaped cultivation (43). Interestingly, some of the uncultured *Bacteroidetes* show a pattern of preferential host distribution. Thus far, however, molecular surveys have been performed only against a limited number of individual hosts, and therefore a more comprehensive examination is necessary to assess the level of host specificity of *Bacteroidetes* species.

Members of the genus *Bifidobacterium* are also recognized as common inhabitants of the animal gut. This bacterial group has regained the attention of the scientific community, due in part to its potential use in probiotics. Bifidobacteria are believed to be involved in the production of short-chain fatty acids which have been linked to the development of healthy guts. The survival of members of this bacterial group outside of the gut is very poor, and as a consequence, its presence in water samples is indicative of recent fecal pollution events. The analysis of the *Bifidobacterium longum* genome has revealed the presence of a genetic background capable of adapting to the conditions normally present in the colonic habitat (144). For example, genomic data have identified some ecologically relevant genes believed to be involved in the catabolism of a variety of oligosaccharides, presumably responsible for the production of glycoprotein-binding fimbriae used for adhesion and persistence in the gut. The most common gut species in adult humans are *Bifidobacterium adolescentis*, *Bifidobacterium catenulatum*, and *B. longum.* However, *Bifidobacterium angulatum*, *Bifidobacterium bifidum*,

Bifidobacterium breve, *Bifidobacterium dentium*, and *Bifidobacterium infantis* also have been isolated often from human feces (68, 78, 114). The density and composition of bifidobacterial species in the human gut changes with age (114). Due to their frequent occurrence in human gut, some *Bifidobacterium* species have been suggested to be good indicators of human pollution (67). However, the presence of bifidobacteria in other animals has not been thoroughly examined, even though PCR-based species assays are available (17). *Bifidobacterium*-specific assays have also been used to discriminate between different sources of fecal pollution (17, 38). It should be noted that species closely related to *Bifidobacterium suis*, *Bifidobacterium globosum*, *Bifidobacterium pseudolongum*, *Bifidobacterium thermophilum*, *Bifidobacterium boum*, and *Bifidobacterium choerinum* have been isolated from pig gut samples, and *B. globosum*, *Bifidobacterium ruminale*, *Bifidobacterium ruminantium*, and *Bifidobacterium merycicum* have been isolated from cows (10, 14, 170). Overall, the numbers of bifidobacteria in pigs and dogs appear to be relatively low (121, 191).

Members of the family *Enterobacteriaceae* are rod-shaped, gram-negative, facultative bacteria capable of fermentation or respiration depending on environmental conditions. The family includes the genera *Citrobacter*, *Enterobacter*, *Erwinia*, *Escherichia*, *Klebsiella*, *Proteus*, *Providencia*, *Salmonella*, *Serratia*, *Shigella*, and *Yersinia*. As a group, they are found in many environments, including the gut, soil, and water. Several members of the *Enterobacteriaceae* are also pathogens to humans, animals, and plants and are among the most studied organisms in modern microbiology. Most *Enterobacteriaceae* are relatively easy to grow, and for many species, like the fecal coliforms, the gut environment may be their primary habitat. While the fecal coliform group is found in the gut systems of practically all animals, their densities are much lower than many strictly anaerobic bacteria. In addition, several studies have shown that fecal coliforms can persist and grow outside of the GI tract, in soils (25, 26, 86) and water and in association with algae (187), bringing into question their use as indicators of fecal contamination. While it has been argued that there is only limited evidence for host specificity among strains of *Escherichia coli* (73) and *Campylobacter jejuni* (85), recent studies suggest that some strains group together in a host-specific manner (91). Strains (or serovars) of *Salmonella enterica* (serovars Dublin [167], Typhi [49], Abortus ovis [131] and Typhimurium [7]) and *Listeria monocytogenes* (190) also appear to group in a host-specific manner. However, specificity is not perfect, and detection in water has been shown to be dependent on shedding from the host, which is generally low within normal populations. Consequently, commensal flora will remain a primary target of interest for MST.

The gram-positive enterococci are also often used as indicators of fecal contamination, especially in marine environments. This is in part due to their frequent association with animal feces and their ability to withstand harsh environmental conditions better than the fecal coliforms. Enterococci have also been associated with plant material (126) and aquatic algae (187), while atypical enterococci have been isolated from seawater and invertebrates (163, 164). The enterococcus group is composed of facultative anaerobes with a fermentative metabolism that preferentially convert carbohydrates into lactic acid. Like other bacteria in the GI tract, enterococci are responsible for the fermentation of nonabsorbed sugars. There are almost 30 recognized species of enterococci, and *Enterococcus faecalis*, *Enterococcus faecium*, *Enterococcus durans*, and *Enterococcus hirae* are believed to be primarily associated with feces (154). Many other enterococcus species (e.g., *Enterococcus casseliflavus*, *Enterococcus gallinarum*, *Enterococcus dispar*, *Enterococcus mundtii*, and *Enterococcus avium*) have been isolated from animal feces; however, their host distribution has not been clearly established. Consequently, additional data are needed in order to assign specific enterococcal species to specific animal hosts. However, there have been several recent reports that some enterococci associate with animal hosts in a species-specific manner (147, 159), suggesting that some members of the genus may be useful in source tracking studies. Several *E. faecalis* and *E. faecium* strains are known to be pathogenic and capable of resisting "last resort" antibiotics like vancomycin. Based on the *E. faecalis* V583 genome sequence, these bacteria are capable of exchanging genetic material at a high frequency, explaining their frequent acquisition of drug resistance (133). Moreover, many genes in the enterococci are also involved in host-cell interaction (90, 168) and microbial cell-to-cell communication (47).

Phages and Viruses

The animal GI tract is also inhabited by a large number of phages and enteric viruses. Coliphages have been extensively studied due to their relevance to genetics, population biology, and bacterial evolution. However, while a number of fecal phages infecting bacteria of the genera *Clostridium*, *Bacteroides*, *Prevotella*, *Enterococcus*, *Ruminococcus*, and *Campylobacter* have also been studied, very little is known about the diversity of viruses in the animal gut, in part due to the overall reliance on culture-dependent techniques. Bacterial culture-based approaches are a relatively inefficient way of studying naturally occurring phages, considering that a significant fraction of environmental phages are specific to difficult-to-culture bacteria. Nevertheless, tissue culture-based assays are needed to study infectivity of enteric viruses, as they are known to be relevant human and animal pathogens. In theory, the diversity of phages in the gut should parallel or exceed bacterial diversity. Indeed, a recent viral metagenome study estimated the presence of 1,200 viral genotypes in human feces alone, most of phage origin (21). Some of the common matches in comparative sequence analyses were with phages infecting nontypical fecal bacteria like *Listeria monocytogenes* and *Burkholderia thailandensis,* raising some interesting possibilities regarding the movement of viral genes between biomes (142). Cann et al. (28) showed that, similar to results found in human studies, equine feces are dominated by a highly diverse, and a yet to be characterized, viral community that is dominated by members of the *Siphoviridae* family.

Protozoa

The diversity of protozoa in feces is likely to be grossly underestimated due to the following: (i) the difficulty of isolating viable cysts in vitro, (ii) biases introduced by historical interests towards clinically relevant strains, and (iii) the lack of adequate culturing technologies. Not surprisingly, research involving gut protozoa has primarily focused on a limited number of genera, including *Cryptosporidium*, *Microsporidium*, *Giardia*, and *Cyclospora* (72). Several of these protozoan species have been implicated in waterborne (34) and foodborne (37) outbreaks, while others that may be responsible for animal diseases also have important economical implications to the food industry (118). Many intestinal protozoa invade gut cells, while others can inhabit the gut as extracellular parasites. The fecal-oral transmission cycle is an integral part of the survival of intestinal protozoa. Evidence for the transmission of several protozoa between multiple host species is widely documented and explains the emergence of these organisms as important zoonotic agents (65). The life cycle of these organisms includes forms that are known to be resistant to a number of environmental factors as well as disinfection treatments used in the drinking water and wastewater industries. Currently, the association between a given protozoan species and particular animal hosts has not been well established, limiting their use in MST studies.

Ecological Role of the Gut Microbiota and Evidence of Host Specificity

Gut microbiota perform important functions for the host, many of which are deemed beneficial from nutritional and immunological standpoints (127). In humans, bacteria are capable of fermenting indigestible dietary residues to short-chain fatty acids relevant to the host physiology. Human gut bacteria also participate in other processes vital to humans, including vitamin synthesis and absorption of calcium, magnesium, and iron (4). From an immunological standpoint, intestinal bacteria have been suggested to play an important role in the development of immunocompetent cells in the gut mucosa. In addition, the resident gut microbiota is also believed to serve as a protection barrier against the colonization of opportunistic pathogens (75). The role of most protozoa and enteric viruses may not be beneficial to the host, although, like pathogenic bacteria, at times they could play relevant roles in the host's population dynamics. For example, the large diversity of phages is clearly relevant to the microbial genetic landscape of the gut bacterial community (21).

Microbial colonization of the human gut occurs shortly after birth, primarily via maternal transmission. Most bacterial groups in the gut persist throughout the entire life of the host. However, variation in the microbial community structure of the gut over time has been documented by using fluorescently labeled probes for 16S rRNA genes and sequence analyses of 16S rRNA gene clone libraries (48, 63). Differences in the bacterial composition have also been observed within parts of the animal gut. For example, the chicken ileum is dominated by *Lactobacillus* spp. (nearly 70%), whereas *Clostridium* species tend to be numerically superior in the cecum (108). In addition, changes in the colonic bacterial community structure have been reported in pigs subjected to different diets (104). Taken together, these data suggest that a combination of anatomical, physiological, and dietary differences have a significant impact on the overall types and proportions of microbial populations that inhabit a particular gut system (111).

Several studies have provided evidence suggesting that different hosts may preferentially harbor certain types of gut microorganisms. This is particularly the case for enteric viruses, as they strictly depend on host cells for proliferation. Due to their narrow host distribution, enteric viruses are good targets for the development of host-specific assays. However, wider application of viral assays used in the future will most likely depend on significantly improving viral particle concentration techniques and molecular assays (60).

Intestinal protozoan species also depend on gut cells for survival. Therefore, it is reasonable to propose that intimate interactions with the host might evolve into the preferential distribution of specific gut protozoan genotypes. Indeed, different levels of host specificity have been reported for protozoan species using phylogenetic analyses of the small subunit rRNA gene (13). However, this relationship may not be strict, as it has also been shown that *Giardia duodenalis* subgenotypes can be shared among different hosts (98).

Studies using genomic fingerprinting techniques suggest that there is differential distribution of bacterial genotypes in some hosts (29, 45). In such cases, however, specific bacterial genes have not been linked to hosts in a specific manner. More direct genetic evidence of host specificity in bacteria has come from studies using *Bifidobacterium* and *Bacteroides* small subunit rRNA gene-based analyses (11), toxin genes and adhesion factors in *E. coli* (81, 93,

94), and surface proteins in enterococci and *Bacteroides* (147, 152). A recent study showed that a 16S rRNA gene-PCR–based *Bacteroides*-specific assay directed against human-borne bacteria could generate false-positive results against those of dogs, chickens, and turkeys (30). Since the validation of the latter assay has been performed with a limited number of isolates from a relatively small number of hosts, it is not possible to conclusively establish the level of host specificity of the studied phylotypes. This is somewhat to be expected with assays that target 16S rRNA genes, as these genes are involved in general protein synthesis and not in host-microbial interactions. While it is necessary to further ascertain the distribution of these markers in targeted and nontargeted hosts, the current 16S rRNA gene-based assays suggest that some phylotypes might indeed be preferentially distributed in certain animals and that some level of coevolution might exist between the host and some members of the gut microbiota.

Additional genetic evidence for host-specific MST markers has come from studies using virulence genes, such as the heat-labile enterotoxin LTII and the heat-stable enterotoxin STII in *E. coli* (32). Taking advantage of the fact that some enterotoxigenic *E. coli* strains are more frequently associated with some animals, Khatib et al. (93, 94) developed LTII and STII PCR-based assays specific to cattle and swine. More recently, Hamilton and coworkers (81) reported the identification of a hybridization-based molecular marker, encoding a putative adhesion factor, showing host specificity for *E. coli* originating from geese and ducks. While these methods require cultivation to identify fecal sources in water samples, the host specificity of these assays further confirms that some bacterial species have developed specific mechanisms to interact with particular hosts and promote their survival (79). The *esp* gene, which encodes a cell wall-associated protein involved in cell adhesion in enterococci, is another example of a bacterial virulence gene that has been shown to display host specificity (147). However, since a preenrichment step is needed to increase the sensitivity of the *E. coli* toxin and enterococci *esp* assays, this suggests that only a subpopulation of these fecal bacterial groups have intimate interactions with the host. More recently, a *Bacteroides thetaiotaomicron* human-specific PCR-based assay was developed using a non-rRNA gene chromosomal region encoding, in part, a hypothetical protein and a putative permease (30). The latter marker was present in 92 and 100% of the human and sewage samples tested, respectively, and absent in cattle, chicken, turkey, horse, swine, and goose samples. It should be noted, however, that 16% of the dogs tested shared the *B. thetaiotaomicron* marker with humans. In general, these studies demonstrate that genes involved in host-microbial interactions may prove to be the best targets for the development of future MST assays.

The Search for New MST Markers: an "omics" Perspective

As previously mentioned, genes encoding 16S rRNA have been used in PCR assays as targets for the development of host-specific markers. However, the cross-amplification with nonspecific fecal samples implies that 16S rRNA genes may not be an ideal target for the development of host-specific assays. Moreover, relying on one gene or one bacterial genus for the development of MST markers may be of limited use, as this approach only uses a small fraction of the entire microbial genetic diversity present in the animal gut. This has prompted some researchers to take a multiphasic approach in source tracking studies (16). However, since a significant fraction of the gut microbial community remains to be discovered, it is reasonable to speculate that relevant genetic differences in functional genes involved in host-microbial interactions are present in yet-to-be-cultured microorganisms. It is necessary to look for alternate methods that do not strictly rely on culture-based approaches in order to discover genes involved in host-microbial interactions (e.g., cell adhesion, cell surface recognition, and avoidance of immunological responses) for which little to no genetic information is available.

Genome sequence analyses, as well as the emergence of other "omic" based technologies including metagenomics, proteomics, and transcriptomics, will enhance our understanding of the molecular and biochemical diversity of the animal GI tract. Linking all aspects of "fecal omics" is a task that will involve emerging technologies, such as microarrays, as well as computational biology and bioinformatics. Such a "systems biology approach" will allow us to understand biochemical processes governing host-microbial interactions and, in turn, will provide the basis to better understand gut physiology and improved host health. From a microbial source tracking standpoint, a systems biology approach will also enhance the number of functional genetic targets that could be used for the development of assays to monitor fecal contamination, identify sources of fecal pollution, and detect fecal pathogens. Such a comprehensive monitoring approach is needed to develop accurate risk assessment models and to evaluate management practices. Like most emerging "omics" areas, researchers involved in "fecal omics" will face the challenge of characterizing an immense number of undiscovered genes and determining which of these genes are unique to specific host-microbial interactions.

Genomics

The availability of complete microbial genome sequences represents a source of information that could help elucidate unique interactions between microbial gut populations and host organisms. Examples of sequenced gut bacterial genomes include those of *Lactobacillus johnsonii*, *B. longum*, *B. fragilis*,

B. thetaiotaomicron, *C. jejuni*, *Helicobacter pylori*, *C. perfringens*, *E. faecalis*, *E. faecium*, *E. coli* O157:H7, *E. coli* K-12, *Salmonella enterica* serovar Typhimurium, *S. enterica*, and *Shigella flexneri*. Several of these bacteria are known to be opportunistic pathogens. Intra- and interspecies genome comparisons of pathogenic and nonpathogenic strains have revealed important aspects of the lifestyles of closely related species, as well as virulence genes that are involved in host-microbial interactions. For example, genome comparisons of *B. thetaiotaomicron* and *B. fragilis* have revealed that the utilization of dietary polysaccharides and the capacity to change surface antigens are two different mechanisms that most likely allow the successful colonization of these species in different parts of the human gut (97). Similar comparisons can be performed with bacterial isolates that show preferential host distribution against cosmopolitan isolates (i.e., those that have been recovered in multiple hosts). Once a pool of genes that are different (or absent) between such strains is identified, studies could then focus on the use of genetic differences to develop host-specific assays via genome subtraction methods (139, 150). It is reasonable to propose that genes involved in host-microbial interactions, like those involved in the production of adhesion proteins, membrane transporters, and transcriptional regulators, will emerge as potential markers from genomic comparisons (81, 84, 93, 94, 112, 152). Additional mechanisms leading to unique interactions are also likely to emerge in the future as the number of sequenced genomes increases.

Genome sequences of hundreds of enteric viruses are also available in public databases (http://www.ncbi.nlm.nih.gov/genomes/VIRUSES/viruses.html). Of significant relevance to fecal source identification are the genomes from the genus *Mastadenovirus* (*Adenoviridae* family), as they infect only mammals. Thus far, the genomes of adenoviruses from bovine, canine, human, ovine, porcine, and simian sources have been sequenced. Whole-genome sequence comparisons between these groups will likely generate several host-specific targets that may be useful for source identification. In some cases, multiple strains from the same subgroup have already been sequenced, which will assist in the identification of conserved sequences that can be used in the development of PCR-based assays.

Similar outcomes are anticipated as a result of protozoan genome-sequencing studies. While genome sequence information for protozoa is not as wide ranging as that for bacteria and viruses, sequence analyses have shown strong host dependence due to the lack of biosynthetic pathways (1). In addition, analyses of the genomes of *Cryptosporidium parvum* and *Cryptosporidium hominis* suggest the presence of proteins potentially relevant to host-protozoan interactions (1, 2). Subtle sequence differences may explain some of the phenotypic differences between these otherwise closely related species (194). Since *C. hominis* is restricted to humans, divergent sequences might be useful in the identification of host-specific loci.

Host cells are an integral part of the host-microbial interaction paradigm. Consequently, genetic information on the host is also necessary to understand the complexity of gut bitrophic interactions. Besides the human genome, the genomes of some of the animals relevant to fecal contamination have been sequenced (http://www.animalgenome.org/) or are in the process of being sequenced (http://www.genome.gov). In addition, the genomes of the chimp (122) and the rat (69) were recently sequenced, providing useful models to detect unique genetic and physiological differences between different mammals and humans (52). This sequence information can also be used to identify genes responsible for maintaining a balance between the occurrence of specific beneficial bacteria in the gut and the host intrinsic immunological response.

Metagenomics

The emergence of high-throughput DNA sequencers and robotics is responsible for the rapid accumulation of genome sequence information in the last decade. Sequencing of pure-culture genomes is becoming a routine exercise in some modern microbiology laboratories. Recently, genomic sequencing has expanded beyond the study of clonal cultures to the analysis of whole microbial community genomes extracted from natural habitats. This emerging discipline is known as metagenomics, although the terms ecogenomics and environmental genomics have also been used. Similar to pure-culture studies, once high-quality genomic DNA is extracted from an environmental sample it is then possible to sequence a large number of relatively small DNA fragments generated by random shotgun cloning. An alternate approach relies on generating larger DNA fragments inserted into bacterial artificial chromosome vectors prior to sequencing (141). In general terms, the objective of metagenomic approaches is the analysis of genomic information extracted directly from environmental communities without relying on culturing techniques (82). Most extractable metagenomic sequences might belong to uncultured microorganisms; therefore, these studies provide a more complete picture of the genetic capacity of natural microbial populations.

Metagenomic analyses have been performed for several environments, including soil (141), marine water (39), and drinking-water biofilms (146). The most extensive metagenomic study was performed with Sargasso Sea samples, where more than a billion base pairs of nonredundant sequences were annotated, and over a million previously unknown genes were identified from an estimated 1,800 genomic species represented in the clone library (179). The sequencing information was primarily derived from numerically dominant species, and it is therefore reasonable to believe that additional novel marine microbial genetic diversity is yet to be discovered. Sequencing the metagenome of highly diverse systems, like pristine soils, will require an even greater level of effort which might not be a viable approach to study complex microbial communities. However, in simpler microbial systems, metagenomic sequencing possibly can provide a good picture of the relevant metabolic pathways and microbial interactions governing the microbial community under study. Metagenomic approaches also can be used to screen for novel bioactive compounds (70, 134, 145), metabolic and biocatalytic reactions of biotechnological value (96, 102, 195), and genes involved in the degradation of xenobiotic compounds (51).

The gut metagenome has been the focus of several studies due to the importance of the gut microbiota to the host's health and its potential value in bioprospecting. In a recent study, Galbraith et al. (66) recovered 81 clones that were unique between two different cows fed similar diets. Using a different genomic subtraction approach and whole-community DNA, Shanks et al. (151) were able to enrich for metagenomic fragments that were different between cow and pig feces. Some DNA fragments presumptively identified as bacterial surface proteins were then used to develop host-specific assays, proving that metagenomic approaches are a viable source of fecal source identification markers. Although the relevancy of these metagenomic fragments to the host physiology is unknown, these results suggest that there are intraspecies differences between microbial communities inhabiting the same gut environment. Sequence analyses of the rumen metagenomic clones revealed a significant number of sequences with no significant match with entries in the publicly available databases. Analysis of metagenomic clones from cecal and colon contents of BALB/c mice also showed the presence of open reading frames with low sequence identity to current database entries (182). These results are not surprising due to the fact that genome sequencing information available for most numerically dominant gut bacteria is scarce.

The study of the gut metagenome has clear implications to human health. This analysis provides a genetic foundation to understand the molecular interactions that result in beneficial products to the host and the mechanisms that promote the invasion and reproduction of infectious agents. From an ecological perspective, metagenomic approaches will help us understand the functional basis for the adaptation of microbial communities in the animal gut and identify the genetic differences between gut microbial communities that are driven by microbial processes relevant to the gut ecosystem. If structural differences between different gut microbial communities are due to more than just changes in the proportions of the numerically dominant microbial populations, then metagenomic approaches will also likely accelerate the discovery of microbial host-specific genes. Such genes represent a rich source of targets that could turn into useful environmental monitoring markers.

Transcriptomics

The goal of transcriptomics is to study the subset of genes that are transcribed or differentially regulated in an organism at a given point in time or under specific growth conditions. Therefore, transcriptomic data serve as a link between genes, gene expression, and biochemical/phenotypic outcomes (124). As hundreds of genes can be simultaneously transcribed in a cell, current surveys involve the use of microarrays to assay the expression of entire genomes. The available genome sequencing data coupled with data mining of gene expression profiles can be used to construct metabolic networks (103). This information can then be used for metabolic engineering (101), for understanding pathways associated with quorum-sensing inhibition (137), bacterial stress, and starvation responses (129), and for the development of molecular diagnostics (175).

Transcriptomics have been used to reveal eukaryotic genes that respond to the gut microbiota. For example, gut-resident microbiota have been shown to affect the expression of genes involved in immune function and water transport in mice. In the digestive tract of zebrafish more than 200 genes were recently found to be regulated by the gut microbiota (138). Some of the responses found in the latter study are also conserved in the mouse intestine and involved in epithelial proliferation, nutrient metabolism, and immune responses. These studies suggest that there is a coordinated biochemical network between the host intestinal cells and the gut microbiota. It is reasonable to speculate that this network is potentially responsible for the persistence of certain bacteria in an otherwise competitive ecosystem. Future studies using comparative transcriptomics will further identify genes that are critical to host-microbial interactions, some of which might exhibit host specificity.

Most microbial genome-scale metabolic studies involving transcriptomics have been performed using pure cultures. However, it is also possible to identify bacterial transcripts after extraction of total RNA from mixed microbial communities. Poretsky et al. (135) recently developed an approach that consisted of the direct retrieval of total RNA from marine and freshwater samples, followed by a subtractive hybridization step to remove rRNA, a DNase treatment to remove coextracted DNA, and a reverse transcriptase PCR step using random primers to generate cDNA. In this study, sequencing analysis of cloned transcripts showed the presence of transcripts linked to sulfur, carbon, and nitrogen biogeochemical processes.

Another transcriptomic approach relies on the addition of a polyadenine tail to environmental RNA, followed by several enzymatic modifications that allow for the introduction of a priming site at the 3′ end of the cDNA (20).

Since total RNA (instead of mRNA) is used in this approach, clone libraries are heavily biased towards rRNA genes. However, this approach should help identify active bacterial populations in the animal gut.

Proteomics

Not all genes that are transcribed in response to an environmental stimulus are expressed as functional proteins. For this reason, the complete molecular characterization of any organism ultimately involves the identification of expressed proteins. In the past, analyses of expressed proteins have involved tedious separation techniques and two-dimensional gel electrophoresis. Recent biotechnological advances have allowed biologists to study the proteome (all proteins expressed in the cell under specific conditions) using automated methods capable of simultaneously dissecting all proteins in a cell or sample (181). Proteomic comparisons of microbial cells exposed to different conditions can facilitate the identification and quantification of enzymes critical to specific biochemical pathways, as well as those involved in pathogenesis and colonization (185). Proteomics analyses can also be used to identify open reading frames that otherwise would be difficult to identify using conventional bioinformatics techniques (92).

Microbial ecologists have recently expanded the use of microbial proteomics, similarly to metagenomics, from solely studying pure cultures to exploring the proteomes of environmental microbial communities. Ram et al. (136) identified over 2,000 proteins expressed in a natural acid mine drainage microbial biofilm community using a shotgun mass spectrometry approach. It is anticipated that microbial community proteomics will soon be used to describe protein expression in other environmental habitats. In the future, studies focusing on the gut microbial proteome will be necessary to completely integrate a systems biology approach to better explain the evolution of host-specific microbial populations in different gut systems and the enzymatic pathways that play a key role in gut-microbial interactions. The understanding of gene expression in situ will be of great value for the development of in vitro cellular models that are needed to accelerate the identification of potential targets for host-specific assay development.

Future Technologies

Future MST technologies most likely will rely on non-library-based methods. Library-independent methods to determine sources of fecal bacteria can be characterized as being growth dependent (e.g., requiring culturing) or growth independent. Moreover, these methods can be qualitative (e.g., merely reporting the absence or presence of a particular host-specific microbe) or quantitative. In our opinion, future developments in MST will focus on library-independent, host-specific markers, analyzed using micro- or macroarrays or PCR-based technologies. Ideally these markers should allow quantitative analysis of the contributors to fecal contamination. The markers should be selectively maintained within the microorganism of interest and would avoid the use of cumbersome library-based methods and statistically intense schemes. Additionally, these markers would not be constrained by time and space and could be useful in all environmental applications. The key question, however, is how do we identify these markers? One approach would be to identify markers via fitness-based assays, but these assays are often time consuming and tedious. Moreover, while certain mutations have been shown to alter fitness (140), they often fail to identify specific markers. Another approach is to use a rational selection process, such as using recognized virulence markers. Scott et al. (147) recently proposed that the *esp* gene from *E. faecium* could serve as a human-host-specific marker for MST. A third approach is to attempt to correlate the presence of specific microbes that are only found in specific hosts. For example, Field and colleagues (11, 12, 41, 42) have used terminal restriction fragment length polymorphism methods to identify 16S rRNA markers that appear to differentiate between organisms that are highly correlated with either humans or other animals. The obvious extension from these efforts is to screen a larger number of markers to identify gene fragments that are highly correlated with host origin for many bacterial species inhabiting the GI tract. The only practical way for this strategy to work, however, is for a large number of markers to be screened simultaneously, and the most efficient tool available for this purpose is DNA microarrays.

Microarrays as tools for screening potential markers

Microarrays are typically constructed on glass slides and are composed of a lattice of spatially registered "spots," where each spot (50 to 250 μm in diameter) represents a single probe sequence. In this case, the probes are complementary to target sequences for microbial genomes. Genomic DNA (or cDNA if expression information is needed) from bacteria is labeled and hybridized to the microarray, and this process is repeated for multiple bacterial isolates. Resulting hybridization data are analyzed to identify the presence or absence of gene fragments between strains of bacteria. In the context of marker identification, this process can be used to identify genetic markers associated with particular groups or lineages of bacteria. For MST applications, the genetic markers are ultimately of interest to the practitioners. The microarrays themselves simply provide a convenient platform for simultaneously screening thousands of markers. Once markers are identified, the microarrays are no longer needed.

Whole-genome microarrays

The information gained from microarray hybridizations is a function of the probes that are used and the targets evaluated. At the simplest level, one could propose to use a whole-genome microarray to compare the genomic content of multiple isolates of a given species of bacteria. The goal would be to identify "polymorphic" genes (present in some strains while absent in others) that are highly correlated with host origin of isolation. Indeed, this approach is incredibly powerful when the goal is to identify genes that are "nonessential" to defining a given species of bacteria, to assess genetic diversity and to identify potential horizontally transferable elements (31, 46, 58, 80, 107, 165). The primary limitation for MST applications, however, is that the only point of reference is the original whole-genome sequence that was used to construct the DNA microarray. Consequently, if genes that are involved in differential fitnesses within different hosts are not represented in the original genome, then they will be missed.

There are several strategies to remedy the limitations of using a single reference strain for comparative genomics. One approach is to incorporate unique gene fragments as they are identified through ongoing whole-genome sequencing projects. Indeed, it is becoming increasingly common to find bacterial genome microarrays composed of "core" genes common to several strains plus unique genes identified for a few strains. While these composite microarrays are very powerful tools, they are still limited by both the number of strains that can be sequenced and the representation of strains relative to different host animals. Understandably, current emphasis involves either unique species or replicate strains important to human health.

Multistrain microarrays

Another approach for microarray analysis relies on the use of random gene fragments rather than a defined set of genes. For example, Cho and Tiedje (33) constructed a microarray composed of 60 to 96 genome fragments from each of four reference genomes of *Pseudomonas* spp. Comparative hybridization of genomic DNA (gDNA) from multiple test strains permitted them to derive taxonomic relationships between different species of *Pseudomonas*. Clearly, this type of microarray incorporates more genetic variation than would be present with a whole-genome microarray, and the random nature of probe selection limits bias in the types of probes represented with the array. From an MST perspective, however, if only a limited number of libraries are constructed, there is some degree of chance involved in selecting the strains to be cloned. More importantly, failure to select a representative, host-specific strain would prevent the method from identifying appropriate markers. In addition, there are questions about how many libraries should be constructed and how many probe sequences should be included from each library. These sample-size considerations are a function of the underlying genetic diversity of the bacteria being sampled (probably unknowable), so logistical constraints will be the deciding factor in construction of the microarray.

Mixed-genome microarrays

If we assume that incorporating more genetic diversity enhances the probability of detecting group-specific genetic markers, then the logical extension would be to use a mixed-genome microarray (19, 27). Like other random library-based microarrays, these arrays are composed of probes that are selected from a shotgun library, but in this case gDNA is pooled from many bacterial isolates prior to construction of the library. The goal is to incorporate as much natural genetic diversity as possible so that subsequent hybridization experiments can be used to partition that variation amongst serovars, clonal variants, lineages, or, potentially, hosts. Similarly to Cho and Tiedje (33), Borucki et al. (19) constructed a shotgun library using 10 different strains of *L. monocytogenes* representing four different serovars. In this and a subsequent paper (27), they showed that the microarray could be used to identify previously recognized phylogenetic lineages. Importantly, several markers were identified that could be used to distinguish between lineages, and this information was combined with other published literature to devise a PCR-based method for serotyping *L. monocytogenes* (18). Consequently, a mixed-genome microarray has the potential to identify multiple markers that are suitable for distinguishing between genetic groups, and it may be possible to extend this strategy for source tracking.

MST Markers

Suppression subtraction hybridization (SSH)-based microarrays

Shotgun cloning has proven to be a useful method for deriving a library for microarray probes, but it is likely that over 85% of the cloned gene fragments will be shared amongst all strains of the same species of bacteria. Consequently, only a relatively small proportion of probes would be potentially useful for marker development. An alternative approach is to enrich the library for genetic differences using a method such as SSH (139). The goal of this method is to amplify unique DNA fragments from the "tester" strain relative to a "driver" strain. DNA fragments that are common to the tester and driver strains are suppressed during the amplification process so that the majority of resulting products represent DNA that is unique to the tester strain. Unique fragments are cloned, and these cloned inserts can serve either as targets or probes for microarray hybridizations. Li et al. (106) used SSH to generate libraries for five species of *Dendrobrium*. They screened library inserts against a membrane-based macroarray composed of gDNA from 72 strains of *Dendrobrium* to identify markers that were unique to each of the five species. In another approach, Soule et al. (157) used SSH to compare two strains of *Flavobacterium psychrophilum*. Reciprocal SSH libraries were used to construct a microarray that was subsequently used to compare genetic relationships between 34 strains of *F. psychrophilum*. The analysis suggests that this species is divided into at least two distinct genetic lineages that are highly correlated with host origin (trout versus salmon). Subsequent genetic and phenotypic analyses have corroborated the hypothesized genetic lineages (158), and this analysis has identified lineage-specific markers.

In the context of MST, SSH methods could be used to generate reciprocal libraries between strains of bacteria isolated from defined hosts. Assuming that host-specific strains are used to construct the libraries, then this would be a very powerful method to identify genetic markers for MST. This is, to some extent, putting the cart before the horse, because you must determine a priori which strains are host specific, or you must conduct a large number of SSH experiments to maximize the probability of including host-specific strains. However, if a large genetically or phenotypically characterized library of known-source strains is available, and strains can be divided by source group, then this would facilitate selection of strains used for SSH (81). A variation on this scheme would be to subtract entire communities from specific hosts (151). For example, gDNA could be extracted from enteric bacteria isolated from either humans or cattle, and SSH could be used to develop a library with unique fragments from each of these communities. While the resulting library is likely to be complex, it may be possible to screen the library against the original community DNA using methods described by Li et al. (106). An SSH approach was recently used to generate goose- and duck-specific markers for *E. coli* that can be used in high-throughout MST studies (81).

Using microarrays to identify MST markers

One potential strategy to identify MST markers using DNA microarrays is outlined in Fig. 1. As with any technique, the initial effort focuses on developing a reference library for screening. In this case, reference strains may be collected from humans, cattle, deer, and geese (A, Fig. 1). In this scenario, multiple isolates of bacteria are pooled to develop a mixed-genome microarray for each host. The organism selected for this analysis could be an easy-to-culture bacterial indicator, such as enterococci or *E. coli*. It is difficult to say how many isolates should be pooled within each library. Clearly, a greater number of isolates provides a better chance of

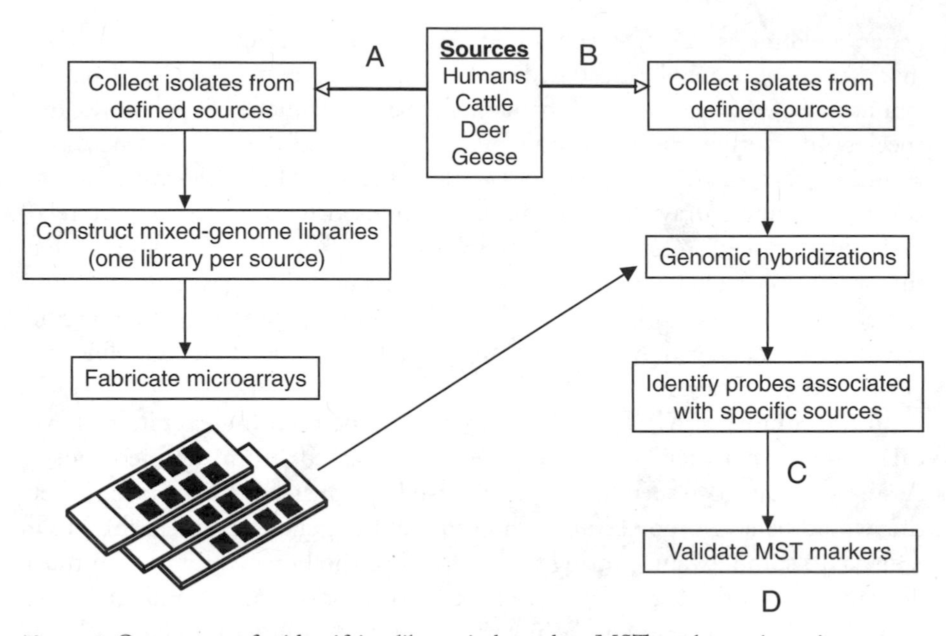

Figure 1 One strategy for identifying library-independent MST markers using microarrays and comparative genomics.

including source-specific strains, but too many isolates may reduce the probability of detecting host-specific markers if the strains harboring these markers represent only a small portion of the total population. It is preferable to limit the number of isolates from each individual sampled in order to reduce any bias that would otherwise be introduced from a single host. Construction of the shotgun libraries and glass slide microarrays involves relatively routine procedures (19, 27). Although there are no guidelines for the number of probes that should be included on the array, this is usually driven by logistical constraints.

The next step is to collect a "large number" of isolates (B, Fig. 1) and hybridize gDNA from these isolates to the microarrays. Probes that are correlated with specific hosts would be candidate markers for MST testing (C, Fig. 1). There are no guidelines for how many isolates should be screened with the microarray, nor can we predict a priori how significant the correlation should be to select candidate markers for further testing (D, Fig. 1). There are, however, likely to be errors due to transient carriage of non-host-specific strains as well as rare host-specific markers that may be difficult to identify if only a few strains are positive with the microarray.

The strategy described above was recently used to identify potential MST markers (159). The microarray is designed from five host-specific libraries, each composed of mixed-genome DNA from >40 isolates of enterococci collected from defined hosts. The microarray is composed of 4,300 gene fragments, and gDNA from over 350 isolates has been hybridized to the microarrays to identify 115 potential markers. Markers are identified based on average hybridization signal for each defined host. Candidate probes can be selected if they, on average, have twofold or greater signal intensity for a specific host relative to "non-specific" hosts. Preliminary validation relies on testing for the presence or absence of the markers (using PCR) in a collection of geographically diverse isolates, and this process has narrowed the number of candidate markers to 15 that distinguish between three host animals (humans, cattle, and deer). Positive control strains show that the markers are related to four different *Enterococcus* species. Thus, it appears that the mixed-genome microarray strategy can successfully identify candidate MST markers. Screening a large number of markers appears justified given that less than 1% of the markers may be suitable for MST applications.

Macroarrays

While microarray technologies have dominated transcriptome studies, and may be applied to MST studies, macroarrays are also a useful alternative to these technologies. Macroarrays are a variation of dot blot technologies initially developed for colony screening of clonal libraries, for large-scale gene-mapping studies, and for analysis of bacterial artificial chromosome libraries. Macroarrays can consist of up to 27,648 clones spotted, in duplicate, onto nylon membranes approximately 8 by 12 cm in size. Spots are usually greater than 300 microns in diameter and can consist of DNA or bacterial colonies. The number of spots on each macroarray is restricted by the diffusion of the probe molecules and the kinetics of hybridization (160). Macroarrays printed on nylon or other membrane support materials often use ^{32}P-labeled DNA or cDNA probes for target identification. Spotting of target cells on membranes is often facilitated by the use of robotic colony picking or DNA transfer systems (81). This is especially useful for MST studies where a large number of target strains may need to be examined using host-species-specific marker genes.

Quantitative PCR

Real-time PCR, often referred to as quantitative PCR (qPCR), is a technology that allows for the direct quantification of DNA or RNA templates from bacteria in soil, sediment, and water samples. In essence, all qPCR methods are based on the quantitative detection of fluorescent reporter molecules, which increase and accumulate with each successive PCR amplification cycle. The technology is based on the premise that the rate (and final amount) of PCR amplification is directly proportional to the amount of template that was originally present in a sample. The most commonly used fluorescent reporter molecules for qPCR are SYBR green I (a dye that binds double-stranded DNA) and sequence-specific probes based on Molecular Beacon or TaqMan technology. The TaqMan technology (ABI) allows quantification of the increase in the amount of PCR product, on a cycle-by-cycle basis, using probes that fluoresce when a quencher region is removed from the growing PCR fragment during the PCR extension cycle. This method is based on inherent $5' \rightarrow 3'$ exonuclease activity present in *Taq* DNA polymerase, which ultimately results in the cleavage of the fluorescently labeled probes during the PCR cycle. Typically, a TaqMan probe has two fluorescent molecules attached, a reporter dye such as 6-carboxyfluorescein (FAM), and a quenching, counter-fluorescent dye, such as 6-carboxytetramethylrhodamine (TAMRA). During the PCR cycle, degradation of the TaqMan probe by *Taq*'s $5' \rightarrow 3'$ exonuclease activity releases the reporter dye from the quenching activity of TAMRA. The increase in fluorescence due to direct cleavage of the probe is proportional to the amount of PCR product formed.

In contrast, molecular beacons (172) are single-stranded oligonucleotide hybridization probes that form a hairpin-shaped, stem-and-loop structure. The loop portion of the molecule is the probe sequence that is complementary to the target DNA or RNA molecule. The stem portion of the beacon is formed by annealing complementary arm sequences to the ends of the probe sequence, and a fluorescent and quenching moiety (a nonfluorescent chromophore) is attached to the end of each arm. By design, the stem keeps these two moieties in close proximity, which results in quenching of fluorescence in free solution. However, when the molecular beacon probe hybridizes to a DNA or RNA strand containing a target sequence, a spontaneous conformational change occurs forcing the stem apart. This results in the movement of the fluorophore and the quencher apart from each other, and restores fluorescence. Since molecular beacons can be made in several different colors by using a wide range of fluorophores (173), they allow for the detection of multiple targets in the same solution or bacterium. However, DABCYL [4-(4′-dimethylaminophenylazo)benzoic acid], the nonfluorescent chromophore, usually serves as the universal quencher for any fluorophore in a molecular beacon. The fluorescence produced by TaqMan or molecular beacon-type probes is typically measured using a PCR thermal cycler coupled to a laser spectrophotometer. Several companies make PCR machines that can monitor fluorescence in 96-well microtiter plates, allowing for high-throughput real-time quantitative analysis of PCR products produced during 35 to 40 cycles of PCR.

Quantitative PCR technologies are now being applied to the analysis of microorganisms germane to food (178, 192), water quality (22, 61, 83), and the environment (55, 100). The main advantage of qPCR over other methods is that the technology is rapid and quantitative, allowing analysis of fecal indicator bacteria in a matter of hours (22), and can be performed on-site using portable instrumentation (178). Currently, a number of primer sets and probes have been developed for the detection of fecal indicator organisms and waterborne pathogens (15, 22, 62, 63, 78, 109, 110, 117, 143). Recently Haugland and colleagues (83) used TaqMan-based probes for the qPCR analysis of *Enterococcus* in Lake Michigan and Lake Erie water samples. They reported that the technique allowed for the accurate and sensitive measurements of enterococci in <3 hours, at a detection limit of 27 cells per sample. Their analyses were facilitated by the use of exogenous positive control DNA from salmon, providing a test for inhibition of PCR. However, since there was not a perfect correlation between the results obtained by using membrane filtration analyses and qPCR, which is expected since not all enterococci in water are viable, these authors suggested that epidemiological analyses need to be performed before qPCR results can be used to directly predict water-associated health risks.

When coupled with host-specific markers, qPCR analyses hold great promise for MST studies. This technology will allow for the rapid detection and quantification of indicator organisms that have specific association with a given host animal. Perhaps the best studied host-specific marker system has been developed for the *Bacteroides-Prevotella* group. Marker genes that discriminate among *Bacteroides* spp. strains originating from human, horse, pig, and ruminant sources of fecal contamination have been developed based on differences in 16S rRNA (12, 30, 42). Recently, Dick and Field (41) demonstrated that 16S rRNA gene-derived marker genes targeting *Bacteroides* strains from human, cow, dog, cat, pig, elk, deer, and gull were useful for qPCR analyses of water samples. While all the primers used were not host species-group-specific, they nevertheless proved useful for the quantitative analysis of enterococci in water. It should be noted, however, that while qPCR holds great promise for MST studies, and will be applied more in the future, further work is needed to define better and more host-specific marker probes and primers. This is in part due to the fact that fecal *Bacteroidales* display both endemic and cosmopolitan distribution among host source groups (43), and it is assumed that similar distribution will be found using marker genes targeting other fecal bacteria. Nevertheless, future applications of qPCR to MST studies look very promising if multiple marker genes can be found to adequately describe specific host-bacterium interactions and future qPCR technology advances to the point where multiple samples can be analyzed in relatively short periods of time.

DNA Fingerprint Analyses and Simple Sequence Repeats (SSR)

DNA fingerprint technologies that allow differentiation among genetically similar bacterial strains have been used in the past in epidemiological studies and have provided the basis for tracking of a wide variety of pathogenic and commensal microorganisms. These technologies appear to provide the necessary discriminating power to differentiate among seemingly near-identical strains within a single genus that cannot be divided into subgroups based on phenotypic methods. At their core, all methods rely on the ability to obtain a snapshot survey of the microbial genome, by using hybridization, electrophoretic, or PCR-based technologies. Thus, the methods can be differentiated from each other based on the means used to visualize genomic differences among closely related strains. However, all the methods produce a pattern of DNA bands (a fingerprint) from an organism and then attempt to group organisms together based on similarities, or differences, in DNA banding patterns. When applied to MST studies, these methods have had varying degrees of success in separating coliform bacteria or enterococci into meaningful groups that correspond to host animals of origin. In the recent past, two major methods have been applied to MST, ribotyping (29, 132) and repetitive extragenic palindromic (rep)-PCR DNA fingerprinting (29, 45, 91, 120). The methods are fundamentally different, as the former technique uses hybridization of rRNA gene probes to restriction enzyme-digested chromosomal DNA to generate banding patterns, and the latter technique relies on PCR amplification of DNA regions between adjacent repetitive DNA elements in the genome (180). While each method has the ability to group strains on the basis of DNA banding patterns, rep-PCR generates more complicated banding patterns and interrogates a larger portion of the microbial genome than does ribotyping. Although the techniques are becoming widely adopted in MST studies, major shortcomings of both techniques are that they require the use of rather extensive libraries of strains from known source hosts, often fail to adequately identify all strains obtained from water sources, and are subject to false-positive and false-negative identifications. In addition, they tend to be labor intensive if not properly set up to capitalize on robotic systems.

Other DNA-fingerprinting methods are also gaining wide acceptance for use in epidemiological and genetic analyses and hold promise for future MST studies. One such DNA-fingerprinting technique involves the use of simple sequence repeat (SSR) elements. These repeated DNA elements were initially discovered in eukaryotes and have been alternately referred to as mini- or microsatellites (88, 89), short tandem repeats, or variable number of tandem repeat (VNTR) loci (128). The VNTR loci are sets of tandemly repeated DNA core sequences, varying from 11 to 60 bp in length. Large numbers of repetitive DNA sequences have been found in eukaryotes, with a few to thousands of copies dispersed throughout the genome. This repetitive DNA is mostly found in extragenic regions and typically consists of either homopolymeric stretches of single-nucleotide repeats of poly(A), poly(G), poly(T), or poly(C), or homogeneous repeats, heterogeneous repeats, or degenerate repeat sequence motifs. Over the last several years, both VNTRs and SSRs, which are highly polymorphic, have been used as molecular targets for analyses in humans (87, 162), as tools for breeding and segregation analyses in plants (197), and for phylogenetic analyses. Since the repeated regions are typically flanked by conserved endonuclease restriction sites, DNA fingerprints can be derived by restriction fragment length polymorphism analyses, or by coupling the PCR technique with primers flanking the conserved regions of VNTR loci (88).

The complete genome sequences of many prokaryotic species have also revealed the presence of large numbers of SSR elements, single-nucleotide polymorphic sites, and genome rearrangements (3, 53, 176, 186). In prokaryotes, many of the repeated elements can be found in intergenic as well as in extragenic loci. The presence of repeated elements in the prokaryotic genome can be exploited for DNA fingerprint analyses, allowing the differentiation of bacteria at the strain or isolate level. Since limited information about each strain, or organism, is needed for production of DNA fingerprints, and many polymorphic sites are present in each genome at the single- and multinucleotide levels, technologies based on SSR hold great promise for future MST studies. This will be facilitated by further sequence analysis of microbial genomes (76). Recently, Diamant and colleagues (40, 156) and Guy et al. (77) reported that large numbers of polymorphic SSRs of repeating mononucleotide motifs (termed MNRs) consisting of at least five repeats are present in the noncoding regions of the *E. coli* genome. These authors showed that the MNRs can be useful in inferring phylogenetic relationships among pathogenic and commensal *E. coli* strains. Moreover, the MNRs may be useful for strain typing and MST studies. A variation on this technology, termed amplified intergenic locus polymorphism analysis, has recently been reported by Somer and coworkers (156). In this particular case, the complete genomic sequence of *E. coli* K-12 was used to design primers for the PCR amplification of randomly repeated genomic targets in this bacterium. The presence, absence, and size variation of PCR amplification products were used as genetic markers for differentiating commensal strains as well as enteropathogenic *E. coli*, enterohemorrhagic *E. coli*, and enterotoxigenic *E. coli* strains. While amplified intergenic locus polymorphism analysis has thus far been used on only a limited number of strains, this technique may have future use in MST studies. However, as was discussed above for rep-PCR and for other fingerprinting methods, library-based methods all suffer from the same shortcomings and limitations. Consequently, for this technology to have a greater impact for MST, new library-independent methods based on repetitive DNA elements will need to be developed.

CONCLUDING REMARKS

Results from the microbial characterization of animal intestinal tracts have suggested that anatomical and physiological differences found in different digestive systems may lead to conditions that affect the structure of gut microbial communities. However, it remains to be determined whether structural differences between different gut microbial communities represent more than just proportional changes in numerically dominant microbial populations or whether they reflect true host-specific differences. If these structural differences translate into the evolution of microbial populations that interact with the host in a species-specific manner, then it seems reasonable to assume that some of the genes responsible for these interactions are good targets for the development of culture- and library-independent methods that will be useful for future MST studies.

Initial attempts to identify host-specific markers focused on using a one bacterial group-one gene target approach, particularly using *Bacteroides* 16S rRNA gene sequencing information. A few studies have shown the application of 16S rRNA gene-based methods in field studies (54). However, recent studies have shown that some 16S rRNA gene-based host-specific markers may also be present in nontargeted fecal sources (30). Thus, future work is needed to further define the specificity of these assays. Similar studies also need to be done to examine the specificity of nonphylogenetic markers that have recently been suggested to be useful as potential markers (81, 93, 94, 147, 151). Once assay specificity has thoroughly been tested, multilaboratory studies should be conducted in order to generate and validate standard operating procedures that will be used in regulatory activities.

In order to more accurately discriminate between different sources of fecal pollution, future MST approaches will most likely require the use of multiple host-specific markers for source identification. In order to achieve this level of discrimination, however, it is necessary to interrogate as many potential genes as possible from microbes inhabiting each gut type. The quest for host-specific markers will lead researchers to further explore the genomes of nontraditional indicator genera like *Bacteroides* and *Bifidobacterium*. Several species from these genera have been shown to interact with the host and, therefore, they should harbor genes involved in host-microbial interactions.

The rich genetic diversity of uncultured microorganisms will be another source for markers. It seems likely that metagenomic approaches as well as genomic-subtraction methods will be part of the path of gene discovery (66, 151). In addition, it is likely that high-throughput technologies (i.e., robotics, microarrays) that allow for the simultaneous detection of multiple targets will play a key role in the discovery of novel targets and in the validation and field application of future host-specific assays.

Future studies will also have to address issues related to the geographical and temporal stability of novel markers, particularly those associated with indicator bacteria. This is important in light of recent reports documenting the persistence and growth of *E. coli* in soils and water (86). The role that sediments play as reservoirs of different fecal bacteria should be critically evaluated. Since differences in the survival rates of fecal bacteria could affect the stability of the markers in different locations (6), it is possible that markers developed in temperate regions and for freshwater systems might not be adequate for identification of fecal sources in tropical countries and in marine waters. If the latter is true, it might be necessary to develop MST assays that are targeted to the specific conditions of the habitat in question. However, currently there are only limited data available on how different the environmental conditions need to be in order to significantly affect the ecology of the chosen fecal source indicator.

Conventional wisdom assumes that most fecal anaerobic bacteria will not survive for extended periods of time outside of their primary habitat in highly oxygenated waters. Poor survivability of anaerobes might be a positive trait when developing indicators to detect recent fecal contamination. While epidemiological data are lacking, it is possible that recent contamination events would show a stronger correlation with risks associated with exposure to polluted waters than to older contamination events. The markers or organisms used for fecal source identification should ultimately provide relevant information pertaining to exposure risks to humans. To establish such correlations, however, it will be necessary to develop quantitative assays targeting several host-specific markers as well as relevant pathogens. This is a very difficult challenge due to the long list of waterborne pathogens and the differences in their ecology. Consequently, it is unlikely that a single marker or organism can be used for risk assessment for all different types of environmental scenarios. Future efforts will also need to incorporate environmental monitoring schemes that simultaneously detect fecal sources and pathogenic agents using multiple markers and, preferably, in a quantitative manner. Such an approach will be necessary to develop better risk assessment models and to develop methods that can validate risk management practices.

REFERENCES

1. **Abrahamsen, M. S., T. J. Templeton, S. Enomoto, J. E. Abrahante, G. Zhu, C. A. Lancto, M. Deng, C. Liu, G. Widmer, S. Tzipori, G. A. Buck, P. Xu, A. T. Bankier, P. H. Dear, B. A. Konfortov, H. F. Spriggs, L. Iyer, V. Anantharaman, L. Aravind, and V. Kapur.** 2004. Complete genome sequence of the Apicomplexan, *Cryptosporidium parvum. Science* **304:**441–445.
2. **Abrahamsen, M. S., V. Kapur, S. Tzipori, and G. A. Buck.** 2004. The genome of *Cryptosporidium hominis. Nature* **431:**1107–1112.
3. **Alland, D., T. S. Whittam, M. B. Murray, M. D. Cave, M. H. Hazbon, K. Dix, M. Kokoris, A. Duesterhoeft, J. A. Eisen, C. M. Fraser, and R. D. Fleischmann.** 2003. Modeling bacterial evolution with comparative-genome-based marker systems: application to *Mycobacterium tuberculosis* evolution and pathogenesis. *J. Bacteriol.* **185:**3392–3399.
4. **Alverdy, J., O. Zaborina, and L. Wu.** 2005. The impact of stress and nutrition on bacterial-host interactions at the intestinal epithelial surface. *Curr. Opin. Clin. Nutr. Metab. Care* 8:205–209.
5. **Amit-Romach, E., D. Sklan, and Z. Uni.** 2004. Microflora ecology of the chicken intestine using 16S ribosomal DNA primers. *Poult. Sci.* **83:**1093–1098.
6. **Anderson, K. L., J. E. Whitlock, and V. J. Harwood.** 2005. Persistence and differential survival of fecal indicator bacteria in subtropical waters and sediments. *Appl. Environ. Microbiol.* **71:**3041–3048.
7. **Andrews-Polymenis, H. L., W. Rabsch, S. Porwollik, M. McClelland, C. Rosetti, L. G. Adams, and A. J. Baumler.** 2004. Host restriction of *Salmonella enterica* serotype Typhimurium pigeon isolates does not correlate with loss of discrete genes. *J. Bacteriol.* **186:**2619–2628.
8. **Arroyo, L. G., S. A. Kruth, B. M. Willey, H. R. Staempfli, D. E. Low, and J. S. Weese.** 2005. PCR ribotyping of *Clostridium difficile* isolates originating from human and animal sources. *J. Med. Microbiol.* **54:**163–166.
9. **Bäckhed, F., R. E. Ley, J. L. Sonnenburg, D. A. Peterson, and J. I. Gordon.** 2005. Host-bacterial mutualism in the human intestine. *Science* **307:**1915–1920.
10. **Beerens, H., B. Hass Brac de la Perriere, and F. Gavini.** 2000. Evaluation of the hygienic quality of raw milk based on the presence of bifidobacteria: the cow as a source of faecal contamination. *Int. J. Food Microbiol.* **54:**163–169.
11. **Bernhard, A. E., and K. G. Field.** 2000. Identification of nonpoint sources of fecal pollution in coastal waters by using host-specific 16S ribosomal DNA genetic markers from fecal anaerobes. *Appl. Environ. Microbiol.* **66:**1587–1594.
12. **Bernhard, A. E., and K. G. Field.** 2000. A PCR assay to discriminate human and ruminant feces on the basis of host differences in *Bacteroides-Prevotella* genes coding for 16S rRNA. *Appl. Environ. Microbiol.* **66:**4571–4574.
13. **Berrilli, F., D. Di Cave, C. De Liberato, A. Franco, P. Scaramozzino, and P. Orecchia.** 2004. Genotype characterisation of *Giardia duodenalis* isolates from domestic and farm animals by SSU-rRNA gene sequencing. *Vet. Parasitol.* **122:**193–199.
14. **Biavati, B., and P. Mattarelli.** 1991. *Bifidobacterium ruminantium* sp. nov. and *Bifidobacterium merycicum* sp. nov. from the rumens of cattle. *Int. J. Syst. Bacteriol.* **41:**163–168.
15. **Blackstone, G. M., J. L. Nordstrom, M. C. L. Vickery, M. D. Bowen, R. F. Meyer, and A. DePaola.** 2003. Detection of pathogenic *Vibrio parahaemolyticus* in oyster enrichments by real-time PCR. *J. Microbiol. Methods* **53:**149–155.

16. **Blanch, A. R., L. Belanche-Munoz, X. Bonjoch, J. Ebdon, C. Gantzer, F. Lucena, J. Ottoson, C. Kourtis, A. Iversen, I. Kuhn, L. Moce, M. Muniesa, J. Schwartzbrod, S. Skraber, G. Papageorgiou, H. D. Taylor, J. Wallis, and J. Jofre.** 2005. Tracking the origin of faecal pollution in surface water: an ongoing project within the European Union research programme. *J. Water Health* **2**:249–260.

17. **Bonjoch, X., E. Ballesté, and A. R. Blanch.** 2004. Multiplex PCR with 16S rRNA gene-targeted primers of *Bifidobacterium* spp. to identify sources of fecal pollution. *Appl. Environ. Microbiol.* **70**:3171–3175.

18. **Borucki, M., and D. R. Call.** 2003. *Listeria monocytogenes* serotype identification using PCR. *J. Clin. Microbiol.* **41**:5537–5540.

19. **Borucki, M., M. Krug, W. Muraoka, and D. Call.** 2003. Discrimination among *Listeria monocytogenes* isolates using a mixed genome DNA microarray. *Vet. Microbiol.* **92**:351–362.

20. **Botero, L. M., S. D'Imperio, M. Burr, T. R. McDermott, M. Young, and D. J. Hassett.** 2005. Poly(A) polymerase modification and reverse transcriptase PCR amplification of environmental RNA. *Appl. Environ. Microbiol.* **71**:1267–1275.

21. **Breitbart, M., I. Hewson, B. Felts, J. M. Mahaffy, J. Nulton, P. Salamon, and F. Rohwer.** 2003. Metagenomic analyses of an uncultured viral community from human feces. *J. Bacteriol.* **185**:6220–6223.

22. **Brinkman, N. E., R. A. Haugland, L. J. Wymer, M. Byappanahalli, R. L. Whitman, and S. J. Vesper.** 2003. Evaluation of a rapid, quantitative real-time PCR method for enumeration of pathogenic *Candida* cells in water. *Appl. Environ. Microbiol.* **69**:1775–1782.

23. **Brooks, S. P., M. McAllister, M. Sandoz, and M. L. Kalmokoff.** 2003. Culture-independent phylogenetic analysis of the faecal flora of the rat. *Can. J. Microbiol.* **49**:589–601.

24. **Bueschel, D. M., B. H. Jost, S. J. Billington, H. T. Trinh, and J. G. Songer.** 2003. Prevalence of cpb2, encoding beta2 toxin, in *Clostridium perfringens* field isolates: correlation of genotype with phenotype. *Vet. Microbiol.* **94**:121–129.

25. **Byappanahalli, M., and R. Fujioka.** 2004. Indigenous soil bacteria and low moisture may limit but allow faecal bacteria to multiply and become a minor population in tropical soils. *Water Sci. Technol.* **50**:27–32.

26. **Byappanahalli, M. N., R. L. Whitman, D. A. Shively, M. J. Sadowsky, and S. Ishii.** 2006. Population structure, persistence, and seasonality of autochthonous *Escherichia coli* in temperate, coastal forest soil from a Great Lakes watershed. *Environ. Microbiol.* **8**:504–513.

27. **Call, D., M. Borucki, and T. Besser.** 2003. Mixed-genome microarrays reveal multiple serotype and lineage-specific differences among strains of *Listeria monocytogenes*. *J. Clin. Microbiol.* **41**:632–639.

28. **Cann, A. J., S. E. Fandrich, and S. Heaphy.** 2005. Analysis of the virus population present in equine faeces indicates the presence of hundreds of uncharacterized virus genomes. *Virus Genes* **30**:151–156.

29. **Carson, C. A., B. L. Shear, M. R. Ellersieck, and J. D. Schnell.** 2003. Comparison of ribotyping and repetitive extragenic palindromic-PCR for identification of fecal *Escherichia coli* from humans and animals. *Appl. Environ. Microbiol.* **69**:1836–1839.

30. **Carson, C. A., J. M. Christiansen, H. Yampara-Iquise, V. W. Benson, C. Baffaut, J. V. Davis, R. R. Broz, W. B. Kurtz, W. M. Rogers, and W. H. Fales.** 2005. Specificity of a

Bacteroides thetaiotaomicron marker for human feces. *Appl. Environ. Microbiol.* **71**:4945–4949.

31. **Chan, K., S. Baker, C. C. Kim, C. S. Detweiler, G. Dougan, and S. Falkow.** 2003. Genomic comparison of *Salmonella enterica* serovars and *Salmonella bongori* by use of an *S. enterica* serovar Typhimurium DNA microarray. *J. Bacteriol.* **185**:553–563.
32. **Chern, E. C., Y. L. Tsai, and B. H. Olson.** 2004. Occurrence of genes associated with enterotoxigenic and enterohemorrhagic *Escherichia coli* in agricultural waste lagoons. *Appl. Environ. Microbiol.* **70**:356–362.
33. **Cho, J., and J. M. Tiedje.** 2001. Bacterial species determination from DNA-DNA hybridization by using genome fragments and DNA microarrays. *Appl. Environ. Microbiol.* **67**:3677–3682.
34. **Craun, G. F., R. L. Calderon, and M. F. Craun.** 2005. Outbreaks associated with recreational water in the United States. *Int. J. Environ. Health Res.* **15**:243–262.
35. **Crump, B. C., G. W. Kling, M. Bahr, and J. E. Hobbie.** 2003. Bacterioplankton community shifts in an arctic lake correlate with seasonal changes in organic matter source. *Appl. Environ. Microbiol.* **69**:2253–2268.
36. **Daly, K., C. S. Stewart, H. J. Flint, and S. P. Shirazi-Beechey.** 2001. Bacterial diversity within the equine large intestine as revealed by molecular analysis of cloned 16S rRNA genes. *FEMS Microbiol. Ecol.* **38**:141–151.
37. **Dawson, D.** 2005. Foodborne protozoan parasites. *Int. J. Food Microbiol.* **103**:207–227.
38. **Delcenserie, V., N. Bechoux, T. Leonard, B. China, and G. Daube.** 2004. Discrimination between *Bifidobacterium* species from human and animal origin by PCR-restriction fragment length polymorphism. *J. Food Prot.* **67**:1284–1288.
39. **Delong, E. F.** 2005. Microbial community genomics in the ocean. *Nat. Rev. Microbiol.* **3**: 459–469.
40. **Diamant, E., Y. Palti, R. Gur-Arie, H. Cohen, E. M. Hallerman, and Y. Kashi.** 2004. Phylogeny and strain typing of *Escherichia coli*, inferred from variation at mononucleotide repeat loci. *Appl. Environ. Microbiol.* **70**:2464–2473.
41. **Dick, L. K., and K. G. Field.** 2004. Rapid estimation of numbers of fecal *Bacteroidetes* by use of a quantitative PCR assay for 16S rRNA genes. *Appl. Environ. Microbiol.* **70**:5695–5697.
42. **Dick, L. K., M. T. Simonich, and K. G. Field.** 2005. Microplate subtractive hybridization to enrich for *Bacteroidales* genetic markers for fecal source identification. *Appl. Environ. Microbiol.* **71**:3179–3183.
43. **Dick, L. K., A. E. Bernhard, T. J. Brodeur, J. W. Santo Domingo, J. M. Simpson, S. P. Walters, and K. G. Field.** 2005. Host distributions of uncultivated fecal *Bacteroidales* bacteria reveal genetic markers for fecal source identification. *Appl. Environ. Microbiol.* **71**:3184–3191.
44. **Dobrindt, U.** 2005. (Patho-) Genomics of *Escherichia coli*. *Int. J. Med. Microbiol.* **295**:357–371.
45. **Dombek, P. E., L. K. Johnson, S. T. Zimmerley, and M. J. Sadowsky.** 2000. Use of repetitive DNA sequences and the PCR to differentiate *Escherichia coli* isolates from human and animal sources. *Appl. Environ. Microbiol.* **66**:2572–2577.
46. **Dorrell, N., J. A. Mangan, K. G. Laing, J. Hinds, D. Linton, H. Al-Ghusein, B. G. Barrell, J. Parkhill, N. G. Stoker, A. V. Karlyshev, P. D. Butcher, and B. W. Wren.** 2001.

Whole genome comparison of *Campylobacter jejuni* human isolates using a low-cost microarray reveals extensive genetic diversity. *Genome Res.* **11**:1706–1715.

47. **Dunny, G. M.** 1990. Genetic functions and cell-cell interactions in the pheromone-inducible plasmid transfer system of *Enterococcus faecalis. Mol. Microbiol.* **4**:689–696.
48. **Eckburg, P. B., E. M. Bik, C. N. Bernstein, E. Purdom, L. Dethlefsen, M. Sargent, S. R. Gill, K. E. Nelson, and D. A. Relman.** 2005. Diversity of the human intestinal microbial flora. *Science* **308**:1635–1638.
49. **Edsall, G. S., S. Gaines, and M. Landy.** 1960. Studies on infection and immunity in experimental typhoid fever. I. Typhoid fever in chimpanzees orally infected with *Salmonella typhosa. J. Exp. Med.* **112**:143–166.
50. **Egert, M., U. Stingl, L. D. Bruun, B. Wagner, A. Brune, and M. W. Friedrich.** 2005. Structure and topology of microbial communities in the major gut compartments of *Melolontha melolontha* larvae (Coleoptera: Scarabaeidae). *Appl. Environ. Microbiol.* **71**:4556–4566.
51. **Eyers, L., I. George, L. Schuler, B. Stenuit, S. N. Agathos, and S. E. Fantroussi.** 2004. Environmental genomics: exploring the unmined richness of microbes to degrade xenobiotics. *Appl. Microbiol. Biotechnol.* **66**:123–130.
52. **Fay, J.** 2006. Human genome: which proteins contribute to human-chimpanzee differences? *Eur. J. Hum. Genet.* **14**:506.
53. **Field, D., and C. Wills.** 1996. Long, polymorphic microsatellites in simple organisms. *Proc. R. Soc. Lond. B Biol. Sci.* **263**:209–215.
54. **Field, K. G., E. C. Chern, L. K. Dick, J. Fuhrman, J. Griffith, P. A. Holden, M. G. LaMontagne, J. Le, B. Olson, and M. T. Simonich.** 2003. A comparative study of culture-independent, library-independent genotypic methods of fecal source tracking. *J. Water Health* **1**:181–194.
55. **Fierer, N., J. A. Jackson, R. Vilgalys, and R. B. Jackson.** 2005. Assessment of soil microbial community structure by use of taxon-specific quantitative PCR assays. *Appl. Environ. Microbiol.* **71**:4117–4120.
56. **Fiksdal, L., J. S. Maki, S. J. LaCroix, and J. T. Staley.** 1985. Survival and detection of *Bacteroides* spp., prospective indicator bacteria. *Appl. Environ. Microbiol.* **49**:148–150.
57. **Finegold, S. M., H. R. Attebery, and V. L. Sutter.** 1974. Effect of diet on human fecal flora: comparison of Japanese and American diets. *Am. J. Clin. Nutr.* **27**:1456–1469.
58. **Fitzgerald, J. R., D. E. Sturdevant, S. M. Mackie, S. R. Gill, and J. M. Musser.** 2001. Evolutionary genomics of *Staphylococcus aureus*: insights into the origin of methicillin-resistant strains and the toxic shock syndrome epidemic. *Proc. Natl. Acad. Sci. USA* **98**:8821–8826.
59. **Fong, T. T., and E. K. Lipp.** 2005. Enteric viruses of humans and animals in aquatic environments: health risks, detection, and potential water quality assessment tools. *Microbiol. Mol. Biol. Rev.* **69**:357–371.
60. **Fong, T.-T., D. W. Griffin, and E. K. Lipp.** 2005. Molecular assays for targeting human and bovine enteric viruses in coastal waters and their application for library-independent source tracking. *Appl. Environ. Microbiol.* **71**:2070–2078.
61. **Foulds, I. V., A. Granacki, C. Xiao, U. J. Krull, A. Castle, and P. A. Horgen.** 2002. Quantification of microcystin-producing cyanobacteria and *E. coli* in water by 5′-nuclease PCR. *J. Appl. Microbiol.* **93**:825–834.

62. **Frahm, E., and U. Obst.** 2003. Application of the fluorogenic probe technique (TaqMan PCR) to the detection of *Enterococcus* spp. and *Escherichia coli* in water samples. *J. Microbiol. Methods* **52:**123–131.

63. **Franks, A. H., H. J. Harmsen, G. C. Raangs, G. J. Jansen, F. Schut, and G. W. Welling.** 1998. Variations of bacterial populations in human feces measured by fluorescent in situ hybridization with group-specific 16S rRNA-targeted oligonucleotide probes. *Appl. Environ. Microbiol.* **64:**3336–3345.

64. **Fujioka, R. S.** 2001. Monitoring coastal marine waters for spore-forming bacteria of faecal and soil origin to determine point from non-point source pollution. *Water Sci. Technol.* **44:**181–188.

65. **Gajadhar, A. A., and J. R. Allen.** 2004. Factors contributing to the public health and economic importance of waterborne zoonotic parasites. *Vet. Parasitol.* **126:**3–14.

66. **Galbraith, E. A., D. A. Antonopoulos, and B. A. White.** 2004. Suppressive subtractive hybridization as a tool for identifying genetic diversity in an environmental metagenome: the rumen as a model. *Environ. Microbiol.* **6:**928–937.

67. **Gavini, F., A. M. Pourcher, C. Neut, D. Monget, C. Romond, C. Oger, and D. Izard.** 1991. Phenotypic differentiation of bifidobacteria of human and animal origins. *Int. J. Syst. Bacteriol.* **41:**548–557.

68. **Germond, J. E., O. Mamin, and B. Mollet.** 2002. Species specific identification of nine human *Bifidobacterium* spp. in feces. *Syst. Appl. Microbiol.* **25:**536–543.

69. **Gibbs, R. A., G. M. Weinstock, M. L. Metzker, D. M. Muzny, E. J. Sodergren, S. Scherer, G. Scott, D. Steffen, K. C. Worley, P. E. Burch, G. Okwuonu, S. Hines, L. Lewis, C. DeRamo, O. Delgado, S. Dugan-Rocha, G. Miner, M. Morgan, A. Hawes, R. Gill, R. A. Holt, M. D. Adams, P. G. Amanatides, H. Baden-Tillson, M. Barnstead, S. Chin, C. A. Evans, S. Ferriera, C. Fosler, A. Glodek, Z. Gu, D. Jennings, C. L. Kraft, T. Nguyen, C. M. Pfannkoch, C. Sitter, G. G. Sutton, J. C. Venter, T. Woodage, D. Smith, H.-M. Lee, E. Gustafson, P. Cahill, A. Kana, L. Doucette-Stamm, K. Weinstock, K. Fechtel, R. B. Weiss, D. M. Dunn, E. D. Green, R. W. Blakesley, G. G. Bouffard, P. J. de Jong, K. Osoegawa, B. Zhu, M. Marra, J. Schein, I. Bosdet, C. Fjell, S. Jones, M. Krzywinski, C. Mathewson, A. Siddiqui, N. Wye, J. McPherson, S. Zhao, C. M. Fraser, J. Shetty, S. Shatsman, K. Geer, Y. Chen, S. Abramzon, W. C. Nierman, P. H. Havlak, R. Chen, K. J. Durbin, A. Egan, Y. Ren, X.-Z. Song, B. Li, Y. Liu, X. Qin, S. Cawley, K. C. Worley, A. J. Cooney, L. M. D'Souza, K. Martin, J. Q. Wu, M. L. Gonzalez-Garay, A. R. Jackson, K. J. Kalafus, M. P. McLeod, A. Milosavljevic, D. Virk, A. Volkov, D. A. Wheeler, Z. Zhang, J. A. Bailey, E. E. Eichler, E. Tuzun, E. Birney, E. Mongin, A. Ureta-Vidal, C. Woodwark, E. Zdobnov, P. Bork, M. Suyama, D. Torrents, M. Alexandersson, B. J. Trask, J. M. Young, H. Huang, H. Wang, H. Xing, S. Daniels, D. Gietzen, J. Schmidt, K. Stevens, U. Vitt, J. Wingrove, F. Camara, M. Mar Alba, J. F. Abril, R. Guigo, A. Smit, I. Dubchak, E. M. Rubin, O. Couronne, A. Poliakov, N. Hubner, D. Ganten, C. Goesele, O. Hummel, T. Kreitler, Y.-A. Lee, J. Monti, H. Schulz, H. Zimdahl, H. Himmelbauer, H. Lehrach, H. J. Jacob, S. Bromberg, J. Gullings-Handley, M. I. Jensen-Seaman, A. E. Kwitek, J. Lazar, D. Pasko, P. J. Tonellato, S. Twigger, C. P. Ponting, J. M. Duarte, S. Rice, L. Goodstadt, S. A. Beatson, R. D. Emes, E. E. Winter, C. Webber, P. Brandt, G. Nyakatura, M. Adetobi, F. Chiaromonte, L. Elnitski, P. Eswara, R. C. Hardison, M. Hou, D. Kolbe, K. Makova, W. Miller, A. Nekrutenko, C. Riemer, S. Schwartz, J. Taylor, S. Yang, Y. Zhang, K. Lindpaintner, T. D. Andrews, M. Caccamo, M. Clamp, L. Clarke,**

V. Curwen, R. Durbin, E. Eyras, S. M. Searle, G. M. Cooper, S. Batzoglou, M. Brudno, A. Sidow, E. A. Stone, B. A. Payseur, G. Bourque, C. Lopez-Otin, X. S. Puente, K. Chakrabarti, S. Chatterji, C. Dewey, L. Pachter, N. Bray, V. B. Yap, A. Caspi, G. Tesler, P. A. Pevzner, D. Haussler, K. M. Roskin, R. Baertsch, H. Clawson, T. S. Furey, A. S. Hinrichs, D. Karolchik, W. J. Kent, K. R. Rosenbloom, H. Trumbower, M. Weirauch, D. N. Cooper, P. D. Stenson, B. Ma, M. Brent, M. Arumugam, D. Shteynberg, R. R. Copley, M. S. Taylor, H. Riethman, U. Mudunuri, J. Peterson, M. Guyer, A. Felsenfeld, S. Old, S. Mockrin, and F. Collins. 2004. Genome sequence of the Brown Norway rat yields insights into mammalian evolution. *Nature* **428**:493–521.

70. **Gillespie, D. E., S. F. Brady, A. D. Bettermann, N. P. Cianciotto, M. R. Liles, M. R. Rondon, J. Clardy, R. M. Goodman, and J. Handelsman.** 2002. Isolation of antibiotics turbomycin A and B from a metagenomic library of soil microbial DNA. *Appl. Environ. Microbiol.* **68**:4301–4306.
71. **Glockner, F. O., E. Zaichikov, N. Belkova, L. Denissova, J. Pernthaler, A. Pernthaler, and R. Amann.** 2000. Comparative 16S rRNA analysis of lake bacterioplankton reveals globally distributed phylogenetic clusters including an abundant group of actinobacteria. *Appl. Environ. Microbiol.* **66**:5053–5065.
72. **Goodgame, R.** 2003. Emerging causes of traveler's diarrhea: *Cryptosporidium*, Cyclospora, Isospora, and Microsporidia. *Curr. Infect. Dis. Rep.* **5**:66–73.
73. **Gordon, D. M.** 2001. Geographical structure and host specificity in bacteria and the implications for tracing the source of coliform contamination. *Microbiology* **147**:1079–1085.
74. **Griffiths, N. J., J. R. Walton, G. B. Edwards, M. Bennett, B. China, J. Mainil, S. Vandevenne, and C. A. Hart.** 1997. The prevalence of *Clostridium perfringens* in the horse. *Rev. Med. Microbiol.* **8**:S52–S54.
75. **Guarner, F., and J. R. Malagelada.** 2003. Gut flora in health and disease. *Lancet* **361**:512–519.
76. **Gur-Arie, R., C. J. Cohen, Y. Eitan, L. Shelef, E. M. Hallerman, and Y. Kashi.** 2000. Simple sequence repeats in *Escherichia coli*: abundance, distribution, composition, and polymorphism. *Genome Res.* **10**:62–71.
77. **Guy, R. A., C. Xiao, and P. A. Horgen.** 2004. Real-time PCR assay for detection and genotype differentiation of *Giardia lamblia* in stool specimens. *J. Clin. Microbiol.* **42**:3317–3320.
78. **Haarman, M., and J. Knol.** 2005. Quantitative real-time PCR assays to identify and quantify fecal *Bifidobacterium* species in infants receiving a prebiotic infant formula. *Appl. Environ. Microbiol.* **71**:2318–2324.
79. **Hajishengallis, G., H. Nawar, R. I. Tapping, M. W. Russell, and T. D. Connell.** 2004. The type II heat-labile enterotoxins LT-IIa and LT-IIb and their respective B pentamers differentially induce and regulate cytokine production in human monocytic cells. *Infect. Immun.* **72**:6351–6358.
80. **Hakenbeck, R., N. Balmelle, B. Weber, C. Gardes, W. Keck, and A. de Saizieu.** 2001. Mosaic genes and mosaic chromosomes: intra- and interspecies genomic variation of *Streptococcus pneumoniae*. *Infect. Immun.* **69**:2477–2486.
81. **Hamilton, M. J., T. Yan, and M. J. Sadowsky.** 2006. Development of goose- and duck-specific DNA markers to determine sources of *Escherichia coli* in waterways. *Appl. Environ. Microbiol.* **72**:4012–4019.

82. **Handelsman, J.** 2004. Metagenomics: application of genomics to uncultured microorganisms. *Microbiol. Mol. Biol. Rev.* **68:**669–685.
83. **Haugland, R. A., S. C. Siefring, L. J. Wymer, K. P. Brenne, and A. P. Dufour.** 2005. Comparison of *Enterococcus* measurements in freshwater at two recreational beaches by quantitative polymerase chain reaction and membrane filter culture analysis. *Water Res.* **39:**559–568.
84. **Hooper, L. V., and J. I. Gordon.** 2001. Glycans as legislators of host-microbial interactions: spanning the spectrum from symbiosis to pathogenicity. *Glycobiology* **11:**1–10.
85. **Hopkins, K. L., M. Desai, J. A. Frost, J. Stanley, and J. M. Logan.** 2004. Fluorescent amplified fragment length polymorphism genotyping of *Campylobacter jejuni* and *Campylobacter coli* strains and its relationship with host specificity, serotyping, and phage typing. *J. Clin. Microbiol.* **42:**229–235.
86. **Ishii, S., W. B. Ksoll, R. E. Hicks, and M. J. Sadowsky.** 2006. Presence and growth of naturalized *Escherichia coli* in temperate soils from Lake Superior watersheds. *Appl. Environ. Microbiol.* **72:**612–621.
87. **Jasinska, A., and W. J. Krzyzosiak.** 2004. Repetitive sequences that shape the human transcriptome. *FEBS Lett.* **567:**136–141.
88. **Jeffreys, A. J., V. Wilson, and S. L. Thein.** 1985. Hypervariable "minisatellite" regions in human DNA. *Nature* **314:**67–73.
89. **Jeffreys, A. J., V. Wilson, R. Newumann, and J. Keyte.** 1988. Amplification of human minisatellites by the polymerase chain reaction: towards DNA fingerprinting of single cells. *Nucleic Acids Res.* **16:**10953–10971.
90. **Johnson, A. P.** 1994. The pathogenicity of enterococci. *J. Antimicrob. Chemother.* **33:**1083–1089.
91. **Johnson, L. K., M. B. Brown, E. A. Carruthers, J. A. Ferguson, P. E. Dombek, and M. J. Sadowsky.** 2004. Sample size, library composition, and genotypic diversity among natural populations of *Escherichia coli* from different animals influence accuracy of determining sources of fecal pollution. *Appl. Environ. Microbiol.* **70:**4478–4485.
92. **Jungblut, P. R., E. C. Müller, J. Mattow, and S. H. E. Kaufmann.** 2001. Proteomics reveals open reading frames in *Mycobacterium tuberculosis* H37Rv not predicted by genomics. *Infect. Immun.* **69:**5905–5907.
93. **Khatib, L. A., Y. L. Tsai, and B. H. Olson.** 2002. A biomarker for the identification of cattle fecal pollution in water using the LTIIa toxin gene from enterotoxigenic *Escherichia coli*. *Appl. Microbiol. Biotechnol.* **59:**97–104.
94. **Khatib, L. A., Y. L. Tsai, and B. H. Olson.** 2003. A biomarker for the identification of swine fecal pollution in water, using the STII toxin gene from enterotoxigenic *Escherichia coli*. *Appl. Microbiol. Biotechnol.* **63:**231–238.
95. **Klietmann, W. F., and K. L. Ruoff.** 2001. Bioterrorism: implications for the clinical microbiologist. *Clin. Microbiol. Rev.* **14:**364–381.
96. **Knietsch, A., T. Waschkowitz, S. Bowien, A. Henne, and R. Daniel.** 2003. Construction and screening of metagenomic libraries derived from enrichment cultures: generation of a gene bank for genes conferring alcohol oxidoreductase activity on *Escherichia coli*. *Appl. Environ. Microbiol.* **69:**1408–1416.
97. **Kuwahara, T., A. Yamashita, H. Hirakawa, H. Nakayama, H. Toh, N. Okada, S. Kuhara, M. Hattori, T. Hayashi, and Y. Ohnishi.** 2004. Genomic analysis of *Bacteroides fragilis*

reveals extensive DNA inversions regulating cell surface adaptation. *Proc. Natl. Acad. Sci. USA* **101**:14919–14924.

98. **Lalle, M., E. Jimenez-Cardosa, S. M. Caccio, and E. Pozio.** 2005. Genotyping of *Giardia duodenalis* from humans and dogs from Mexico using a beta-giardin nested polymerase chain reaction assay. *J. Parasitol.* **91**:203–205.
99. **Larue, R., Z. Yu, V. A. Parisi, A. R. Egan, and M. Morrison.** 2005. Novel microbial diversity adherent to plant biomass in the herbivore gastrointestinal tract, as revealed by ribosomal intergenic spacer analysis and *rrs* gene sequencing. *Environ. Microbiol.* **7**:530–543.
100. **Lebuhn, M., M. Effenberger, G. Garces, A. Gronauer, and P. A. Wilderer.** 2004. Evaluating real-time PCR for the quantification of distinct pathogens and indicator organisms in environmental samples. *Water Sci. Technol.* **50**:263–270.
101. **Lee, S. J., D. Y. Lee, T. Y. Kim, B. H. Kim, J. Lee, and S. Y. Lee.** 2005. Metabolic engineering of *Escherichia coli* for enhanced production of succinic acid, based on genome comparison and in silico gene knockout simulation. *Appl. Environ. Microbiol.* **71**:7880–7887.
102. **Lee, S. W., K. Won, H. K. Lim, J. C. Kim, G. J. Choi, and K. Y. Cho.** 2004. Screening for novel lipolytic enzymes from uncultured soil microorganisms. *Appl. Microbiol. Biotechnol.* **65**:720–726.
103. **Lengeler, J. W.** 2000. Metabolic networks: a signal-oriented approach to cellular models. *Biol. Chem.* **381**:911–920.
104. **Leser, T. D., R. H. Lindecrona, T. K. Jensen, B. B. Jensen, and K. Møller.** 2000. Changes in bacterial community structure in the colon of pigs fed different experimental diets and after infection with *Brachyspira hyodysenteriae*. *Appl. Environ. Microbiol.* **66**:3290–3296.
105. **Leser, T. D., J. Z. Amenuvor, T. K. Jensen, R. H. Lindecrona, M. Boye, and K. Møller.** 2002. Culture-independent analysis of gut bacteria: the pig gastrointestinal tract microbiota revisited. *Appl. Environ. Microbiol.* **68**:673–690.
106. **Li, T. X., J. Wang, Y. Bai, X. Sun, and Z. Lu.** 2004. A novel method for screening species-specific gDNA probes for species identification. *Nucleic Acids Res.* **32**:e45.
107. **Liang, X., X. Q. Pham, M. V. Olson, and S. Lory.** 2001. Identification of a genomic island present in the majority of pathogenic isolates of *Pseudomonas aeruginosa*. *J. Bacteriol.* **183**:843–853.
108. **Lu, J., U. Idris, B. Harmon, C. Hofacre, J. J. Maurer, and M. D. Lee.** 2003. Diversity and succession of the intestinal bacterial community of the maturing broiler chicken. *Appl. Environ. Microbiol.* **69**:6816–6824.
109. **Ludwig, W., and K. H. Schleifer.** 2000. How quantitative is quantitative PCR with respect to cell counts? *Syst. Appl. Microbiol.* **23**:556–562.
110. **Lyon, W. J.** 2001. TaqMan PCR for detection of *Vibrio cholerae* O1, O139, non-O1, and non-O139 in pure cultures, raw oysters, and synthetic seawater. *Appl. Environ. Microbiol.* **67**:4685–4693.
111. **Mai, V.** 2004. Dietary modification of the intestinal microbiota. *Nutr. Rev.* **62**:235–242.
112. **Maroncle, N., D. Balestrino, C. Rich, and C. Forestier.** 2002. Identification of *Klebsiella pneumoniae* genes involved in intestinal colonization and adhesion using signature-tagged mutagenesis. *Infect. Immun.* **70**:4729–4734.

113. **Marteau, P., P. Pochart, J. Doré, C. Béra-Maillet, A. Bernalier, and G. Corthier.** 2001. Comparative study of bacterial groups within the human cecal and fecal microbiota. *Appl. Environ. Microbiol.* **67:**4939–4942.

114. **Matsuki, T., K. Watanabe, R. Tanaka, M. Fukuda, and H. Oyaizu.** 1999. Distribution of bifidobacterial species in human intestinal microflora examined with 16S rRNA-gene-targeted species-specific primers. *Appl. Environ. Microbiol.* **65:**4506–4512.

115. **Matsuki, T., K. Watanabe, J. Fujimoto, Y. Miyamoto, T. Takada, K. Matsumoto, H. Oyaizu, and R. Tanaka.** 2002. Development of 16S rRNA-gene-targeted group-specific primers for the detection and identification of predominant bacteria in human feces. *Appl. Environ. Microbiol.* **68:**5445–5451.

116. **Matsuki, T., K. Watanabe, and R. Tanaka.** 2003. Genus- and species-specific PCR primers for the detection and identification of bifidobacteria. *Curr. Issues Intest. Microbiol.* **4:**61–69.

117. **Matsuki, T., K. Watanabe, J. Fujimoto, T. Takada, and R. Tanaka.** 2004. Use of 16S rRNA gene-targeted group-specific primers for real-time PCR analysis of predominant bacteria in human feces. *Appl. Environ. Microbiol.* **70:**7220–7228.

118. **McDougald, L. R.** 1998. Intestinal protozoa important to poultry. *Poult. Sci.* **77:**1156–1158.

119. **McLellan, S. L.** 2004. Genetic diversity of *Escherichia coli* isolated from urban rivers and beach water. *Appl. Environ. Microbiol.* **70:**4658–4665.

120. **McLellan, S. L., A. D. Daniels, and A. K. Salmore.** 2003. Genetic characterization of *Escherichia coli* populations from host sources of fecal pollution by using DNA fingerprinting. *Appl. Environ. Microbiol.* **69:**2587–2594.

121. **Mikkelsen, L. L., C. Bendixen, M. Jakobsen, and B. B. Jensen.** 2003. Enumeration of bifidobacteria in gastrointestinal samples from piglets. *Appl. Environ. Microbiol.* **69:**654–658.

122. **Mikkelsen, T. S., L. W. Hillier, E. E. Eichler, M. C. Zody, D. B. Jaffe, S.-P. Yang, W. Enard, I. Hellmann, K. Lindblad-Toh, T. K. Altheide, N. Archidiacono, P. Bork, J. Butler, J. L. Chang, Z. Cheng, A. T. Chinwalla, P. deJong, K. D. Delehaunty, C. C. Fronick, L. L. Fulton, Y. Gilad, G. Glusman, S. Gnerre, T. A. Graves, T. Hayakawa, K. E. Hayden, X. Huang, H. Ji, W. J. Kent, M.-C. King, E. J. Kulbokas III, M. K. Lee, G. Liu, C. Lopez-Otin, K. D. Makova, O. Man, E. R. Mardis, E. Mauceli, T. L. Miner, W. E. Nash, J. O. Nelson, S. Pääbo, N. J. Patterson, C. S. Pohl, K. S. Pollard, K. Prüfer, X. S. Puente, D. Reich, M. Rocchi, K. Rosenbloom, M. Ruvolo, D. J. Richter, S. F. Schaffner, A. F. A. Smit, S. M. Smith, M. Suyama, J. Taylor, D. Torrents, E. Tuzun, A. Varki, G. Velasco, M. Ventura, J. W. Wallis, M. C. Wendl, R. K. Wilson, E. S. Lander, and R. H. Waterston (The Chimpanzee Sequencing and Analysis Consortium).** 2005. Initial sequence of the chimpanzee genome and comparison with the human genome. *Nature* **437:**69–87.

123. **Moore, W. E. C., and L. V. Holdeman.** 1974. Human fecal flora: the normal flora of 20 Japanese-Hawaiians. *Appl. Microbiol.* **27:**961–979.

124. **Mostowy, S., C. Cleto, D. R. Sherman, and M. A. Behr.** 2004. The *Mycobacterium tuberculosis* complex transcriptome of attenuation. *Tuberculosis* **84:**197–204.

125. **Moune, S., P. Caumette, R. Matheron, and J. C. Willison.** 2003. Molecular sequence analysis of prokaryotic diversity in the anoxic sediments underlying cyanobacterial mats of two hypersaline ponds in Mediterranean salterns. *FEMS Microbiol. Ecol.* **44:**117–130.

126. **Muller, T., A. Ulrich, E. M. Ott, and M. Muller.** 2001. Identification of plant-associated enterococci. *J. Appl. Microbiol.* **91**:268–278.

127. **Mutch, D. M., R. Simmering, D. Donnicola, G. Fotopoulos, J. A. Holzwarth, G. Williamson, and I. Corthesy-Theulaz.** 2004. Impact of commensal microbiota on murine gastrointestinal tract gene ontologies. *Physiol. Genomics* **19**:22–31.

128. **Nakamura, Y., M. Leppert, P. O'Connell, R. Wolff, T. Holm, M. Culver, C. Martin, E. Fujimoto, M. Hoff, E. Kumlin, and R. White.** 1987. Variable number tandem repeat (VNTR) markers for human gene mapping. *Science* **235**:1616–1622.

129. **Nystrom, T.** 2001. Not quite dead enough: on bacterial life, culturability, senescence, and death. *Arch. Microbiol.* **176**:159–164.

130. **O'Sullivan, L. A., A. J. Weightman, and J. C. Fry.** 2002. New degenerate *Cytophaga-Flexibacter-Bacteroides*-specific 16S ribosomal DNA-targeted oligonucleotide probes reveal high bacterial diversity in River Taff epilithon. *Appl. Environ. Microbiol.* **68**:201–210.

131. **Pardon, P., R. Scanchis, J. Marly, F. Lantier, M. Pepin, and M. Popoff.** 1988. Ovine salmonellosis caused by *Salmonella abortus ovis*. *Ann. Rech. Vet.* **19**:221–235. (In French.)

132. **Parveen, S., K. M. Portier, K. Robinson, L. Edmiston, and M. L. Tamplin.** 1999. Discriminant analysis of ribotype profiles of *Escherichia coli* for differentiating human and nonhuman sources of fecal pollution. *Appl. Environ. Microbiol.* **65**:3142–3147.

133. **Paulsen, I. T., L. Banerjei, G. S. Myers, K. E. Nelson, R. Seshadri, T. D. Read, D. E. Fouts, J. A. Eisen, S. R. Gill, J. F. Heidelberg, H. Tettelin, R. J. Dodson, L. Umayam, L. Brinkac, M. Beanan, S. Daugherty, R. T. DeBoy, S. Durkin, J. Kolonay, R. Madupu, W. Nelson, J. Vamathevan, B. Tran, J. Upton, T. Hansen, J. Shetty, H. Khouri, T. Utterback, D. Radune, K. A. Ketchum, B. A. Dougherty, and C. M. Fraser.** 2003. Role of mobile DNA in the evolution of vancomycin-resistant *Enterococcus faecalis*. *Science* **299**:2071–2074.

134. **Piel, J., D. Butzke, N. Fusetani, D. Hui, M. Platzer, G. Wen, and S. Matsunaga.** 2005. Exploring the chemistry of uncultivated bacterial symbionts: antitumor polyketides of the pederin family. *J. Nat. Prod.* **68**:472–479.

135. **Poretsky, R. S., N. Bano, A. Buchan, G. LeCleir, J. Kleikemper, M. Pickering, W. M. Pate, M.-A. Moran, and J. T. Hollibaugh.** 2005. Analysis of microbial gene transcripts in environmental samples. *Appl. Environ. Microbiol.* **71**:4121–4126.

136. **Ram, R. J., N. C. VerBerkmoes, M. P. Thelen, G. W. Tyson, B. J. Baker, R. C. Blake II, M. Shah, R. L. Hettich, and J. F. Banfield.** 2005. Community proteomics of a natural microbial biofilm. *Science* **308**:1915–1920.

137. **Rasmussen, T. B., M. E. Skindersoe, T. Bjarnsholt, R. K. Phipps, K. B. Christensen, P. O. Jensen, J. B. Andersen, B. Koch, T. O. Larsen, M. Hentzer, L. Eberl, N. Hoiby, and M. Givskov.** 2005. Identity and effects of quorum-sensing inhibitors produced by *Penicillium* species. *Microbiology* **151**:1325–1340.

138. **Rawls, J. F., B. S. Samuel, and J. I. Gordon.** 2004. Gnotobiotic zebrafish reveal evolutionarily conserved responses to the gut microbiota. *Proc. Natl. Acad. Sci. USA* **101**:4596–4601.

139. **Rebrikov, D. V., S. M. Desai, P. D. Siebert, and S. A. Lukyanov.** 2004. Suppression subtractive hybridization. *Methods Mol. Biol.* **258**:107–134.

140. **Remold, S. K., and R. E. Lenski.** 2001. Contribution of individual random mutations to genotype-by-environment interactions in *Escherichia coli*. *Proc. Natl. Acad. Sci. USA* **98**:11388–11393.

141. **Rondon, M. R., P. R. August, A. D. Bettermann, S. F. Brady, T. H. Grossman, M. R. Liles, K. A. Loiacono, B. A. Lynch, I. A. MacNeil, C. Minor, C. L. Tiong, M. Gilman, M. S. Osburne, J. Clardy, J. Handelsman, and R. M. Goodman.** 2000. Cloning the soil metagenome: a strategy for accessing the genetic and functional diversity of uncultured microorganisms. *Appl. Environ. Microbiol.* **66:**2541–2547.

142. **Sano, E., S. Carlson, L. Wegley, and F. Rohwer.** 2004. Movement of viruses between biomes. *Appl. Environ. Microbiol.* **70:**5842–5846.

143. **Santo Domingo, J. W., S. C. Siefring, and R. A. Haugland.** 2003. Real-time PCR method to detect *Enterococcus faecalis* in water. *Biotechnol. Lett.* **25:**261–265.

144. **Schell, M. A., M. Karmirantzou, B. Snel, D. Vilanova, B. Berger, G. Pessi, M. C. Zwahlen, F. Desiere, P. Bork, M. Delley, R. D. Pridmore, and F. Arigoni.** 2002. The genome sequence of *Bifidobacterium longum* reflects its adaptation to the human gastrointestinal tract. *Proc. Natl. Acad. Sci. USA* **29:**14422–14427.

145. **Schirmer, A., R. Gadkari, C. D. Reeves, F. Ibrahim, E. F. DeLong, and C. Richard Hutchinson.** 2005. Metagenomic analysis reveals diverse polyketide synthase gene clusters in microorganisms associated with the marine sponge *Discodermia dissolute*. *Appl. Environ. Microbiol.* **71:**4840–4849.

146. **Schmeisser, C., C. Stöckigt, C. Raasch, J. Wingender, K. N. Timmis, D. F. Wenderoth, H.-C. Flemming, H. Liesegang, R. A. Schmitz, K.-E. Jaeger, and W. R. Streit.** 2003. Metagenome survey of biofilms in drinking-water networks. *Appl. Environ. Microbiol.* **69:**7298–7309.

147. **Scott, T. M., T. M. Jenkins, J. Lukasik, and J. B. Rose.** 2005. Potential use of a host associated molecular marker in *Enterococcus faecium* as an index of human fecal pollution. *Environ. Sci. Technol.* **19:**145–152.

148. **Seksik, P., L. Rigottier-Gois, G. Gramet, M. Sutren, P. Pochart, P. Marteau, R. Jian, and J. Dore.** 2003. Alterations of the dominant faecal bacterial groups in patients with Crohn's disease of the colon. *Gut* **52:**237–242.

149. **Sghir, A., G. Gramet, A. Suau, V. Rochet, P. Pochart, and J. Dore.** 2000. Quantification of bacterial groups within human fecal flora by oligonucleotide probe hybridization. *Appl. Environ. Microbiol.* **66:**2263–2266.

150. **Shanks, O. C., J. W. Santo Domingo, and J. E. Graham.** Use of competitive DNA hybridization to identify differences in the genomes of bacteria. *J. Microbiol. Methods* **66:**321–330.

151. **Shanks, O. C., J. W. Santo Domingo, and J. E. Graham.** 2006. Competitive metagenomic DNA hybridization identifies host-specific microbial genetic markers in cattle fecal samples. *Appl. Environ. Microbiol.* **72:**4054–4060.

152. **Simpson, J. M., J. W. Santo Domingo, and D. J. Reasoner.** 2002. Microbial source tracking: state of the science. *Environ. Sci. Technol.* **36:**5279–5288.

153. **Simpson, J. M., J. W. Santo Domingo, and D. J. Reasoner.** 2004. Assessment of equine fecal contamination: the search for alternative bacterial source-tracking targets. *FEMS Microbiol. Ecol.* **47:**65–75.

154. **Sinton, L. W., A. M. Donnison, and C. M. Hastie.** 1993. Faecal streptococci as faecal pollution indicators: a review. Part II: sanitary significance, survival, and use. *N. Z. J. Mar. Freshwater Res.* **27:**117–137.

155. **Sjoling, S., and D. A. Cowan.** 2003. High 16S rDNA bacterial diversity in glacial meltwater lake sediment, Bratina Island, Antarctica. *Extremophiles* **7:**275–282.

156. **Somer, L., Y. Danin-Poleg, E. Diamant, R. Gur-Arie, Y. Palti, and Y. Kashi.** 2005. Amplified intergenic locus polymorphism as a basis for bacterial typing of *Listeria* spp. and *Escherichia coli*. *Appl. Environ. Microbiol.* **71**:3144–3152.
157. **Soule, M., K. Cain, S. LaFrentz, and D. R. Call.** 2005. Combining suppression subtractive hybridization and microarrays to map the intraspecies phylogeny of *Flavobacterium psychrophilum*. *Infect. Immun.* **73**:3799–3802.
158. **Soule, M., S. LaFrentz, K. Cain, S. LaPatra, and D. R. Call.** 2005. Polymorphisms in 16S rRNA genes of *Flavobacterium psychrophilum* correlate with elastin hydrolysis and tetracycline resistance. *Dis. Aquat. Organ.* **65**:209–216.
159. **Soule, M., E. Kuhn, F. Loge, J. Gay, and D. R. Call.** 2006. Using DNA microarrays to identify library-independent markers for bacterial source tracking. *Appl. Environ. Microbiol.* **72**:1843–1851.
160. **Southern, E., K. Mir, and M. Shchepinov.** 1999. Molecular interactions on microarrays. *Nat. Genet.* **21**:5–9.
161. **Stewart, J. R., J. Vinjé, S. J. G. Oudejans, G. I. Scott, and M. D. Sobsey.** 2006. Sequence variation among group III F-specific RNA coliphages from water samples and swine lagoons. *Appl. Environ. Microbiol.* **72**:1226–1230.
162. **Sutherland, G. R., and R. I. Richards.** 1995. Simple tandem DNA repeats and human genetic disease. *Proc. Natl. Acad. Sci. USA* **92**:3636–3641.
163. **Švec, P., L. A. Devriese, I. Sedlacek, M. Baele, M. Vancanneyt, F. Haesebrouck, J. Swings, and J. Doskar.** 2002. Characterization of yellow-pigmented and motile enterococci isolated from intestines of the garden snail *Helix aspersa*. *J. Appl. Microbiol.* **92**:951–957.
164. **Švec, P., M. Vancanneyt, L. A. Devriese, S. M. Naser, C. Snauwaert, K. Lefebvre, B. Hoste, and J. Swings.** 2005. *Enterococcus aquimarinus* sp. nov., isolated from sea water. *Int. J. Syst. Evol. Microbiol.* **55**:2183–2187.
165. **Taboada, E. N., R. R. Acedillo, C. D. Carrillo, W. A. Findlay, D. T. Medeiros, O. L. Mykytczuk, M. J. Roberts, C. A. Valencia, J. M. Farber, and J. H. Nash.** 2004. Large-scale comparative genomics meta-analysis of *Campylobacter jejuni* isolates reveals low level of genome plasticity. *J. Clin. Microbiol.* **42**:4566–4576.
166. **Tajima, K., R. I. Aminov, T. Nagamine, K. Ogata, M. Nakamura, H. Matsui, and Y. Benno.** 1999. Rumen bacterial diversity as determined by sequence analysis of 16S rDNA libraries. *FEMS Microbiol. Ecol.* **29**:159–169.
167. **Taylor, D. N., J. M. Bied, J. S. Munro, and R. A. Feldman.** 1982. *Salmonella dublin* infections in the United States, 1979–1980. *J. Infect. Dis.* **146**:322–327.
168. **Tendolkar, P. M., A. S. Baghdayan, and N. Shankar.** 2005. The N-terminal domain of enterococcal surface protein, Esp, is sufficient for Esp-mediated biofilm enhancement in *Enterococcus faecalis*. *J. Bacteriol.* **187**:6213–6222.
169. **Todorov, J. R., A. Y. Chistoserdov, and J. Y. Aller.** 2000. Molecular analysis of microbial communities in mobile deltaic muds of Southeastern Papua New Guinea. *FEMS Microbiol. Ecol.* **33**:147–155.
170. **Trovatelli, L. D., and D. Matteuzzi.** 1976. Presence of bifidobacteria in the rumen of calves fed different rations. *Appl. Environ. Microbiol.* **32**:470–473.
171. **Trusova, M. Y., and M. I. Gladyshev.** 2002. Phylogenetic diversity of winter bacterioplankton of eutrophic Siberian reservoirs as revealed by 16S rRNA gene sequence. *Microb. Ecol.* **44**:252–259.

172. **Tyagi, S., and F. R. Kramer.** 1996. Molecular beacons: probes that fluoresce upon hybridization. *Nat. Biotechnol.* **14:**303–308.

173. **Tyagi, S., D. P. Bratu, and F. R. Kramer.** 1998. Multicolor molecular beacons for allele discrimination. *Nat. Biotechnol.* **16:**49–53.

174. **U.S. Environmental Protection Agency.** 2005. *Microbial Source Tracking Guide Document.* EPA-600/R-05/064. Office of Research and Development, U.S. Environmental Protection Agency, Washington, D.C.

175. **van Belkum, A.** 2003. Molecular diagnostics in medical microbiology: yesterday, today and tomorrow. *Curr. Opin. Pharmacol.* **3:**497–501.

176. **van Belkum, A., S. Scherer, L. van Alphen, and H. Verbrugh.** 1998. Short-sequence DNA repeats in prokaryotic genomes. *Microbiol. Mol. Biol. Rev.* **62:**275–293.

177. **van Belkum, A., M. Struelens, A. de Visser, H. Verbrugh, and M. Tibayrenc.** 2001. Role of genomic typing in taxonomy, evolutionary genetics, and microbial epidemiology. *Clin. Microbiol. Rev.* **14:**547–560.

178. **van Kessel, J. S., J. S. Karns, and M. L. Perdue.** 2003. Using a portable real-time PCR assay to detect *Salmonella* in raw milk. *J. Food Prot.* **66:**1762–1767.

179. **Venter, J. C., K. Remington, J. F. Heidelberg, A. L. Halpern, D. Rusch, J. A. Eisen, D. Wu, I. Paulsen, K. E. Nelson, W. Nelson, D. E. Fouts, S. Levy, A. H. Knap, M. W. Lomas, K. Nealson, O. White, J. Peterson, J. Hoffman, R. Parsons, H. Baden-Tillson, C. Pfannkoch, Y.-H. Rogers, and H. O. Smith.** 2004. Environmental genome shotgun sequencing of the Sargasso Sea. *Science* **304:**66–74.

180. **Versalovic, J., M. Schneider, F. J. de Bruijn, and J. R. Lupski.** 1994. Genomic fingerprinting of bacteria using repetitive sequence-based polymerase chain reaction. *Methods Mol. Cell. Biol.* **5:**25–40.

181. **Volker, U., and M. Hecker.** 2005. From genomics via proteomics to cellular physiology of the Gram-positive model organism *Bacillus subtilis. Cell. Microbiol.* **7:**1077–1085.

182. **Walter, J., M. Mangold, and G. W. Tannock.** 2005. Construction, analysis, and β-glucanase screening of a bacterial artificial chromosome library from the large-bowel microbiota of mice. *Appl. Environ. Microbiol.* **71:**2347–2354.

183. **Wang, M., S. Ahrné, B. Jeppsson, and G. Molin.** 2005. Comparison of bacterial diversity along the human intestinal tract by direct cloning and sequencing of 16S rRNA genes. *FEMS Microbiol. Ecol.* **54:**219–231.

184. **Wang, X., S. P. Heazlewood, D. O. Krause, and T. H. Florin.** 2003. Molecular characterization of the microbial species that colonize human ileal and colonic mucosa by using 16S rDNA sequence analysis. *J. Appl. Microbiol.* **95:**508–520.

185. **Washburn, M. P., and J. R. Yates III.** 2000. Analysis of the microbial proteome. *Curr. Opin. Microbiol.* **3:**292–297.

186. **Wei, J., M. B. Goldberg, V. Burland, M. M. Venkatesan, W. Deng, G. Fournier, G. F. Mayhew, G. Plunkett III, D. J. Rose, A. Darling, B. Mau, N. T. Perna, S. M. Payne, L. J. Runyen-Janecky, S. Zhou, D. C. Schwartz, and F. R. Blattner.** 2003. Complete genome sequence and comparative genomics of *Shigella flexneri* serotype 2a strain 2457T. *Infect. Immun.* **71:**2775–2786.

187. **Whitman, R. L., D. A. Shively, H. Pawlik, M. B. Nevers, and M. N. Byappanahalli.** 2003. Occurrence of *Escherichia coli* and enterococci in *Cladophora* (Chlorophyta) in nearshore water and beach sand of Lake Michigan. *Appl. Environ. Microbiol.* **69:**4714–4719.

188. **Whitman, W. B., D. C. Coleman, and W. J. Wiebe.** 1998. Prokaryotes: the unseen majority. *Proc. Natl. Acad. Sci. USA* **95:**6578–6583.

189. **Whittam, T. S.** 1996. Genetic variation and evolutionary processes in natural populations of *Escherichia coli*, p. 2708–2720. *In* F. C. Neidhardt, R. C. Curtiss III, J. L. Ingraham, E. C. C. Lin, K. B. Low, B. Magasanik, W. S. Reznikoff, M. Riley, M. Schaechter, and H. E. Umbarger (ed.), *Escherichia coli and Salmonella: cellular and molecular biology*, 2nd ed. American Society for Microbiology, Washington, D.C.

190. **Wiedmann, M., J. L. Bruce, C. Keating, A. E. Johnson, P. L. McDonough, and C. A. Batt.** 1997. Ribotypes and virulence gene polymorphisms suggest three distinct *Listeria monocytogenes* lineages with differences in pathogenic potential. *Infect. Immun.* **65:**2707–2716.

191. **Willard, M. D., R. B. Simpson, N. D. Cohen, and J. S. Clancy.** 2000. Effects of dietary fructooligosaccharide on selected bacterial populations in feces of dogs. *Am. J. Vet. Res.* **61:**820–825.

192. **Wolffs, P., B. Norling, and P. Radstrom.** 2005. Risk assessment of false-positive quantitative real-time PCR results in food, due to detection of DNA originating from dead cells. *J. Microbiol. Methods* **60:**315–323.

193. **Xu, J., M. K. Bjursell, J. Himrod, S. Deng, L. K. Carmichael, H. C. Chiang, L. V. Hooper, and J. L. Gordon.** 2003. A genomic view of the human-*Bacteroides thetaiotaomicron* symbiosis. *Science* **299:**2074–2076.

194. **Xu, P., G. Widmer, Y. Wang, L. S. Ozaki, J. M. Alves, M. G. Serrano, D. Puiu, P. Manque, D. Akiyoshi, A. J. Mackey, W. R. Pearson, P. H. Dear, A. T. Bankier, D. L. Peterson, M. S. Abrahamsen, V. Kapur, S. Tzipori, and G. A. Buck.** 2004. The genome of *Cryptosporidium hominis*. *Nature* **431:**1107–1112.

195. **Yun, J., and S. Ryu.** 2005. Screening for novel enzymes from metagenome and SIGEX, as a way to improve it. *Microb. Cell Fact.* **4:**8.

196. **Zhu, X. Y., T. Zhong, Y. Pandya, and R. D. Joerger.** 2002. 16S rRNA-based analysis of microbiota from the cecum of broiler chickens. *Appl. Environ. Microbiol.* **68:**124–137.

197. **Zhu, Y.-L, Q.-J. Song, D. L. Hyten, C. P. Van Tassell, L. K. Matukumalli, D. R. Grimm, S. M. Hyatt, E. W. Fickus, N. D. Young, and P. B. Cregan.** 2003. Single nucleotide polymorphisms (SNPs) in soybean. *Genetics* **163:**1123–1134.

Index

G

H